Salim Chikhi · Amine Bermak ·
Abdelmalek Amine · Allaoua Chaoui ·
Djamel Eddine Saidouni ·
Mohamed Khireddine Kholladi
Editors

Artificial Intelligence and Internet of Things

Proceedings of the 8th International Symposium, MISC 2024, Tamanghasset, Algeria, December 1–3, 2024

Springer

Editors
Salim Chikhi
Faculty of New Information
and Communication Technologies
University of Constantine2
Constantine, Algeria

Abdelmalek Amine
Faculty of Technology
University of Saida
Saida, Algeria

Djamel Eddine Saidouni
Faculty of New Information
and Communication Technologies
University of Constantine2
Constantine, Algeria

Amine Bermak
Hamad Bin Khalifa University
Doha, Qatar

Allaoua Chaoui
Faculty of New Information
and Communication Technologies
University of Constantine2
Constantine, Algeria

Mohamed Khireddine Kholladi
Faculty of Sciente and Technology,
Department of Mathematics and Computer
Science
University of El Oued
El-Oued, Algeria

ISSN 2367-3370 ISSN 2367-3389 (electronic)
Lecture Notes in Networks and Systems
ISBN 978-3-031-82111-0 ISBN 978-3-031-82112-7 (eBook)
https://doi.org/10.1007/978-3-031-82112-7

This Springer imprint is published by the registered company Springer Nature Switzerland AG
The registered company address is: Gewerbestrasse 11, 6330 Cham, Switzerland

Lecture Notes in Networks and Systems 1267

The series "Lecture Notes in Networks and Systems" publishes the latest developments in Networks and Systems—quickly, informally and with high quality. Original research reported in proceedings and post-proceedings represents the core of LNNS.

Volumes published in LNNS embrace all aspects and subfields of, as well as new challenges in, Networks and Systems.

The series contains proceedings and edited volumes in systems and networks, spanning the areas of Cyber-Physical Systems, Autonomous Systems, Sensor Networks, Control Systems, Energy Systems, Automotive Systems, Biological Systems, Vehicular Networking and Connected Vehicles, Aerospace Systems, Automation, Manufacturing, Smart Grids, Nonlinear Systems, Power Systems, Robotics, Social Systems, Economic Systems and other. Of particular value to both the contributors and the readership are the short publication timeframe and the world-wide distribution and exposure which enable both a wide and rapid dissemination of research output.

The series covers the theory, applications, and perspectives on the state of the art and future developments relevant to systems and networks, decision making, control, complex processes and related areas, as embedded in the fields of interdisciplinary and applied sciences, engineering, computer science, physics, economics, social, and life sciences, as well as the paradigms and methodologies behind them.

Indexed by SCOPUS, EI Compendex, INSPEC, WTI Frankfurt eG, zbMATH, SCImago.

All books published in the series are submitted for consideration in Web of Science.

For proposals from Asia please contact Aninda Bose (aninda.bose@springer.com).

Preface

In a constantly changing world, where technology shapes our daily lives, scientific research plays a crucial role in creating innovative and sustainable solutions. This book brings together a collection of research articles that explore diverse themes ranging from artificial intelligence and the Internet of things to data science and the impact of these technologies in the field of health.

This volume contains research papers accepted and presented at the 8th International Symposium on Modelling and Implementation of Complex Systems (MISC 2024), held on 1–3 December 2024 in University of Tamanghasset, Algeria. The contributions presented here reflect the authors' commitment to addressing contemporary challenges while highlighting technological and methodological advances. The articles address issues ranging from anomaly detection in satellite systems to glioma classification using deep learning techniques to intrusion detection in IoT networks. Each illustrates how emerging technologies can be used to improve daily life, optimise existing systems, and pave the way for new applications.

We have organised this book into four main sections, each of which addresses different dimensions of the topics covered. The first section focuses on the design and implementation of IoT and cloud computing systems, while the second section highlights the applications of artificial intelligence in various fields. The third section explores the applications of data science, and finally, the fourth section is dedicated to artificial intelligence in the healthcare sector.

The integration of ideas and research contained in this book represents a significant contribution to the scientific and technical community. We hope that it will serve as a valuable resource for researchers, students, and professionals wishing to deepen their knowledge and draw inspiration from the latest advances in these fascinating fields.

We would like to thank all the authors for their hard work and dedication to research. Their contributions enrich our understanding of today's technological challenges and pave the way for innovative solutions for tomorrow. We also thank Dr. Akram Kout and Prof. Nabil Belala for managing Microsoft's Conference Management Toolkit for MISC 2024 from submissions to proceedings elaboration. We are grateful to the program committee and organizing committee members for their contribution in the success of the symposium. We hope you enjoy reading and exploring the ideas and discoveries presented in this book.

October 2024 Salim Chikhi

Organization

The 8th International Symposium on Modelling and Implementation of Complex Systems (MISC 2024) was Co-organized by University of Constantine 2 – Abdelhamid Mehri, University of Tamanghasset and the Center for Development of Advanced Technologies (CDTA Algiers), and took place in Tamanghasset, Algeria (December 1–3, 2024).

Honorary Chairs

Abdelghani Choucha	Rector of University of Tamanghasset, Algeria
Youcef Lakhdar Hamina	Rector of University of Constantine 2, Algeria

General Chairs

Salim Chikhi	University Constantine 2, Algeria
Mohamed Traiche	Center for Development of Advanced Technologies, Algeria

Steering Committee

Allaoua Chaoui	University of Constantine 2, Algeria
Salim Chikhi	University of Constantine 2, Algeria
Mohamed-Khireddine Kholladi	University of El Oued, Algeria
Djamel Eddine Saïdouni	University of Constantine 2, Algeria
Mohamed Bachir Yagoubi	University of Laghouat, Algeria
Moumen Hamouma	University of Batna 2, Algeria
Belabbes Yagoubi	University Center of Tindouf, Algeria

Organizing Committee Chairs

Arsalane Chouaib Guidoum	University of Tamanghasset, Algeria
Abdelkrim Bouramoul	University of Constantine 2, Algeria

Organizing Committee

Adil Chekati	University of Constantine 2, Algeria
Badreddine Miles	University of Constantine 1, Algeria
Said Labed	University of Constantine 2, Algeria
Radouane Nouara	University of Constantine 2, Algeria
Mourad Bouzenada	University of Constantine 2, Algeria

Publication Chairs

Nabil Belala	University of Constantine 2, Algeria
Ahmed-Chawki Chaouche	University of Constantine 2, Algeria
Akram Kout	University of Setif 1, Algeria

Program Committee Chairs

Allaoua Chaoui	University of Constantine 2, Algeria
Ali Lemouari	University of Tamanghasset, Algeria
Djamel Eddine Saïdouni	University of Constantine 2, Algeria

Program Committee

Messaoud Abbas	University of El Oued, Algeria
Isma Akli	CDTA, Algeria
Amar Balla	ESI, Algeria
Mounir Beggas	University of El Oued, Algeria
Abdesselem Beghriche	University of Setif 1, Algeria
Ali Behloul	University of Batna 2, Algeria
Faiza Belala	University of Constantine 2, Algeria
Nabil Belala	University of Constantine 2, Algeria
Ghalem Belalem	University of Oran 1, Algeria
Meriem Belguidoum	University of Constantine 2, Algeria
Hacene Belhadef	University of Constantine 2, Algeria
Tahar Bendouma	University of Laghouat, Algeria
Messaoud Bengherabi	CDTA, Algeria
Djamel Benmerzoug	University of Constantine 2, Algeria
Mohamed Benmohammed	University of Constantine 2, Algeria
Imene Bensalem	ESCF Constantine, Algeria
Bilal Bouaita	University of Setif 1, Algeria

Mustapha Bouakkaz	University of Laghouat, Algeria
Karim Bouamrane	University of Oran 1, Algeria
Zizette Boufaida	University of Constantine 2, Algeria
Kamel Boukhalfa	USTHB, Algeria
Abdelkrim Bouramoul	University of Constantine 2, Algeria
Rachid Chalal	ESI, Algeria
Allaoua Chaoui	University of Constantine 2, Algeria
Adil Chekati	University of Constantine 2, Algeria
Saloua Chettibi	University of Jijel, Algeria
Salim Chikhi	University of Constantine 2, Algeria
Mohamed Skander Daas	University of Constantine 1, Algeria
Gregorio Díaz-Descalzo	University of Castilla-La Mancha, Spain
Karim Djemame	University of Leeds, UK
Mahieddine Djoudi	University of Poitiers, France
Cédric Eichler	INSA Centre Val de Loire, France
Raida El Mansouri	University of Constantine 2, Algeria
Farou Brahim Farou	University of Guelma, Algeria
Mohamed Gharzouli	University of Constantine 2, Algeria
Nassira Ghoualmi-Zine	University of Annaba, Algeria
Nadia Hocine	University of Mostaghanem, Algeria
Okba Kazar	University of Biskra, Algeria
Elhillali Kerkouche	University of Jijel, Algeria
Khaled Khalfaoui	University of Jijel, Algeria
Ilham Kitouni	University of Constantine 2, Algeria
Zineddine Kouahla	University of Guelma, Algeria
Mohamed Nadjib Kouahla	University of Guelma, Algeria
Akram Kout	University of Setif 1, Algeria
Abdesslem Layeb	University of Constantine 2, Algeria
Ali Lemouari	University of Jijel, Algeria
Smaine Mazouzi	University of Skikda, Algeria
Kamal Melkemi	University of Batna 2, Algeria
Salah Merniz	University of Constantine 2, Algeria
Chaker Mezioud	University of Constantine 2, Algeria
Mohamed Chaouki Babahenini	University of Biskra, Algeria
Abdelouahab Moussaoui	University of Setif 1, Algeria
Mustapha Aouache	CDTA, Algeria
Nassir Harrag	University of Setif 1, Algeria
Mathieu Roche	CIRAD of Montpellier, France
Baroudi Rouba	University of Mostaghanem, Algeria
Rachid Seghir	University of Batna 2, Algeria
Hichem Talbi	University of Constantine 2, Algeria
Nacer Eddine Zarour	University of Constantine 2, Algeria

Nadia Zeghib	University of Constantine 2, Algeria
Abdelhafid Zitouni	University of Constantine 2, Algeria

Co-editors

Salim Chikhi	University of Constantine 2, Algeria
Abdelmalek Amine	University of Saïda, Algeria
Allaoua Chaoui	University of Constantine 2, Algeria
Mohamed Khireddine Kholladi	University of El Oued, Algeria
Djamel Eddine Saïdouni	University of Constantine 2, Algeria

Contents

Design and Implementation of IoT Systems and Cloud Computing

Supervised ML-Based Multi-criteria for Detecting Malicious Nodes in FANET 3
Said Neciri, Noureddine Chaib, and Messaoud Benguenane

A Semi-decentralized Congestion-Free Multi-path Energy Routing for P2P Energy Trading Systems 14
Sara Hebal, Saad Harous, Djamila Mechta, and Lemia Louail

Privacy-Preserving Intrusion Detection Using FedSplit Learning 29
Malika Abid, Mohammed Kamel Benkaddour, and Mohamed Benouis

Unveiling Human Activity Patterns in Smart Cities Through a CNN-LSTM Approach 43
Imane Ameur, Mohamed El Amine Ameur, Mazene Ameur, and Houssem Eddine Dagha

Emerging Systems and Technologies a Comparative Study 53
Amel Houha and Abdelkamel Tari

An Approach for UML Sequence Diagram Models/MetaModels Co-evolution Using Transformation Rules 69
Asma Chouikh, Khaled Khalfaoui, ELHillali Kerkouche, and Allaoua Chaoui

Artificial Intelligence and its Applications

Advancing Speech Emotion Recognition: A Comparative Study and Enhanced Performance in English and Algerian Dialect Using Uni and Multimodal Transformer 87
Sabrina Khemis, Abdelouahab Moussaoui, and Skander Hamdi

Material Property Prediction with CNN-LSTM Hybrid Models and Periodic Table as Input Representation 102
Saad Boudabia, Youcef Mahdjoub, Ahmed Draoui, Fatima Zohra Bensalem, and Kheira Laiourate

Attentional Seq2Seq Model for Arabic Opinion Question Generation 112
Leila Ouahrani and Djamal Bennouar

Evaluation of Machine Learning Methods for Anomaly Detection in Satellite Telemetry: Advantages and Limitations 127
Achraf Djerida

Data Science and its Applications

Assessing Landslide Susceptibility Using Machine Learning and Remote Sensing Data: A Case Study of Southeastern Constantine, Algeria 143
Zakaria Matougui, Mohamed Yacine Daksi, and Mehdi Dib

Data-Driven Bandwidth Selection in Kernel Density Estimation via Unsupervised Learning 156
Arsalane Chouaib Guidoum and Amel Henider

Autonomous Data Tiering Using Reinforcement Learning for 3-Tier Hierarchical Storage 171
Nedjah Oussama, Mokhtari Omar, Boumahdi Fatima, and Mancer Yasmine

Urban Change Detection from Satellite Images: Leveraging Edge AI for Advanced Insights 182
Meriem Baziz, Anfal Yousra Boucetta, Leila Hamdad, and Imed Allal

Artificial Intelligence in Healthcare

Enhanced Glioma Classification Through Multi-modal Deep Learning: Integrating Histopathological and MRI Data 205
Linda Ait Mohammed, Fatiha Alim-Ferhat, and Farid Talbi

Automated Medical Labels Detection and Text Extraction Using Tesseract 212
Imène Ait Abderrahim and Youcef Benrached

Mobile AR for Learning Skeleton Anatomy in Algerian Undergraduates 226
Mohamed Amine Guerroudji, Samir Benbelkacem, Leila Boudine, Kahina Amara, Samya Taright, Nesrine Toubaline, Imene Issolah, Rayane Gacem, Nadia Zenati, Fatima-Zohra Touia, Mostefa Masmoudi, and Ali Benkacimi

MobileViT-Based Approach for Arrhythmia Classification Using ECG Scalograms 240
Hadjer Bechinia, Djamel Benmerzoug, and Nawres Khlifa

A Combined MaskRCNN-Watershed Model for Colorectal Gland Segmentation from Histological Images 254
Ferialle Lahreche, Slimane Oulad-Naoui, Ballou Mansoura, Asma Harrouzi, and Abdelouahab Moussaoui

CASF: Conflict Areas Survivor Framework 267
Nardjes Bouchemal, Mounira Bouzahzah, and Saber Ziad Barki

Author Index 279

Design and Implementation of IoT Systems and Cloud Computing

Supervised ML-Based Multi-criteria for Detecting Malicious Nodes in FANET

Said Neciri[1(✉)], Noureddine Chaib[2,3], and Messaoud Benguenane[3]

[1] Computer Science and Mathematics Laboratory (LIM), Laghouat University, Laghouat, Algeria
s.neciri@lagh-univ.dz

[2] Amar Telidji University of Laghouat, Laghouat, Algeria
n.chaib@lagh-univ.dz

[3] Department of Computer Science and IT Kasdi Merbah University of Ouargla, Ouargla, Algeria

Abstract. As unmanned aerial vehicles (UAVs) are rapidly being adopted in various industries, there is a need to formulate concepts for safe operability. In this paper, the problem of identifying malicious UAVs is tackled by employing a multi-criteria decision making (MCDM) approach which is based on supervised machine learning. The approach looks to enhance the efficiency and precision of UAVs classification by measuring several performance parameters such as the fall rate, delivery rate, and loss rate, as well as residual energy, among others. MCDM helps in giving a decision on the classification of a given UAV in the dataset as either normal or malignant. After this, the features of the dataset are standardized to undertake inter-comparison. The user-trained classifiers include Random Forest, Decision trees, Naïve Bayes, and Support Vector Machine models where cross-validation sequences are employed to test for overfitting. The results support the effectiveness of Random Forest with classification accuracy reaching 99.9% in execution time of about 0.45 s. These results also indicate the possibility of improving the operational security of Flying Ad-Hoc Network (FANETs) by enhancing the capability of malicious nodes identification. This framework employs multi criteria decision making (MCDM) enabled by machine learning to tackle the problem of malicious identification clearly.

Keywords: FANET · UAV · Security · Machine learning · MCDM

1 Introduction

The quick rise of Unmanned Aerial Vehicles (UAVs) has brought developmental advancements to various industries such as surveillance, agriculture, logistics and disaster relief. However, spotting suspicious actions within the systems of UAVs still remains a big challenge. Determining drop attacks where a UAV is made to discard packets as an attack, is necessary in enhancing the trust and reliability of UAV networks, especially in Fanets. On the other hand, drop attacks are

S. Chikhi et al. (Eds.): MISC 2024, LNNS 1267, pp. 3–13, 2025.
https://doi.org/10.1007/978-3-031-82112-7_1

very detrimental with regard to the Quality of service (QoS) in FANET systems due to increased packet dropping rates and consequent poor throughput and performance of the networks [3]. Common detection techniques such as anomaly based and signature based have also been shown to be insufficient considerations because of the flexible and distributed nature of FANETs [10]. Anomaly detection is simply the detection of change where the present conduct is raised against a pre-established level of performance and any difference is considered a potential threat. Even though the methods are efficient, they tend to raise the amount of false positive rates in situations that are dynamic as those in FANETs [6]. Signature based detection is predominantly based on attack patterns that have been known however it has difficulties in the detection of unknown or emerging attacks [5]. Given such stipulations, there is urgent need for such detection algorithms that are able to withstand any forms of changes and any deployments as are characteristic of FANETs. In this paper, we propose a novel approach that integrates MCDM methods with machine learning algorithms to solve the problem of prevention of drop attacks in FANETs. The MCDM framework uses the TOPSIS approach to ranking order of preferences. Within the MCDM framework, TOPSIS is used to analyze data sets such as Drop Ratio, Delivered Ratio, Loss Ratio, Remaining Energy and categorize them accordingly. This labeled dataset is used to build models, such as Naive Bayes, for classification purposes.

In summary, the key contributions of this paper include:

- Introducing an MCDM-based approach using TOPSIS to label UAVs for drop attack detection.
- Implementing and evaluating various supervised ML algorithms to classify UAVs based on the labeled data.
- Demonstrating the effectiveness of the proposed model through performance analysis.

This study highlights the potential of combining MCDM and ML to address the critical issue of drop attacks in FANETs, offering a promising solution for enhancing UAV network security and reliability.

2 Literature Review

Urgent action is necessary to identify hostile drones due to the threats they may pose to FANET security. Researchers have proposed several approaches to address this problem. One approach that uses deep learning and machine learning models for drone detection emphasizes the importance of a hybrid approach for better results. This method includes vision-based, radio-frequency-based, and audio-based approaches [13]. Using multi-agent reinforcement learning (RL) to fine-tune radar parameters and positions could be a way to tackle the problems of swarm UAV detection, tracking, and classification. This method outperforms previous AI algorithms [8]. A computer vision-based surveillance system that uses supervised machine intelligence models, such as neural networks, has been able to tell the difference between dangerous UAVs and other objects in the air with an impressive classification accuracy rate of over 95% [14]. Another novel method is the combined UAV detection model (CUDM), which uses regular surveillance equipment to efficiently and accurately identify suspicious targets, such as harmful UAVs, in real-time [4]. New studies on FANETs have mostly been on hybrid deep learning models that can stop Distributed Denial of Service (DDoS) attacks and find strange things [11]. Neciri et al. proposed a supervised machine learning approach to detect drop attacks on Unmanned Aerial Vehicle Ad-hoc Networks (UANETs). K-Nearest Neighbors (KNN), Logistic Regression (LR), and Decision Tree (DT) are some of the methods used by this system to identify dangerous or harmless UAVs. The method's high classification accuracy allowed it to reliably identify malicious nodes, which in turn boosted security in UAV networks [12]. According to [1], the generalized likelihood ratio test (MaDe) is suggested by the authors as a means to detect malicious UAVs. This simple data integrity method was able to detect these UAVs in performance testing. A novel framework has also been proposed in [9], which integrates hybrid handcrafted and deep features to detect and localize malicious drones from their sound and image information, showing improved performance compared to other related methods. Wang et al. [15] presented a deep reinforcement learning approach to detect hostile attacks in UAV aerial computer networks. They demonstrate how to safeguard UAV services by utilizing a deep reinforcement learning-based intrusion detection case study. The authors use Arduino and RF communication devices to set up encrypted communication in order to identify drones that are either friendly or enemy when they enter airspace [7]. By establishing a secure encrypted connection, we confirmed the drone's identity. Brown et al. [2] propose a method of passively tracking high-capability malicious UAVs to their final destination using a swarm of surveillance UAVs. The UAV swarm members proactively position themselves to enhance their chances of identifying and pursuing the hostile UAV, and they continuously transfer the tracking responsibility from one pursuing UAV to another within the swarm. This proactive stance is based on the current hostile UAV trajectory that the members of the networked swarm share.

3 Methodology

This section outlines the methodology for implementing a supervised machine learning-based multi-criteria decision-making (MCDM) approach to detect malicious UAVs. The proposed framework combines key performance metrics with advanced machine-learning techniques to achieve robust and accurate UAV classification.

3.1 Overall Framework

The overall architecture of the proposed solution for secure communication of UAV swarm is depicted in Fig. 1. The process involves evaluating UAVs based on multiple criteria using MCDM and classifying them using supervised machine learning algorithms.

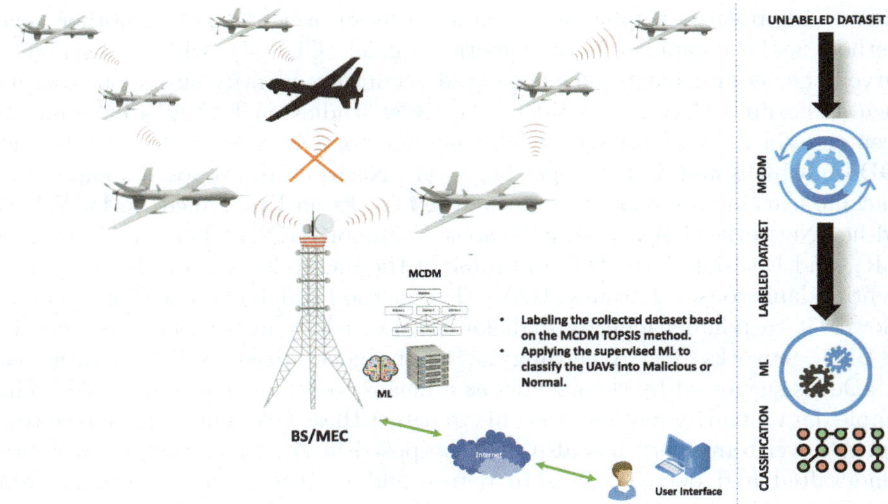

Fig. 1. System Model

3.2 Design Components

UAV Swarm: The UAV swarm consists of multiple UAVs equipped with various sensors and communication devices to collect and transmit data in real time. Each UAV in the swarm performs specific tasks such as surveillance, data collection, and relay communication.

Edge Computing Servers: Edge computing servers are deployed to process the data collected by the UAVs. These servers run MCDM and machine learning algorithms to evaluate the performance metrics of each UAV and classify them accordingly. Edge computing helps reduce latency and enhances the efficiency of data processing.

User Interface: The user interface allows operators to monitor the status of the UAV swarm, receive alerts about potential malicious UAVs, and manage the overall operation of the swarm. The interface provides real-time updates on UAV performance metrics and classification results.

Communication Network: A robust communication network ensures seamless data transmission between the UAVs, edge servers, and the user interface. This network supports real-time monitoring and control of the UAV swarm, enabling prompt detection and response to any malicious activities.

3.3 Proposed Model

Our approach's dataset labeling process is based on TOPSIS, where higher scores indicate better alignment with the desired characteristics defined by the weighted criteria. This method provides a comprehensive framework for evaluating and labeling UAVs by balancing multiple performance metrics, integrating both the benefits and costs associated with each UAV's performance. The key performance criteria used include drop ratio, delivered ratio, loss ratio, and residual energy, with a particular emphasis on drop ratio to effectively identify potential drop attacks. The following are the key steps for labeling the dataset using TOPSIS (Fig. 2).

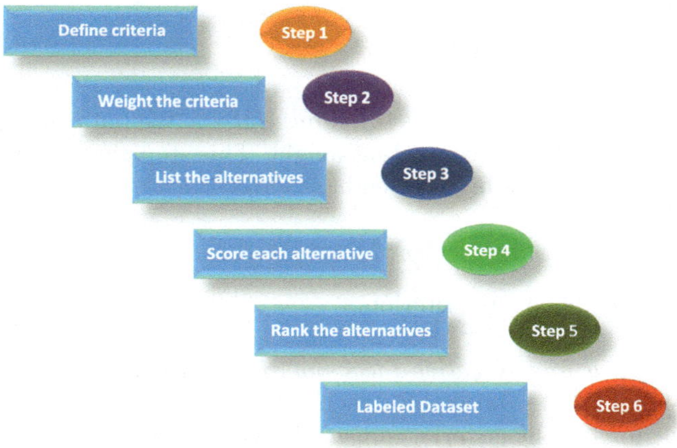

Fig. 2. Multi-criteria Decision-Making steps

Define Criteria and Alternatives: List the performance metrics (criteria) and the UAVs (alternatives) to be evaluated. For this specific problem, the criteria include drop ratio, delivered ratio, loss ratio, and residual energy.

Assign Weights to Criteria: The weights of the TOPSIS criterion were determined according to their significance in identifying drop attacks within UAV networks. The procedure entailed allocating greater significance to the drop ratio relative to other criteria because of its critical role in detecting potential harmful activities. The designated weights were as follows: Drop Ratio = 0.417, Delivery Ratio = 0.167, Loss Ratio = 0.167 and Residual Energy equals = 0.25. The TOPSIS framework guarantees an appropriate equilibrium of the criteria when the cumulative sum of these weights equals 1. The technique would exhibit a certain sensitivity to modifications in these weights, especially when adjusting the weights of essential criteria such as the drop ratio. Given that the drop ratio carries the greatest weight, its reduction would likely impair the system's capacity to identify UAVs engaged in drop attacks, while its augmentation would further accentuate this characteristic. Each criterion, especially the more significant ones like the drop ratio, determines the classification outcomes.

Construct the Decision Matrix: We create a decision matrix where each row i represents an alternative and each column j represents a criterion. The matrix entries u_{ij} are the values of each alternative for each criterion.

Normalize the Decision Matrix: We normalize the decision matrix to make the criteria comparable. The normalized value N_{ij} for each criterion u_{ij} is calculated as:

$$N_{ij} = \frac{u_{ij}}{\sqrt{\sum_{k=1}^{m} u_{ik}^2}}$$

where m is the number of UAVs.

Calculate the Weighted Normalized Decision Matrix: We multiply each normalized criterion value by its corresponding weight. This step gives more importance to the drop ratio due to its higher weight.

$$v_{ij} = W_j * N_{ij}$$

where w_j is the weight of the criterion.

Determine the Ideal and Negative-Ideal Solutions: We identify the ideal (A^*) and negative-ideal (A^-) solutions for each criterion.
$A^* = \{\max v_{ij}$ for benefit criteria, $\min v_{ij}$ for cost criteria$\}$
$A^- = \{\min v_{ij}$ for benefit criteria, $\max v_{ij}$ for cost criteria$\}$
Since a higher drop ratio indicates potential drop attacks, it should be treated as a cost criterion.

Calculate the Separation Measures: We compute the separation measures from the ideal S^* and negative-ideal S^- solutions for each UAV:

$$S_i^* = \sqrt{\sum_{j=1}^{n} (v_{ij} - A_j^*)^2}$$

$$S_i^- = \sqrt{\sum_{j=1}^{n} (v_{ij} - A_j^-)^2}$$

Calculate the Relative Closeness to the Ideal Solution: We calculate the relative closeness (Ci*) of each UAV to the ideal solution:

$$C_i^* = \frac{S_i^-}{S_i^* + S_i^-}$$

A higher Ci* value indicates a UAV closer to the ideal solution, thus more likely to be normal.

Label the UAVs: We label each UAV based on its *Ci** value. UAVs with lower Ci* values (indicating higher drop ratios) can be labeled as malicious, while those with higher Ci* values can be labeled as normal.

Dataset Description: Table 1 shows a synthetic dataset created from collation of useful characteristics such as Drop Ratio, Delivered Ratio, Loss Ratio, and Residual Energy. The parameters enable an analysis of the examination of each UAV's performance by concentrating on how hostile actions can be detected. A UAV that has a high drop ratio, low delivery ratio, significant loss ratio can perform a drop attack when it resorts to dropping of packets with the intention of bringing down the network. In addition, The dataset includes two more column: One is to compute the TOPSIS score and another for labels involve "Normal" or "Malicious". Furthermore, the dataset was divided into 80% o for training and the remaining 20% was used for testing. Then, the TOPSIS score obtained becomes higher for the UAVs classified as normal and lower for the ones classified as malicious.

Table 1. Dataset Snapshot

UAVID	Drop	Dlivered	Loss	Residual Energy	TOPSIS Score	Label	
uav1	0.12	0.71	0.04	61.55	0.442141	normal	
uav2	0.29	0.86	0.01	75.94	0.562727	malicious	
uav3	0.22	0.78	0.02	76.22	0.525269	normal	
uav4	0.18	0.83	0.05	79.12	0.652659	malicious	
uav5	0.06	0.93	0,04	81,78	0.377338	normal	
uav6	0.06	0.76	0.01	89.28	0.202514	normal	
uav7	0.03	0.8	0.02	75.49	0.160778	normal	
...	...	...	...	...	...	...	...
...	...	...	...	...	...	...	...

4 Performance Evaluation

This section outlines the system model for detecting malicious UAVs, describes the selected dataset, summarizes the performance metrics used, and compares the performance of various classifiers. The evaluation focuses on identifying the most effective model based on accuracy, precision, recall, F1 score, and execution time.

4.1 Dataset Description

Table 1 illustrates a synthetic dataset produced by collecting diverse data, including Drop Ratio, Delivered Ratio, Loss Ratio, and Residual Energy. This dataset is utilized to assess, analyze, and detect the drop attack. We added two columns to the dataset: one to compute the TOPSIS score using the collected metrics and another to label the dataset as "normal" or "malicious" based on MCDM TOPSIS. The dataset is divided into two parts: eighty percent for training purposes and twenty percent for model performance evaluation and validation.

4.2 Performance Metrics

Performance measures in machine learning are crucial for assessing and contrasting the efficiency of models, especially in detecting malicious UAVs. The common performance metrics used include:

Accuracy: The percentage of correctly classified instances (both true positives and true negatives) out of the total instances evaluated.

$$\text{Accuracy} = \frac{\text{True Positives} + \text{True Negatives}}{\text{Total Population}}$$

Precision: The ratio of true positives to the sum of true and false positives, reflecting the model's accuracy in predicting the positive class.

$$\text{Precision} = \frac{\text{True Positives}}{\text{True Positives} + \text{False Positives}}$$

Recall: The ratio of true positives to the sum of true positives and false negatives, measuring the model's ability to identify positive instances accurately.

$$\text{Recall} = \frac{\text{True Positives}}{\text{True Positives} + \text{False Negatives}}$$

F1 Score: The harmonic mean of precision and recall, providing a balance between the two, particularly useful in cases of imbalanced class distributions.

$$\text{F1 Score} = 2 \times \frac{\text{Precision} \times \text{Recall}}{\text{Precision} + \text{Recall}}$$

where:
-True positives (TP) are correctly predicted positive observations.
-True Negatives (TN) are correctly predicted negative observations.
-False Positives (FP) are incorrectly predicted positive observations.
-False Negatives (FN) are incorrectly predicted negative observations.

4.3 Comparative Analysis

In the context of detecting malicious UAVs, we compare our technique's performance metrics using various machine learning algorithms. Accuracy is a key measure for assessing classifier performance, providing an overall indication of how successfully algorithms classify UAVs as "normal" or "malicious". We implemented several established machine learning algorithms, such as Random Forest, Decision Tree, Naive Bayes, and SVM, each initialized with specific hyperparameters. Table 2 shows the overall performance of the various classification processes. Random Forest demonstrated the best results in all categories, followed by Decision Tree. Naive Bayes performed moderately, while SVM was the least effective. Random Forest has achieved high accuracy, precision, recall, and F1 scores, indicating strong performance in detecting malicious UAVs. Decision Tree: Also performed well, though slightly less effective than Random Forest. Naive Bayes showed moderate performance, with lower scores compared to Random Forest and Decision Tree. SVM had the lowest performance metrics, suggesting it is less suitable for this specific detection task. The comparative analysis highlights the effectiveness of our proposed approach, particularly with the Random Forest algorithm, in accurately detecting malicious UAVs based on the provided dataset. The results underscore the capability of our framework to optimize UAV service delivery and enhance overall network security (Fig. 3).

Table 2. The performance classification processes.

Algo	Accuracy	Precision	Recall	F1-Score	E-time
RF	0.99	0.99	0.99	0.98	0.513
DT	0.988	0.988	0.988	0.988	0.002
NB	0.973	0.974	0.973	0.973	0.002
SVM	0.95	0.99	0.9	0.94	0.002

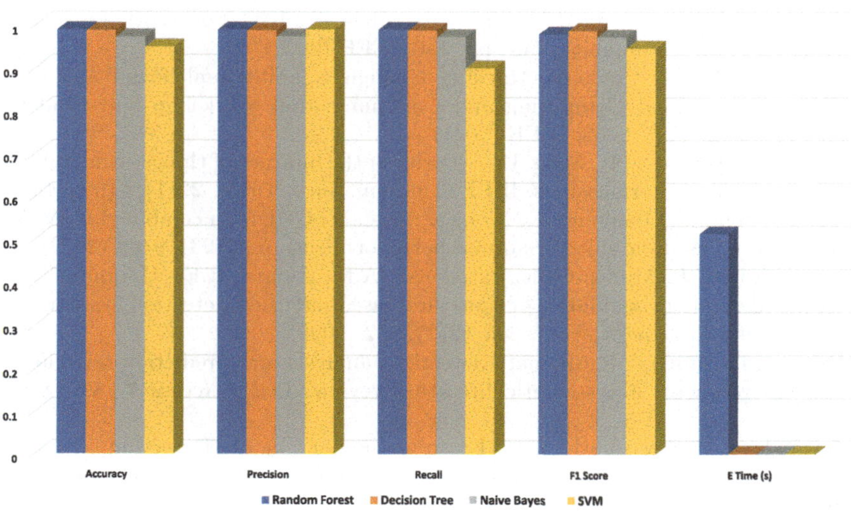

Fig. 3. Comparative Study of Classifiers

5 Conclusions And Future Works

The paper addresses concerning how to use machine learning (ML) and multi-criteria decision-making (MCDM) methods, especially TOPSIS, together to find FANET malicious nodes. This project aims to enhance the decision-making process for categorizing datasets into normal or dangerous nodes, thereafter implementing supervised machine learning algorithms to categorize UAVs based on these classifications. Using MCDM TOPSIS labels with machine learning classifiers for model training makes finding UAVs much easier, as shown by data from experiments. The approach has shown enhanced accuracy, precision, recall, and F1 score metrics. We can establish edge computing infrastructure to evaluate real-time data from UAV swarms, enabling swift detection of detrimental activities. Expanding edge servers is essential for accommodating larger UAV networks, and model retraining will be required to adjust to new data and threats. Nonetheless, practical implementation encounters constraints, like ensuring secure communication between UAVs and edge servers, managing heightened processing requirements, and minimizing false positives in fluctuating conditions. Addressing these difficulties will necessitate enhancing energy efficiency, augmenting data security, and perfecting detection algorithms. Future endeavors will concentrate on improving scalability, creating more flexible machine learning models, and investigating hybrid detection approaches for mitigation.

References

1. Abhishek, N.V., Aman, M.N., Lim, T.J., Sikdar, B.: MaDe: malicious aerial vehicle detection using generalized likelihood ratio test. In: ICC 2022-IEEE International Conference on Communications, pp. 1–6. IEEE (2022)
2. Brown, J., Raj, N.: Predictive tracking of a high capability malicious UAV. In: 2021 IEEE 11th Annual Computing and Communication Workshop and Conference (CCWC), pp. 0747–0752. IEEE (2021)
3. Butun, I., Österberg, P., Song, H.: Security of the internet of things: vulnerabilities, attacks, and countermeasures. IEEE Commun. Surv. Tutor. **22**(1), 616–644 (2019)
4. Cai, H., Song, Z., Jianlong, X., Xiong, Z., Xie, Y.: CUDM: a combined UAV detection model based on video abnormal behavior. Sensors **22**(23), 9469 (2022)
5. Díaz-Verdejo, J., Muñoz-Calle, J., Alonso, A.E., Alonso, R.E., Madinabeitia, G.: On the detection capabilities of signature-based intrusion detection systems in the context of web attacks. Appl. Sci. **12**(2), 852 (2022)
6. Fahim, M., Sillitti, A.: Anomaly detection, analysis and prediction techniques in IoT environment: a systematic literature review. IEEE Access **7**, 81664–81681 (2019)
7. Jayapalan, A., Savarinathan, P., Karuppasamy, T., et al.: Detection of malicious UAV in airspace. In: 2021 International Conference on Computer Communication and Informatics (ICCCI), pp. 1–4. IEEE (2021)
8. Khawaja, W., Yaqoob, Q., Guvenc, I.: RL-based detection, tracking, and classification of malicious UAV swarms through airborne cognitive multibeam multifunction phased array radar. Drones **7**(7), 470 (2023)

9. McCoy, J., Rawal, A., Rawat, D.B., Sadler, B.M.: Ensemble deep learning for sustainable multimodal UAV classification. IEEE Trans. Intell. Transp. Syst. **24**(12), 15425–15434 (2022)
10. Nair, A.S., Thampi, S.M.: Flying ad hoc networks: security, authentication protocols, and future directions. In: Internet of Things and Secure Smart Environments, pp. 223–272. Chapman and Hall/CRC (2020)
11. Neciri, S., Chaib, N.: Hybrid deep learning for anomaly detection in FANETs: a defense against DDoS attacks. Int. J. Intell. Syst. Appl. Eng. **12**(4), 3799–3809 (2024)
12. Neciri, S., Chaib, N., Djeddi, C.: Supervised machine learning for detecting drop attack in UAV ad-hoc network. In: International Conference on Emerging Intelligent Systems for Sustainable Development (ICEIS 2024), pp. 286–297. Atlantis Press (2024)
13. Shahbazian, R., Trubitsyna, I.: Human sensing by using radio frequency signals: a survey on occupancy and activity detection. IEEE Access **11**, 40878–40904 (2023)
14. Shakhatreh, H., et al.: Unmanned aerial vehicles (UAVs): a survey on civil applications and key research challenges. IEEE Access **7**, 48572–48634 (2019)
15. Wang, W., Lv, Z., Xiaozhen, L., Zhang, Y., Xiao, L.: Distributed reinforcement learning based framework for energy-efficient UAV relay against jamming. Intell. Converg. Netw. **2**(2), 150–162 (2021)

A Semi-decentralized Congestion-Free Multi-path Energy Routing for P2P Energy Trading Systems

Sara Hebal[1](✉), Saad Harous[2], Djamila Mechta[1], and Lemia Louail[3]

[1] LRSD Laboratory, Department of Computer Science, Sétif 1 University Ferhat Abbas, 19000 Sétif, Algeria
{sara.hebal,mechtadjamila}@univ-setif.dz

[2] Department of Computer Science, College of Computing and Informatics, University of Sharjah, Sharjah, UAE
harous@sharjah.ac.ae

[3] Université de Lorraine, CNRS, CRAN, 54000 Nancy, France
lemia.louail@univ-lorraine.fr

Abstract. The increased emphasis on clean energy has accelerated the integration of renewable energy sources into electrical grids, resulting in the rise of peer-to-peer energy trading systems. For optimal power dispatch and management the Energy Internet concept advocates using energy routers. Efficient energy routing algorithms are crucial for successful energy transfer between trading pairs in peer-to-peer energy trading across the complex Energy Internet. This research treats the energy routing problem as a non-convex, non-linear optimization problem using a combination technique of graph theory and non-convex nonlinear programming. It efficiently generates a multi-path solution for a trading pair, minimizing energy transmission loss while considering power flow direction constraints and accurately computing power losses during transmission which has not been considered in existing energy routing algorithms. A semi-decentralized approach with a priority concept is proposed for managing simultaneous power transmissions from multiple energy trading pairs, to avoid congestion and shift computational workloads to individual routers and the network system operator. The approach's performance is assessed through numerical simulations and comparisons with existing energy routing algorithms to showcase its novelty in power loss reduction, congestion management, and adaptability.

Keywords: Energy Internet · Energy routing · Graph-theory · Minimum energy loss · Non-convex non-linear optimization · Peer-to-peer energy trading · Sequential quadratic programming

1 Introduction

The power sector is transitioning from conventional centralized systems to Smart Grids (SGs) due to increasing demand for clean, low-carbon energy and advancements in renewable energy technologies [1]. The intermittent, and unpredictable

S. Chikhi et al. (Eds.): MISC 2024, LNNS 1267, pp. 14–28, 2025.
https://doi.org/10.1007/978-3-031-82112-7_2

nature of renewable energy sources (RESs) makes it challenging to integrate them into the grid real-time operation [2]. To promote RESs use in SGs, peer-to-peer energy trading (P2PET) has been introduced to enable direct power trading between consumers and prosumers [3]. The power grid is a top-down structure that transmits power in a unidirectional way, uncontrolled injection of P2PET power in the grid from multiple sources in multiple paths can create reverse power flow, network voltage and power lines capacity violation, and increases network energy losses as power lines losses are inevitable during energy transmission in P2PET [4]. Thus, deploying end-to-end power exchange in the power grid is challenging giving rise to the energy routing problem. To promote P2PET over the power system, Energy Internet (EI) has recently arisen, where distributed energy devices comprising loads, generating units, and energy storage systems (ESSs) are interconnected through a mesh network of energy routers (ERs) with bidirectional power and communication flow [5]. ERs are the EI's backbone, responsible for routing and converting energy, exchanging information, regulating voltage and frequency, scheduling, and controlling processes[6]. How to efficiently deliver energy flows between trading pairs while avoiding the issues above and reducing energy transmission losses is the main aspect of the EI and it is known as energy routing [4]. A successful transfer of energy flows between trading pairs in P2PET across the complex EI relies heavily on implementing an efficient energy routing algorithm.

The interest of designing energy-routing algorithms for EI has recently increased. Researchers in [7,8] proposed using Electric Vehicles (EVs) with their ESSs to store and transmit energy between RESs and consumers, where battery-powered charging stations serve as ERs. This method has drawbacks, involving an intensive charging and discharging process that shortens the EV's ESSs lifespan and increases energy losses, ESSs' limited capacity promotes demand for high-capacity ESSs and charging stations raising system costs, traffic issues, and longer transmission times compared to using power lines. To solve these issues, authors in [9] proposed an energy routing algorithm for energy local area networks. It uses an exhaustive search to find the energy minimum loss path. To avoid congestion problems, a pruning process is employed to eliminate the power lines and ERs that can not support the transmitted power before calculating the minimum loss path. However, defining the minimum energy loss path after evaluating all the possible paths is complex and time-consuming, particularly in large networks. The study in [10] uses the same pruning process to prevent congestion where the energy minimum loss path is selected by using an ACO-based energy routing algorithm. The energy is transmitted through a single path with enough capacity (virtual circuit mode). Ignoring power lines with limited capacity by the pruning process leads to more restrictive solutions and potentially congested networks. Authors in [11], proposed using the Dijkstra algorithm to select the energy minimum loss path between a trading pair in EI. The energy packet mode is employed to transmit power when the minimum loss path can not transmit the total requested power. Power is divided, with the minimum loss path transporting as much as its capacity allows, while the remaining power is routed through other potentially less efficient paths. The study in [12] employed the Dijkstra algorithm to determine all the feasible paths between a trading pair using distance as a metric. The minimum loss path is selected after comparing the power loss of each feasible path which is time-consuming, especially in large networks. The Depth First Search (DFS) algorithm has been utilized in several research articles to tackle the energy routing issue. In [13, 14], the DFS algorithm is employed to compute all feasible paths between the trading pair and based on each path's power loss, the minimum loss path is defined. In [15], authors used the DFS algorithms to determine all the feasible paths between the trading pair and then applied the non-linear programming to split the power on different paths, minimising the total energy transmission loss of the selected paths.

Energy routing algorithms described in [12, 13] are centralized where a network system operator (NSO) determines energy transmission paths for trading pairs to minimize grid power loss. These algorithms provide a global optimal solution for congestion relief, even in simultaneous energy transmissions by multiple trading pairs. However, they require significant computational resources and time, especially in larger systems. Some trading pairs may incur larger losses, as they must renounce high-efficiency transmission paths to minimise total network losses for the other pairs' benefit. Whereas, decentralized energy routing algorithms, described in [9–11], allow energy trading pairs to calculate their paths according to their preferences, reducing the computational burden on the NSO. They manage congestion using virtual circuits or energy packet mode, but their ability to provide congestion-free energy transmission in simultaneous power transmission scenarios is questionable. Energy packet mode optimizes power line use and can reduce energy transmission loss. The discussed centralized and decentralized energy routing algorithms are single-path algorithms.

Researchers recently have shifted from single-path centralized and decentralized algorithms to multi-path semi-decentralized ones to improve energy routing efficiency [14, 15]. Existing energy routing algorithms in literature have not accurately calculated power losses during transmission, as they assume constant transmitted power throughout the path to simplify the computation and path search process, leading to higher power losses than actual losses as power diminishes along the path (see Fig. 1). P2PET systems transport energy across grid distribution lines, which are designed for unidirectional power flow and are unable to transfer electrical energy in two opposing directions simultaneously. This physical limit, known as power flow direction constraint (PFD), is not considered in the discussed energy routing algorithms. The power line in Fig. 1b between ERs 2 and 5 transmits power flows in opposite directions potentially causing reverse power flow, line overloading, voltage fluctuations, and equipment failures affecting the power grid's performance, efficiency, and reliability. Thus, for an appropriate operation of the EI, the power line's PFD constraint must be considered in energy routing.

The energy routing problem is formulated in this paper as a non-convex, nonlinear optimization problem. To solve it, we proposed a novel semi-decentralized multi-path energy routing approach that computes power losses accurately dur-

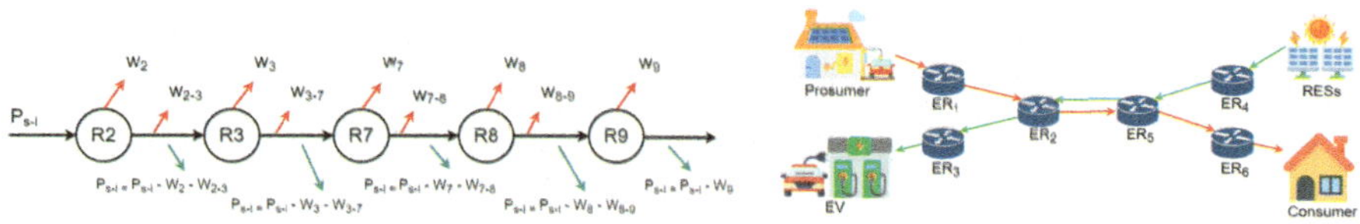

(a) (b)

Fig. 1. Graphical depiction of (a) power transmission losses along the path. (b) simultaneous bidirectional power flow in a power line PFD case.

ing path selection while respecting grid physical constraints, including the PFD constraint. The proposed approach involves decentralized local path discovery at the ER level, where trading pairs autonomously choose transmission paths. If conflicts arise, the global path conflict resolution phase is invoked at the NSO level, constructing the centralized part of the approach.

The rest of the paper is structured as follows: Sect. 2 introduces the novel energy routing problem formulation. The proposed semi-decentralized multi-path energy routing approach is detailed in Sect. 3. Section 4 discusses the evaluation and numerical simulation of the energy routing approach, focusing on scenarios involving networks of varying sizes and simultaneous energy transmission time. Section 5 concludes the paper.

2 Energy Routing Problem Formulation and System Model

EI is a mesh network where many paths can transmit energy between trading pairs. The mentioned works used a single energy transmission path solution to transmit energy. It provides congestion-free energy transmission paths and limits power line selection to those with sufficient capacity for the traded power. This may cause disputes between trading pairs over access to these power lines, as lower-capacity lines are ignored. Instead, traded power can be split into multiple energy packets and transmitted through different paths. Thus, building upon the energy routing problem formulation in [10,11], we introduce an enhanced mathematical formulation of the problem incorporating additional physical constraints. The multi-path energy routing optimization problem is defined by Eq. (1).

$$min\left(TL_{Total}^{s\to l}\right) \tag{1}$$

where $TL_{Total}^{s\to l}$ represents the total transmission loss of transmitting power amount between a trading pair $(s-l)$ through N_p path (Eq. (2)).

$$TL_{Total}^{s\to l} = \sum_{k=1}^{N_p}\left(W_{Total}^{p_k:s\to l}\right) \tag{2}$$

The energy transmission loss of a path p ($W_{Total}^{p:s\to l}$) is the sum of power losses of ERs (w_i) and power lines (w_{ij}) that construct the path. eff_i represents the

ER_i conversion efficiency, r_{ij} and V_{ij} are the resistance and voltage of power line L_{ij}, while P_{ij} is the pre-existing power on it.

$$W_{Total}^{p:s \to l} = \sum_{R_i \in p} w_i + \sum_{L_{i,j} \in p} w_{i,j} \tag{3}$$

$$w_i = (1 - eff_i) \times P_{s-l} \tag{4}$$

$$w_{ij} = \frac{r_{ij}}{V_{ij}^2} \times \left[(P_{s-l} + P_{ij})^2 - P_{ij}^2 \right] \tag{5}$$

The multi-path energy routing must adhere to the following constraints:

$$TP_{s-l} = \sum_{k=1}^{N_p} P_{s-l}^k \tag{6}$$

$$TL_{Total}^{s \to l} < TP_{s-l} \tag{7}$$

$$W_{Total}^{p_k:s \to l} < P_{s-l}^k \tag{8}$$

$$\sum_{(s-l) \in N} \sum_{k=1}^{Np} P_{s-l}^k + P_{ij} \leq P_{(L_{i,j})}^{Tc}, \quad \forall L_{i,j} \in E \tag{9}$$

$$\sum_{(s-l) \in N} \sum_{k=1}^{Np} P_{s-l}^k + P_{ij} \leq min(P_{(R_i)}^{Tc}, P_{(R_j)}^{Tc}), \quad \forall R_i, R_j \in V \tag{10}$$

$$P_{ij}^{total} = \sum_{(s-l) \in N} \sum_{k=1}^{Np} P_{s-l}^k + P_{ij}, \quad \forall L_{i,j} \in E \tag{11}$$

$$P_{ij}^{total} \times P_{ji}^{total} = 0, \quad \forall L_{i,j} \in E \tag{12}$$

$$0 \leq P_{s-l}^k \leq TP_{s-l} \tag{13}$$

TP_{s-l} is the total power transmitted from energy source s to load l. The sum of the transmitted power through a path k between s and l (P_{s-l}^k) should equal the overall power initially flowing from s (Eq. (6)). The combined transmission loss across these paths must be lower than TP_{s-l} (Eq. (7)), and the power loss of each path must be lower than the transmitted power through it (Eq. (8)). A power line L_{ij} can be used by multiple paths to deliver power to the same trading pair $s - l$ (self-path overlapping) or multiple trading pairs simultaneously (path overlapping). Equations (9) and (10) ensure that aggregate power transferred across a power line meets the maximum capacity of the power line $P_{(L_{i,j})}^{Tc}$ and ERs in the line's extremities $P_{(R_i)}^{Tc}$. P_{ij}^{Total} depicts the total power sent over the power line L_{ij}. Constraint (12) ensures that a power line L_{ij} may only transmit power in one direction at a time.

Multi-path energy routing aims to minimize total energy transmission loss by selecting the best paths with the power amount sent in each path. Power line loss increases quadratically with the amount of power flowing through it (Eq. (5)), creating a non-linear optimization problem. PFD constraint (Eq. (12)) increases the complexity of the problem by defining a non-convex set, resulting in a non-convex nonlinear optimization problem.

3 Proposed Energy Routing Algorithm

The multi-path energy routing problem is a non-convex, non-linear optimization problem. Non-convexity offers numerous local optima but doesn't guarantee a global optimum, making it challenging to determine if a solution is a global optimum. To solve this problem, a semi-decentralized energy routing approach is proposed, with routing decisions made locally at the ER level, coordinated by an NSO. We presume the existence of a P2PET communication platform where trading pairs are created and power system characteristics are shared. It allows ERs to submit their status information, including capacities, conversion efficiency, and connected power lines and ERs, providing a comprehensive view of the network structure and status information. The proposed approach involves local path discovery and global path conflict resolution, as depicted in Fig. 2.

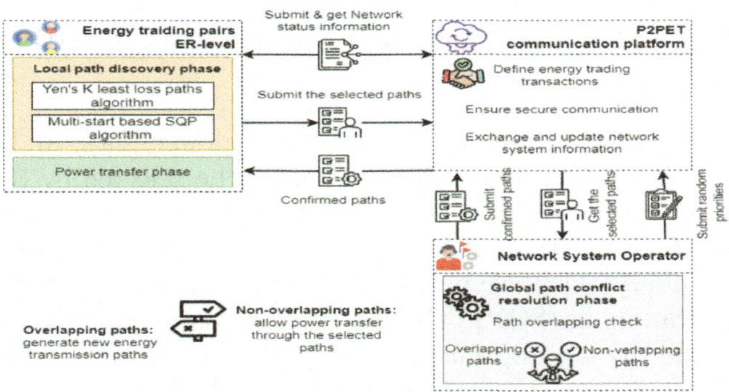

Fig. 2. The semi-decentralized multi-path energy routing approach overview.

3.1 Local Path Discovery Phase

Executed at the consumer-associated ER. Instead of finding all possible paths between a trading pair, which increases the search space for the power dispatch algorithm, we proposed the use of a Modified Yen's k energy least loss paths algorithm (Fig. 3) to determine the k most promising paths between them. The algorithm assumes that each ER and power line will transmit the total power and invokes a modified bidirectional Dijkstra algorithm (Fig. 4) to select the k-least loss paths. The bidirectional search from source and load sides simultaneously, reduces processing time, especially in large networks. A filtering phase excludes

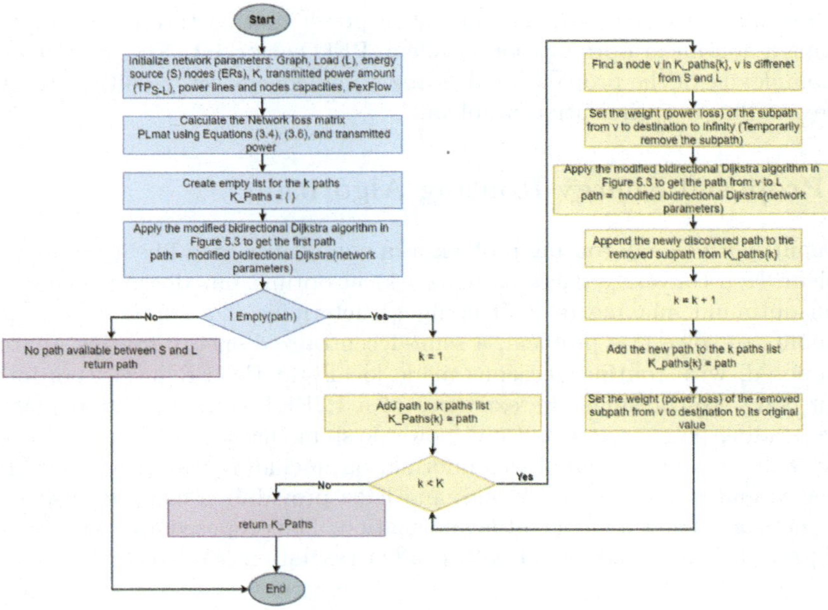

Fig. 3. Flowchart of the Yen's k least loss paths algorithm.

power lines with opposite power flow, saturated power lines, and ERs. Power lines with small capacities are considered and can be selected by the algorithm.

To determine the best paths to select from this set with the amount of power to send in each path to optimize energy transmission loss while respecting the constraints (6)–(13), we proposed the use of multi-start point Sequential Quadratic Programming (SQP). The multi-start points enhance the possibility of moving outside the local optimal by identifying an improved solution nearest to the global optimal solution. As illustrated in Algorithm 1, the multi-start point SQP is an iterative algorithm, for faster execution, it can be parallelized if the network's ERs are powerful. The $XBest$ vector represents the power to be transmitted through each path in the k paths set; if $XBest(i) > 0$ the path i is selected to transmit $XBest(i)$ kW; otherwise, it is not selected. TL_{Total}^{XBest} represents the total transmission loss of the selected paths in $XBest$. As power diminishes along the transmission path, the multi-start points SQP algorithm uses Algorithm 2 for accurate transmission loss calculations. At the end of this phase, each ER submits to the P2PET communication platform the selected paths, along with the amount of power to send on each one.

3.2 Global Path Conflict Resolution Phase

The NSO extracts ER routing decisions from the communication platform. If no path overlaps, it sends approval to the trading pairs, allowing them to start

Algorithm 1: Multi-Start Point SQP Algorithm

Input: KPaths, KPathsCap, NetworkParameters
Output: $XBest, TL_{Total}^{XBest}$

`/* Initialize a random solution */`

$XBest \leftarrow$ generate a random solution in bounds of the k paths capacities
$TL_{Total}^{XBest} \leftarrow 0$

`/*` **Determine the power loss of each selected path in** $XBest$ `*/`

for $i = 1$ ***to*** K **do**
  $W^P \leftarrow$ compute loss of path i in the KPaths using Algotithm 2
  $TL_{Total}^{XBest} \leftarrow TL_{Total}^{XBest} + W^P$
  Update Network Prameters

while *stopping condition is not satisfied* **do**
  $X_0 \leftarrow$ generate a random solution in bounds of the k paths capacities
  $(TL_{Total}, X) \leftarrow$ Apply SQP on X_0

  `/* Update` $XBest$ `*/`

  if $TL_{Total} < TL_{Total}^{XBest}$ **then**
    $XBest \leftarrow X$
    $TL_{Total}^{XBest} \leftarrow TL_{Total}$

Algorithm 2: Precise path power loss Algorithm

Input: P, Path, NetworkParameters
Output: W_{Total}^{p}
$W_{Total}^{p} \leftarrow 0$
for $i = 1$ ***to*** $(length(Path) - 1)$ **do**
  $w_{Path(i)} \leftarrow$ calculate $ER_{Path(i)}$ power loss using Equation (4), P, and NetworkParameters

  `/* Update transmitted power` P `to enter the next power line */`

  $P \leftarrow P - w_{Path(i)}$
  $w_{Path(i,i+1)} \leftarrow$ calculate $L_{Path(i,i+1)}$ power loss using (5), P, and NetworkParameters

  `/* Update transmitted power` P `to enter the` $ER_{Path(i+1)}$ `*/`

  $P \leftarrow P - w_{Path(i,i+1)}$
  $W_{Total}^{p} \leftarrow W_{Total}^{p} + w_{Path(i)} + w_{Path(i,i+1)}$

`/* Calculate the power loss of the last ER in the path */`

$w_{Path(end)} \leftarrow$ calculate $ER_{Path(end)}$ power loss using Equation (4), P, and NetworkParameters
$W_{Total}^{p} \leftarrow W_{Total}^{p} + w_{Path(end)}$

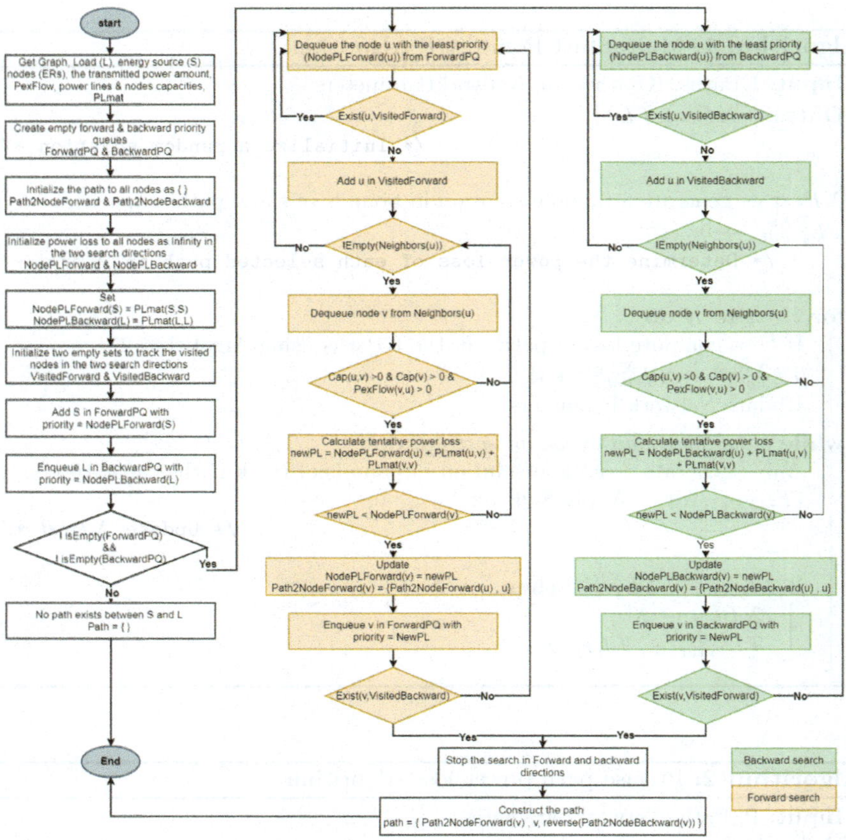

Fig. 4. Flowchart of the Modified Bidirectional Dijkstra.

the energy transfer. Otherwise, it orders conflicted trading pairs based on their priority, assigns the conflicted path to the higher-ranked pair and decreases its ranking. If two pairs have the same ranking, one is randomly chosen. For remaining trading pairs that did not take their selected paths, the NSO constructs new paths for each. Two methods are proposed for generating new paths:

- In the first method, as pre-existing power in the power line affects its transmission loss, the order of path allocation impacts the network's total loss. To reduce this loss, the NSO selects the ideal sequence in which each pair is accommodated (affected new paths). It creates a combination set of trading pairs and allocates paths according to their order in the set It determines the total transmission loss of each set and chooses the set with the lowest transmission loss. The allocation of new paths for trading pairs based on total network loss may favor some pairs, increasing power loss for others. For fairness, the total power loss of all affected pairs will be divided between them.

Equation (14) shows the power loss charged to each trading pair.

$$PL_{pair} = \frac{\sum_{(s-l)\in N} TL_{Total}^{s\rightarrow l}}{\sum_{(s-l)\in N} \sum_{k=1}^{N_p} P_{s-l}} \times TP_{s-l} \tag{14}$$

- In the second method, the NSO takes the trading pairs randomly, one by one, and attributes new paths using the proposed local path discovery phase. The power losses charged to each pair are calculated using Eq. (14).

The first method reduces network energy transmission loss, benefiting trading pairs and the utility grid, as in P2PET, the grid compensates for the loss and bills trading pairs. However, it takes time and increases with network size and number of pairs. The second method is better when dealing with a large number of trading pairs. We assume that NSO assigns random priorities to all network participants, distributing them at the start of the day. The trading pair's priority is the average of its consumer's and producer's priorities. When a trading pair gets a conflicted path, its priority decreases, allowing other pairs to take desired paths in the next path-overlapping cases preventing some pairs from dominating power lines. Energy trading pairs must obtain NSO approval before initiating a power transfer; this ensures system stability and risk management.

4 Numerical Simulation and Result Analysis

The proposed semi-decentralized multipath energy routing approach is evaluated through MATLAB simulations using the same networks, ERs, and power line parameters as in [15] for consistency. The approach is compared to existing energy routing algorithms based on Exhaustive search [9], Dijkstra [11], DFS [13], and DFS with non-linear programming [15] to demonstrate its superior performance.

4.1 Single Predetermined Source-Load Pair

The study assumes a single source-load trading pair in each network, located at its extremes to increase power transmission paths. A 10 kW power amount should be transmitted from ER_2 to ER_9 and from ER_{14} to ER_{29}, with pre-existing power initially set to zero. Table 1 and Figs. 5, 6 illustrate the simulation outcomes showing that exhaustive search, Dijkstra, and DFS algorithms provide a single-path solution for transmitting power, while non-linear programming and the proposed approach provide a multi-path solution using the same paths. However, the proposed approach provides different power distribution across chosen paths resulting in total transmission loss reduction by 24.15% and 10.34% in Figs. 5 and 6 respectively, compared to the solution found by exhaustive search, Dijkstra, and DFS algorithms and better energy transmission loss by 8.56% and 8.13% compared to the solution provided by non-linear programming.

Assuming the utility grid charges trading pairs a 0.25 $ for 1 kW power loss. The compensation fees for the trading pair ($ER_9 - ER_2$) decrease from 0.43 $

Table 1. Simulation results for single predetermined source-load pair case.

EI Network	Single-path energy routing			Multi-path energy routing	
	Exhaustive search (kW) [9]	Dijkstra (kW) [11]	DFS (kW) [13]	non-linear programming with DFS (kW) [15]	Proposed Approach (kW)
Figure 5	1.706	1.706	1.706	1.415	1.294
Figure 6	1.691	1.691	1.691	1.655	1.516

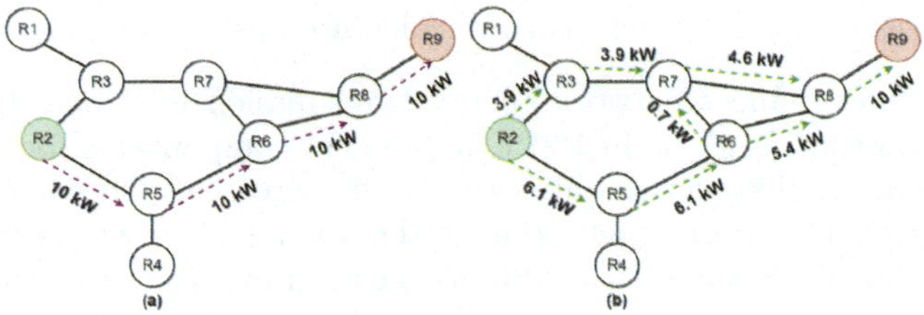

Fig. 5. The least loss paths selected by (a) Exhaustive search, Dijkstra and DFS algorithms, (b) non-linear programming, and the proposed approach for pair ($ER_9 - ER_2$).

when using DFS, Dijkstra, or exhaustive search to 0.36 \$ when using nonlinear programming. However, applying the suggested approach with accurate power loss further reduces compensation fees to 0.32 \$. This ensures that prosumers pay only their actual loss, and prevents the utility grid from injecting excess energy to compensate for losses, reducing unnecessary power circulation, grid infrastructure pressure, inefficiencies and costs for all parties involved. According to Fig. 7 the suggested approach outperforms the previously mentioned algorithms in minimizing power transmission loss.

Grid lines always transfer some degree of demand, indicating that the preexisting power on certain lines differs from zero. We assumed that a 5 kW power is transmitted from ER_6 to ER_4 through the path $6 \rightarrow 5 \rightarrow 4$ and from ER_6 to ER_3 through the path $6 \rightarrow 4 \rightarrow 3$. The selected paths by exhaustive search, Dijkstra, DFS, and non-linear programming-based energy routing algorithms adhere to the ERs and power lines capacity constraints, but create PFD conflict on the power lines L_{5-6} in Fig. 8 and L_{4-6} in Fig. 9. This creates reverse power flow causing voltage fluctuations, overheating, and potential power line damage, causing grid instability (power transmission failure). On the contrary, the proposed approach generates congestion-free energy least loss paths (1.807 kW in Fig. 8, and 1.642 kW in Fig. 9) that adhere to the network capacity and PFD constraints. The proposed approach outperforms cited algorithms in providing congestion-free least loss paths, accurate power loss, and adhering to grid physical constraints, including PFD constraints.

4.2 Multiple Predetermined Source-Load Pair

Using the same EI with increased capacity of ER_{18} and power line $L_{15,18}$ to 25 kW, it is assumed that the following energy transactions are executed simulta-

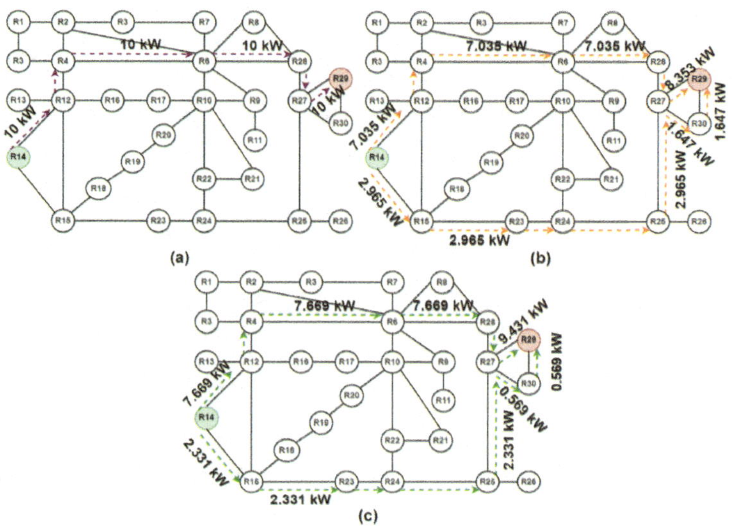

Fig. 6. The least loss paths selected by (a) Exhaustive search, Dijkstra and DFS algorithms, (b) non-linear programming with DFS, and (c) the proposed approach for the trading pair ($ER_{14} - ER_{29}$).

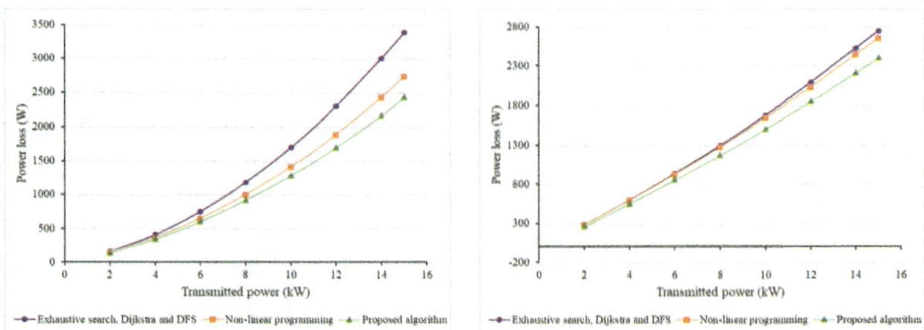

(a) Trading pair ($ER_2 - ER_9$) (b) Trading pair ($ER_{14} - ER_{29}$)

Fig. 7. Energy transmission loss comparison of the proposed approach with existing algorithms concerning transmitted power in different EI networks.

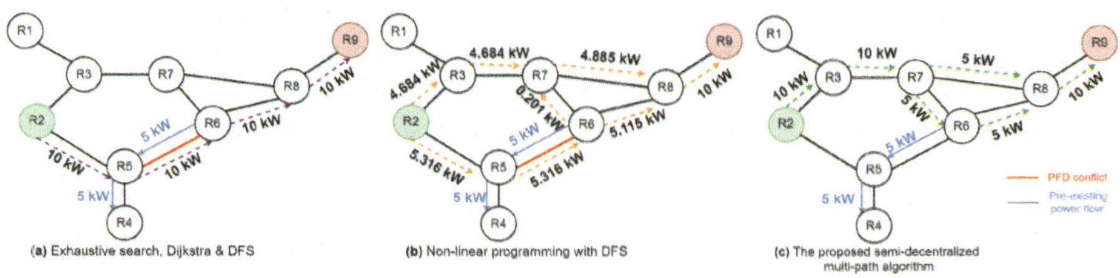

Fig. 8. The selected least loss paths for trading pair (ER_2, ER_9) with pre-existing power flow.

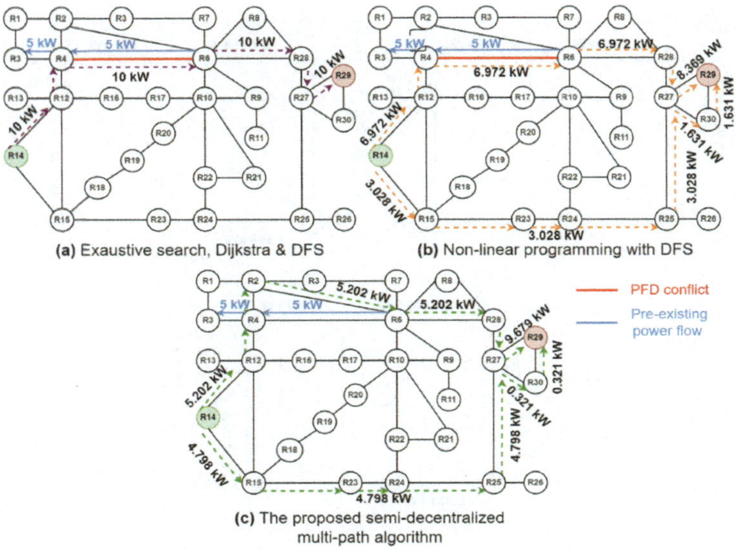

Fig. 9. The selected least loss paths for trading pair (ER_{14}, ER_{29}) with pre-existing power flow.

neously $\{(ER_{12}, ER_3), 30\,\text{kW}\}$, $\{(ER_2, ER_{18}), 25\,\text{kW}\}$, $\{(ER_{13}, ER_6), 25\,\text{kW}\}$, $\{(ER_{30}, ER_{28}), 10\,\text{kW}\}$, and $\{(ER_{25}, ER_{23}), 10\,\text{kW}\}$. As shown in Fig. 10.a, the selected paths by the decentralized exhaustive search and Dijkstra-based algorithms for pairs (ER_{12}, ER_3), (ER_2, ER_{18}), and (ER_{13}, ER_6) overlap at power line $L_{4,12}$, causing congestion and PFD conflict. The centralized DFS-based algorithm avoids congestion on line $L_{4,12}$ by splitting transmitted power into two energy packets and transmitting them through two different paths as shown in Fig. 10.b, creating PFD conflict in power line $L_{4,12}$. The non-linear programming-based algorithm uses a coalition cooperative game (Fig. 10d) to resolve the overlapping path of the pairs' decentralized decisions in Fig. 10c. The new paths avoid line congestion but create a PFD conflict in power lines L_{4-12} and L_{6-10}. On the contrary, each trading pair use the proposed local discovery phase and submits its routing decision to the NSO (Fig. 10e). The NSO invokes the global path conflict resolution phase (Fig. 10f), generates an approval for pairs (ER_{30}, ER_{28}) and (ER_{25}, ER_{23}). It assigns the overlapping path to the highest-priority pair assuming that it is (ER_{13}, ER_6) and decreases its priority. It builds new paths for the remaining pairs to reduce overall transmission loss while adhering to the grid's physical limits, including the PFD constraint. The new paths successfully eradicated the congestion and PFD conflict. As shown in Table 2, the execution time of the proposed approach is reduced by 45.61% and 82.23% compared to the algorithms in [15] and in [13] respectively.

Table 2. A comparison of computational time between the different algorithms

Algorithm	Decentralized part	Centralized part	Total computational time
Proposed semi-decentralized approach	9.52	43.94	53.46
Semi-decentralized in [15]	28.30	70	98.30
Centralized in [13]	–	–	301

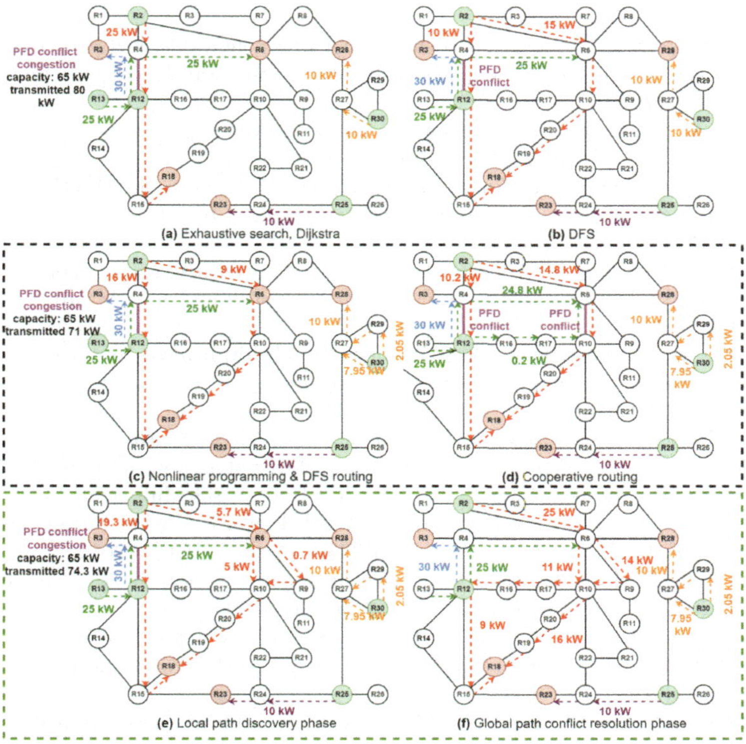

Fig. 10. The selected least loss paths for trading pair (ER_{14}, ER_{29}) with pre-existing power flow.

5 Conclusion

This paper presents a semi-decentralized multi-path energy routing approach, combining the advantages of fully centralized and decentralized energy routing algorithms. The approach determines the least-loss energy transmission paths for a trading pair. It uses modified Yen's k least loss path and multi-start SQP algorithms for path enumeration and power dispatch. It accurately computes power losses during path selection while respecting physical power system constraints, including the PFD constraint. The algorithm benefits trading pairs and the utility grid by reducing transmission losses, improving grid power line utilization, and alleviating line congestion. The deployment of the NSO keeps the P2PET under control and solves path conflict issues during simultaneous power transmission. Simulation findings confirm the algorithm's efficacy, making it a potential option for developing P2PET in EI.

References

1. Kataray, T., et al.: Integration of smart grid with renewable energy sources: opportunities and challenges-a comprehensive review. Sustain. Energy Technol. Assess. **58**, 103363 (2023)
2. Khalid, M.: Smart grids and renewable energy systems: perspectives and grid integration challenges. Energ. Strat. Rev. **51**, 101299 (2024)
3. Khorasany, M., Dorri, A., Razzaghi, R., Jurdak, R.: Lightweight blockchain framework for location-aware peer-to-peer energy trading. Int. J. Electr. Power Energy Syst. **127**, 106610 (2021)
4. Hebal, S., Mechta, D., Harous, S., Louail, L.: A comparative study of energy routing algorithms to optimize energy transmission in energy internet. J. Supercomput. 1–57 (2024)
5. Wu, Y., Wu, Y., Guerrero, J.M., Vasquez, J.C.: Digitalization and decentralization driving transactive energy internet: key technologies and infrastructures. Int. J. Electr. Power Energy Syst. **126**, 106593 (2021)
6. Umar, B.M., Jibril, Y., Jimoh, B., Kunya, A.B., Aliyu, S.: Energy router applications in the electric power system. J. Appl. Mater. Technol. **4**(2), 56–68 (2023)
7. Li, S., Yi, P., Huang, Z., Xie, T., Zhu, T.: Energy scheduling and allocation in electric vehicles energy internet. In: 2016 IEEE Power & Energy Society Innovative Smart Grid Technologies Conference (ISGT), pp. 1–5. IEEE (2016)
8. Sui, Y., Yi, P., Liu, X., Zhu, T.: Energy transport station deployment in electric vehicles energy internet. IEEE Access **7**, 97986–97995 (2019)
9. Wang, R., Jiande, W., Qian, Z., Lin, Z., He, X.: A graph theory based energy routing algorithm in energy local area network. IEEE Trans. Industr. Inf. **13**(6), 3275–3285 (2017)
10. Hebal, S., Mechta, D., Harous, S., Dhriyyef, M.: Hybrid energy routing approach for energy internet. Energies **14**(9), 2579 (2021)
11. Guo, H., Wang, F., Li, L., Zhang, L., Luo, J.: A minimum loss routing algorithm based on real-time transaction in energy internet. IEEE Trans. Industr. Inf. **15**(12), 6446–6456 (2019)
12. Koottappillil, D.P., Naidoo, R.M., Mbungu, N.T., Bansal, R.C.: Distribution of renewable energy through the energy internet: a routing algorithm for energy routers. Energy Rep. **8**, 355–363 (2022)
13. Razi, R., Pham, M.-C., Hably, A., Bacha, S., Tran, Q.-T., Iman-Eini, H.: A novel graph-based routing algorithm in residential multimicrogrid systems. IEEE Trans. Industr. Inf. **17**(3), 1774–1784 (2020)
14. El Hafdaoui, H., Khallaayoun, A.: Internet of energy (IOE) adoption for a secure semi-decentralized renewable energy distribution. Sustain. Energy Technol. Assess. **57**, 103307 (2023)
15. Jiang, X., Sun, C., Cao, L., Law, N.-F., Loo, K.H.: Semi-decentralized energy routing algorithm for minimum-loss transmission in community energy internet. Int. J. Electr. Power Energy Syst. **135**, 107547 (2022)

Privacy-Preserving Intrusion Detection Using FedSplit Learning

Malika Abid[1(✉)], Mohammed Kamel Benkaddour[1], and Mohamed Benouis[2]

[1] Department of Computer Science and Information Technology, Artificial Intelligence and Information Technology Laboratory (LINATI), University of Kasdi Merbah, 30000 Ouargla, Algeria
{abid.malika,Benkaddour.kamel}@univ-ouargla.dz

[2] Human-Centered Artificial Intelligence, University of Augsburg, Augsburg, Germany
mohamed.benouis@uni-a.de

Abstract. The rapid development of networking technologies has chang-ed how we live and allows us to grow as people and society work together. However, as we depend increasingly on these technologies, significant security worries exist, especially about people hacking into network data and intrusion. This study introduces a Federated split Intrusion Detection System (FIDS) that utilizes federated learning techniques to effectively identify, categorize, and handle network intrusions while reducing the occurrence of false positives in non-identically and independently distributed (non-IID) data sets. The system employs a Convolutional Neural Network (CNN) combined with a Long Short-Term Memory (LSTM) model, which has been trained using the CICIDS2017 dataset to reach a high level of accuracy in detection. Federated split learning offers robust privacy preservation and computational efficiency benefits; this strategy decreases the susceptibility to reverse engineering attacks, provides protection against such assaults, and enhances privacy. The system proposed in this work utilizes federated split learning techniques to protect sensitive data and ensure privacy. It is a robust and privacy-focused solution for detecting intrusions in IoT networks.

Keywords: Federated learning · IDS · CNN-LSTM · CICIDS2017 · FIDS · non-IID

1 Introduction

The growing use of artificial intelligence and machine learning has transformed several sectors, facilitating the implementation of automation and enhancing productivity. However, these sophisticated technologies can create fresh weaknesses, such as AI-driven assaults and data manipulation. In light of the ongoing changes in the digital environment, it is crucial to acknowledge and tackle these developing security issues by creating protective, solid measures to prevent possible risks.

S. Chikhi et al. (Eds.): MISC 2024, LNNS 1267, pp. 29–42, 2025.
https://doi.org/10.1007/978-3-031-82112-7_3

Intrusion detection systems (IDS) have advanced to include machine learning (ML) and deep learning techniques, resulting in improved detection capabilities compared to conventional methods. These systems can detect and classify harmful behaviors in network traffic in real-time. Federated Learning (FL) is a viable method for deploying intrusion detection systems (IDS) in the Internet of Things (IoT) ecosystem[1]. Federated Learning (FL) allows several agents to collectively train a shared model without disclosing their training data. This approach effectively tackles privacy concerns and minimizes computing requirements.

However, traditional FL methods are vulnerable to man-in-the-middle attacks, where an adversary can intercept and alter the model parameters transmitted between clients and servers. We suggest an innovative approach called Split-Fed Learning(SFL) to reduce this potential danger. Split Federated Learning combines the advantages of Split Learning (SL) and Federated Learning (FL) to provide strong protection against such assaults[2]. SFL combines FL and SL, overcoming the limitations of both methods while allowing for privacy preservation, communication efficiency optimization, and model resilience reinforcement.

The primary objective of this project is to design and develop a Federal Intrusion Detection System (FIDS) that can accurately identify and respond to network intrusions in non-IID data distributions while minimizing false positive rates. To achieve this, the system integrates a hybrid model and feature selection, which enables it to handle complex and diverse network traffic patterns effectively.

The system aims to safeguard confidential information and maintain privacy using federated learning methods. The FedIDS system integrates SFL, a novel approach that combines FL and SL, to enhance the precision and robustness of intrusion detection. This approach also reduces the vulnerability to reverse engineering attacks on the student model.

To further minimize communication threats, the system employs traffic analysis and other security measures to prevent man-in-the-middle attacks and other forms of data breaches. By eliminating communication threats and vulnerabilities, the FedIDS system provides an additional layer of defense against attacks, thereby enhancing privacy and protecting sensitive data.

2 Related Work

Centralized IDS have been extensively researched, with various machine learning and deep learning techniques being employed to improve detection accuracy and reduce false positives. For instance, S. Krsteski et al. proposed a Random Forest (RF) based IDS that achieved an accuracy of 96.6% on the CICIDS2017 dataset [3]. Similarly, A. Guezzaz developed a Decision Tree (DT) based IDS that demonstrated a detection rate 99.42% on the NSL-KDD dataset [4].

More recently, deep learning techniques, such as the Convolutional Neural Network (CNN) based IDS proposed by Saba et al. [5] on the UNSW-NB15 dataset, have been explored. Additionally, hybrid models have been proposed, such as the CNN-Long Short-Term Memory (LSTM) model developed by [6,7] that demonstrated a detection rate of more than 90% on the CICIDS2017, N-BaIoT, UNSW-NB15, and WSN-DS dataset. Other approaches have focused on feature extraction and selection, such as [8], which proposed a feature selection method based on mutual information on the KDD Cup 1999 dataset. These works demonstrate the ongoing efforts to develop more effective and efficient centralized IDS to detect and prevent network attacks and anomalies.

Federated Learning (FL) has been increasingly explored in Intrusion Detection Systems (IDS) since its introduction by Google in 2016 and McMahan et al. in 2017 [9]. The integration of FL with IDS has shown promising results, with studies demonstrating the effectiveness of Convolutional Neural Networks (CNN) and Deep Neural Networks with the NSL-KDD dataset [10,11], Deep Belief Networks (DBN) with the TON-IoT dataset [12], and hybrid models such as CNN-GRU by Z. Jin et al. [13] and CNN-BiLSTM by Javeed and all [14]. A GRU-SVM-based federated learning approach is proposed to detect assaults on wireless edge devices. The needed federation rounds have been decreased by 70% despite the more excellent performance; the model has not been evaluated for an IoT network [15]. Ammar Alazab et al. [16] employed a client selection technique with federated averaging (EA) on the NSL-KDD dataset to enhance the effectiveness of federated learning for intrusion detection systems. Qin et al. [17] propose a novel three-tier algorithm, Hier-SFL, which synergistically integrates federated and split learning to efficiently train deep learning models for traffic classification in a client-edge-cloud architecture. These models have been tested on both Independent and Identically Distributed (IID) and non-IID data, showcasing the potential of FL in improving IDS performance.

3 Method and Architecture

3.1 CICIDS 2017 Dataset

The CICIDS2017 dataset is an extensive and authentic dataset designed for intrusion detection systems (IDS). It includes both harmless and malicious network traffic. The dataset was produced using a testbed with several operating systems, network devices, and protocols. The dataset comprises many types of attacks, including Brute Force FTP, Brute Force SSH, DoS, Heartbleed, Web Attack, Infiltration, Botnet, and DDoS. These assaults were carried out throughout both morning and afternoon periods on Tuesday, Wednesday, Thursday, and Friday. The collection contains benign traffic created using the B-Profile technology, which analyzes the abstract behavior of human interactions to produce lifelike background traffic [18].

3.2 Data Pre-processing

We consolidated the CICIDS2017 dataset, which contained various variables (numerical, nominal, etc.) distributed across eight separate files, into a single file containing 3,119,345 occurrences. We preprocessed the data by performing numerous operations. These included deleting duplicate entries and missing values, eliminating constants and their related features, and categorizing labels into seven distinct categories: standard, Dos/DDos, PortScan, brute force, web assault, botnet, and infiltration. We used Principal Component Analysis (PCA) to reduce the number of features, resulting in a final set of 25 features from the initial 77. Ultimately, the data underwent scaling. After preprocessing, we divided the dataset into three subsets: The dataset consists of three subsets: the training set, which accounts for 60% of the data; the validation set, which represents 20% of the data; and the testing set, which contributes another 20% of the data. The training set trained the model, and the validation set fine-tuned its hyperparameters and evaluated its performance during the training phase. The testing set assessed the model's capacity to generalize, preventing overfitting and ensuring that the model acquired underlying patterns instead of just remembering the training data (Fig. 1).

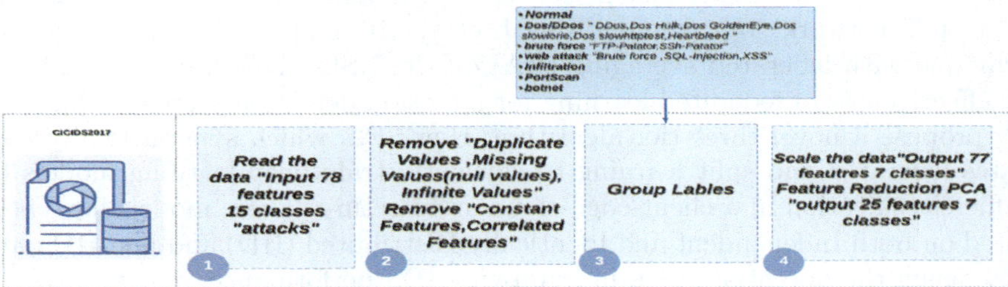

Fig. 1. Data pre-processing

3.3 Evaluation Method

The proposed models are evaluated using evaluation metrics, including precision, recall, accuracy, and F1 score, commonly used in various fields, including threat detection. The following sections provide a detailed description of each [19]:

- Accuracy = (TP+TN)/(TP+TN+FP+FN).
- Precision = (TP)/(TP+FP)
- Recall = (TP)/(TP+FN)
- F1 score = (2 × Precision × Recall)/(Precision + Recall)

Where **TP** represents the number of instances accurately classified as standard classification, and **TN**corresponds to the number of cases with aberrant characteristics correctly designated as abnormal classification. On the other hand, **FP** represents the number of instances with strange features that were inaccurately classified as usual. In contrast, **FN** represents the number of regular cases misclassified as aberrant classes.

3.4 Data Distribution

Federated learning classifies data distribution into independent and identically distributed (IID) or non-independent and non-identically dispersed (non-IID) [20]. The IID data distribution guarantees that the data is independent and the same across clients, without any correlation. Non-IID data distribution refers to a situation where the data is not independent and similar among clients. This may occur owing to several causes, such as disparities in client populations, variations in data-gathering techniques, or changes in usage habits.

Dirichlet Distribution: The Dirichlet distribution [21] is used to represent the dispersal of client data in a distributed manner. We assume that a Dirichlet distribution governs each client's data, signifying the proportion of each category to generate datasets for each participant, where Dir(α) represents the Dirichlet distribution with the concentration parameter α. To ensure that the datasets are non-uniform (Non-IID), we selectα values that are dispersed and remote from one another to evaluate a mini-distribution. To incorporate variability into the data generation procedure, we select $\alpha = 0{,}5$, which is not the default value.

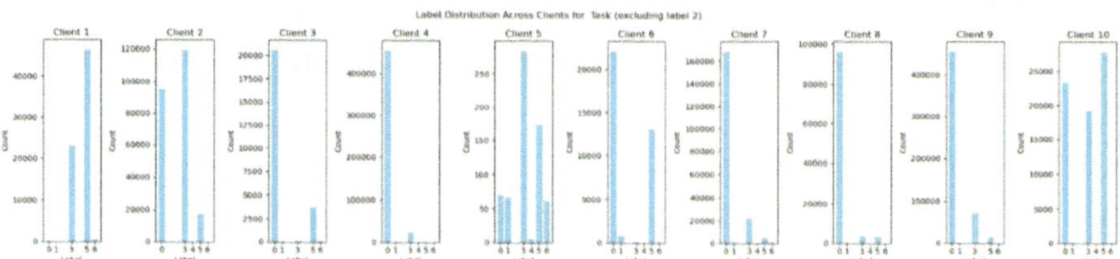

Fig. 2. The Non-IID data

As illustrated in Fig. 2, we developed a list of ten clients with labels of various dimensions. Client 5 has six classes, while customers 3 and 4 have only two.

3.5 Proposed Model

Our work is divided into three parts. The first part involves local training of a CNN-LSTM model, which follows the same process as centralized learning. The second part involves federated learning, specifically FedCNN-LSTM, designed to ensure data privacy and minimize the need for ample storage and high computational power. Finally, the third part involves federated split learning, known as FedSplitCNN-LSTM, invented to address several key challenges, including ensuring data and model privacy and minimizing communication costs.

Centralized Traing. After completing studies including both CNN and LSTM models, we have proposed a hybrid CNN-LSTM model. Our CNN model architecture comprises two convolutional layers with max pooling, a flattening layer and two dense layers. On the other hand, our LSTM model consisted of a solitary LSTM layer of 50 units, followed by a dense layer containing seven units, each representing one of the seven classes in our classification task.

The hybrid architecture CNN-LSTM is proposed in this work to process sequential data with spatial hierarchies. It uses the strengths of convolutional neural networks (CNNs) and long short-term memory (LSTM) networks to extract spatial features and temporal dependencies. The CNN extracts relevant spatial features, while the LSTM models temporal relationships between these features. This combination improves model performance by capturing spatial and temporal patterns in the data, leading to more accurate and robust predictions. The CNNL-STM architecture combines the strengths of both architectures to enhance the model's performance. This model takes in data with 25 features and runs it through two convolutional layers with ReLU activation and max pooling. This removes spatial dimensions and pulls out local features. We then feed the output into an LSTM layer, which captures long-term dependencies in the data. We then pass the LSTM output through two fully connected layers with ReLU activation, resulting in a final production with seven classes. We design the model's architecture to handle sequential data and extract local and global features, making it suitable for time series forecasting and sequence prediction tasks (Fig. 3).

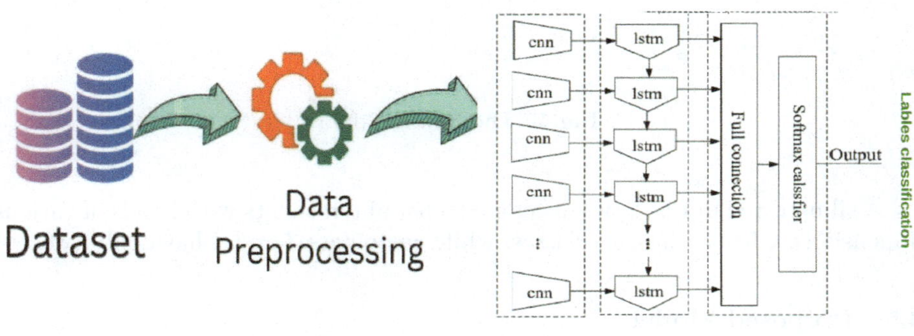

Fig. 3. Centralized Training Architecture CNN-LSTM

Federated Learning Model. In the federated learning process known as FedCNN-LSTM, the clients work together to build a standard CNN-LSTM model without exchanging their own data. Each client starts by initializing a local CNN-LSTM model with identical architecture. Every client trains its local model using its data and transmits the model updates to a central server during each iteration. To generate a fresh global model, the server consolidates the updates by using a federated averaging technique, known as FedAvg [22]. Each client receives the international model, allowing them to update their local models based on it. We iterate this technique for numerous rounds until the model reaches convergence (Fig. 4).

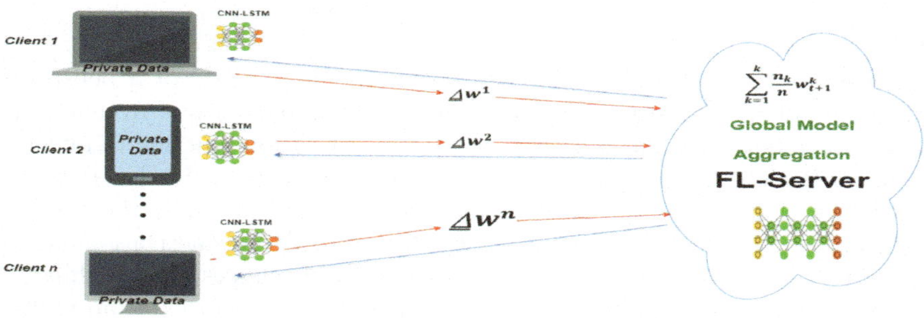

Fig. 4. Federated Learning Architecture

Federated Split Learning Architecture. In FedSliptCNN-LSTM's federated split learning process, the client In the federated split learning setting, we divide the model into two components: a client-side model and a server-side model (Fig. 5).

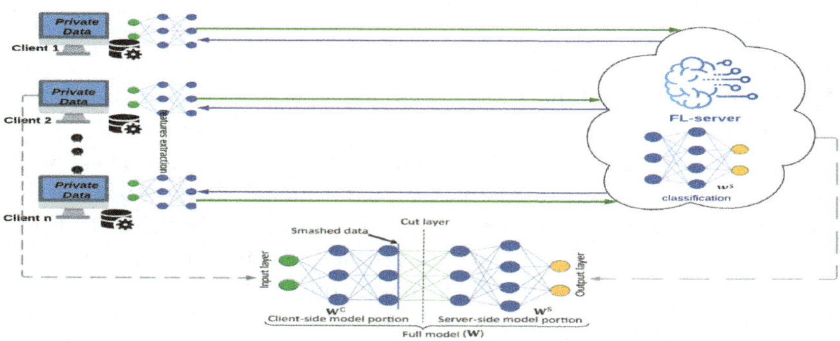

Fig. 5. Federated Split Learning Architecture

Client-Side CNN-LSTM. The client-side CNN-LSTM model consists of a convolutional neural network (CNN) and a long short-term memory (LSTM) layer. The CNN component comprises two convolutional layers, each with a kernel size of 1, 10, and 20 filters, respectively. A ReLU activation function and a max pooling layer with a kernel size of 2 follow each convolutional layer. The output of the CNN component is then flattened and provided into an LSTM layer with an input size of 20, a hidden size of hidden1 size, and a single layer. The LSTM layer is responsible for modelling the temporal dependencies in the input data, and its last hidden state is output, which is the final representation of the input data.

Server-Side CNN-LSTM. The server-side model consists of a dense layer with a size of hidden2size that takes the concatenated output from multiple client-side models as input, followed by a ReLU activation function to introduce non-linearity. The output is then provided into another dense layer with an output size that makes the final prediction or classification. If the task is a classification problem, the output is passed through a softmax activation function to produce a probability distribution over the possible classes. This architecture enables the server-side model to aggregate the knowledge learned from multiple clients and make accurate predictions or classifications.

Model Update Mechanism in Split Learning The model update mechanism in split learning is a crucial component that allows client-side and server-side models to learn from each other and improve performance. It updates model parameters on both models using gradients computed from the loss function. The process includes a forward pass, a loss computation, a backward pass, and the update of parameters using optimization algorithms, according to Adam. The client and server share the model parameters until they reach convergence or a stopping criterion, which offers benefits like improved privacy, reduced communication overhead, and faster training.

4 Experiment and Results

In this report, we share the results of our model's assessment. These were the hyperparameters of the centralized CNN-LSTM model that we used. It had 30 epochs, the Adam optimization function, and a learning rate 0.001. In FedCNN-LSTM and FedSliptCNN-LSTM, we established a framework consisting of 10 clients and performed 30 rounds. Each round consisted of 15 epochs. We aggregated the model using the FedAVG method.

4.1 The Outcomes of the Proposed Model

Initially, we present the results of the centralized CNN, LSTM, and CNN-LSTM models, which were trained using a sparse categorical cross-entropy loss function, an Adam optimizer, and a learning rate 0.001 over 30 epochs. The performance metrics for each model are summarized in Table 1 and Fig. 6.

Table 1. The centralized model performance

	Accuracy	Recull	Precision	f1-score
CNN	0.99	0.95	0.86	0.98
LSTM	0.99	0.993	0.87	0.97
CNN-LSTM	0.99	0.92	0.92	0.97

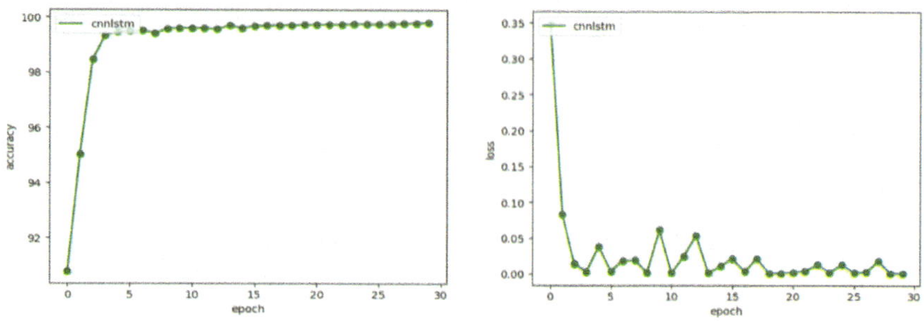

Fig. 6. Centralized learning's results

The CNN-LSTM model is considered a more suitable choice for our proposed system due to its superior predictive capabilities and better accuracy compared to the CNN model. The hybrid CNN-LSTM architecture, which has both convolutional and recurrent neural network parts, is good at finding patterns in data that spans space and time, which makes it easier to make predictions. Despite the CNN model's similar accuracy, the CNN-LSTM model shows slightly better precision and f1 score compared to the CNN model. The hybrid architecture also offers more flexibility and adaptability to future data or problem domain changes, facilitating easier integration of new features or modifications. Thus, the CNN-LSTM model is considered a more promising choice for our system.

We present the results of the FedCNN-LSTM model's accuracy on both IID (independent and identically distributed) and non-IID (non-independent and identically distributed) data distributions. We also evaluate the performance of the FedSplitCNN-LSTM model on non-IID data distribution. The accuracy plot, shown in Fig. 7, illustrates the model's performance. Furthermore, we provide additional insights into the model's performance on non-IID data through the ROC (Receiver Operating Characteristic) curve and precision-recall curve, as depicted in Fig. 8.

We compare the centralized, federated, and federated split learning models (Fig. 9), including their accuracy, F1 score, recall, and precision. This comparison provides a comprehensive view of the performance of each model, allowing for a more informed decision on which approach to use based on the specific requirements (Fig. 10).

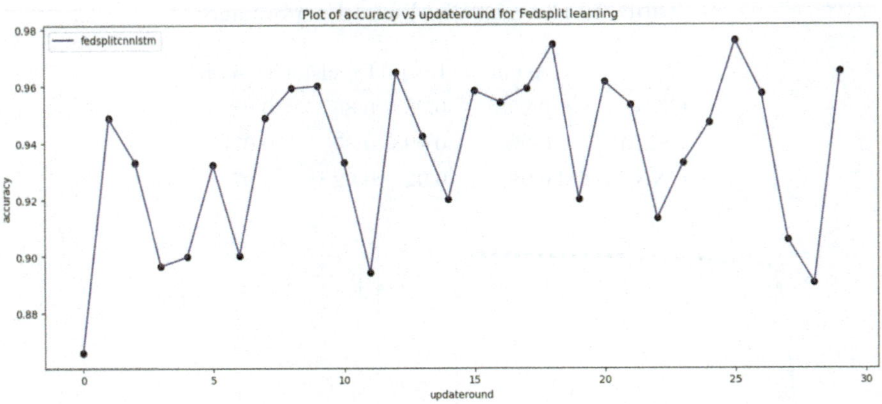

Fig. 7. Federated split learning's model results.

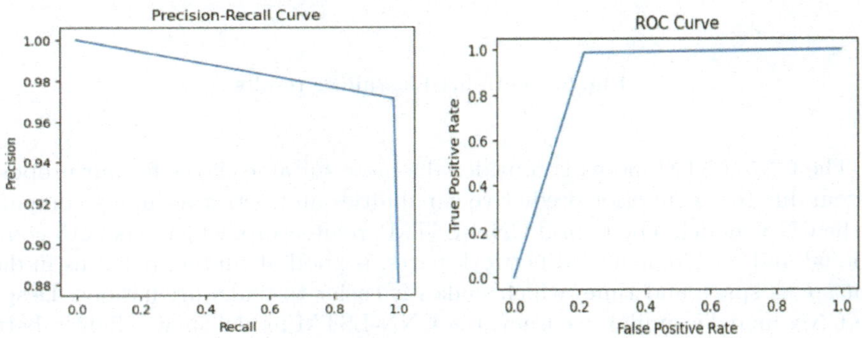

Fig. 8. Performance of federated split learning's model

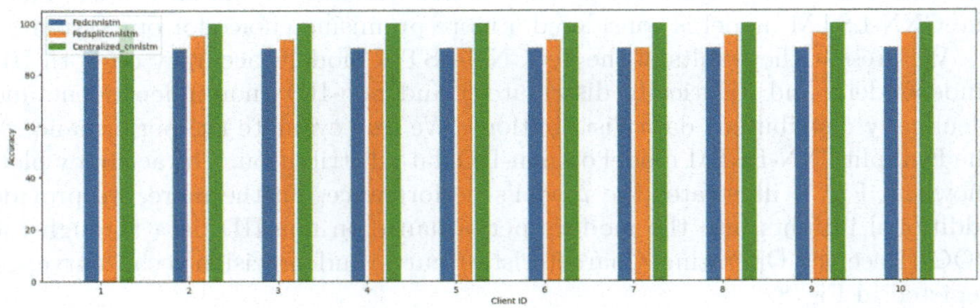

Fig. 9. Comparison between centralized, FedCNN-LSTM and FedSplitCNN-LSTM accuracy results on Non-IID Distributions.

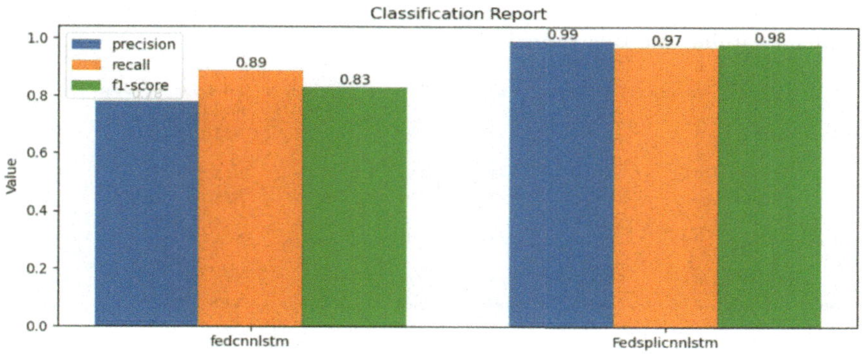

Fig. 10. FedCNN-LSTM model vs. FedSplitCNN-LSTM in Non-IID

Our results show that the centralized model achieves the best performance but at the cost of requiring ample storage and high computational power and compromising privacy and security by collecting and processing sensitive customer data. On the other hand, the FedCNN-LSTM model sacrifices some performance to preserve data privacy and security, but it still requires significant computational power and communication overhead. The FedSplitCNN-LSTM model, on the other hand, finds a good balance between performance and privacy. It offers a better solution that works with non-IID data distributions and needs less computing power from clients and communication overhead. By keeping individual customer data confidential during training and only communicating model updates, FedSplitCNN-LSTM ensures the privacy and security of sensitive data.

4.2 Performance Evaluation and Comparison

We assessed the efficacy of established techniques by using the CICIDS 2017 dataset. Specifically, we examined two federated learning strategies in a Non-IID scenario, Fed-ANIDS (autoencoder 'AE') [23]and SFedsi [24], and centralized models, such as DBN [25], MLP [26], and CNN-LSTM [27]. The evaluation included the target number of classes, features, accuracy, precision, recall, and f1-score. We also considered security issues with both the data and the model. As shown in Table 2, federated learning maintained the privacy of the data, while federated split learning maintained the model's privacy.

5 Discussion

This research conducted a performance comparison of centralized, federated learning, and federated split learning models in a non-IID environment. The findings demonstrate that the FedSplitCNN-LSTM model surpasses the federated learning models accuracy, precision, recall, and F1-score.

Table 2. Contribution comparison of the proposed Federated Split Learning against Federated Learning and Centralized Models

	Approach	Accuracy	Precision	Recall	F1-score	Privacy of data	Privacy of model	Classes	Features
Centralized Model	DBN [17]	N\A	0.88	0.99	0.94	N/A	N/A	6	25
	MLP [18]	0.99	0.99	0.99	0,99	N/A	N/A	7	60
	CNN-LSTM [19]	0.99	0.99	0.99	0.99	N/A	N/A	5	77
Federated Model	AE [23]	0.93	N/A	N/A	0.92	YES	N/A	2	N/A
	CFedSI [25]	0.83	N/A	N/A	N/A	YES	N/A	5	256
Our Study	**FedSplit CNN-LSTM**	**0.96**	**0.99**	**0.97**	**0.98**	**YES**	**YES**	**7**	**25**

The centralized models, such as DBN, MLP, and CNN-LSTM, demonstrated excellent accuracy and F1-scores. However, these models need a centralized dataset, which may not be practical in real-world situations where data is distributed and confidential. On the other hand, the federated learning models, such as FedAvg, could preserve data privacy, albeit their performance was subpar compared to the centralized models.

The FedSplitCNN-LSTM, a federated split learning model, demonstrated superior accuracy and F1-score compared to other models while preserving data privacy and model privacy. This indicates that federated split learning is an up-and-coming method for decentralized machine learning, mainly when the data is non-independent and identically distributed (Non-IID).

The findings also underscore the significance of taking data and model privacy into account in decentralized machine learning. The centralized models could not maintain data privacy, whereas the federated and split learning models could. Furthermore, the federated split learning model could preserve model privacy, a critical factor in situations where the intellectual property of the model must be safeguarded.

6 Conclusions

This study introduces a Federal Intrusion Detection System (FedIDS) that effectively identifies, categorizes, and reacts to network intrusions while reducing the occurrence of false positive results in non-independent and Identically Distributed (non-IID) data distributions. The system utilizes a CNN-LSTM model trained using the CICIDS2017 dataset. It incorporates FedSplit Learning as an alternative federated learning fashion to meet the utility privacy tradeoff.

The suggested FIDS system has been extensively tested and has shown a high level of accuracy in detecting intrusions while also having a low rate of false positives. This makes it an up-and-coming option for detecting unauthorized access in IoT networks. FedSplit Learning integration mitigates the susceptibility to reverse engineering assaults on the student model and offers an extra layer of protection against such attacks, hence bolstering privacy.

Overall, FIDS is a robust and privacy-conscious solution for detecting intrusions in IoT networks. The tool's capacity to precisely identify and react to network incursions while upholding privacy and security makes it a significant asset for safeguarding sensitive information in intricate and ever-changing network settings. Subsequent research might prioritize enhancing the system's efficiency and capacity to handle larger workloads while investigating its possible implementations in other fields.

As future work, we plan to extend FIDS in other IoT applications, such as industrial control systems, smart homes, or autonomous cars, to evaluate its suitability and efficiency in various settings. Another line of research work is to investigate the impact of different federated learning algorithms, such as FedProx or FedMOON, on the performance and privacy of FIDS. In addition, integrating blockchain technology into decentralized, federated learning might improve the levels of transparency and privacy in FIDS, providing a robust and secure solution for IoT networks.

References

1. Agrawal, S., et al.: Federated learning for intrusion detection system: concepts, challenges and future directions (n.d.)
2. Zheng, J., Chen, Y., Lai, Q.: PPSFL: privacy-preserving split federated learning for heterogeneous data in edge-based internet of things (2024)
3. Krsteski, S., Tashkovska, M., Sazdov, B., Radojichikj, L., Cholakoska, A., Efnusheva, D.: Intrusion detection with supervised and unsupervised learning using PyCaret over CICIDS 2017 dataset. In: Computer Science On-line Conference, pp. 125–132. Springer (2023)
4. Guezzaz, A., Benkirane, S., Azrour, M., Khurram, S.: A reliable network intrusion detection approach using decision tree with enhanced data quality, pp. 1–15 (2021)
5. Saba, T., Rehman, A., Sadad, T., Kolivand, H., Bahaj, S.A.: Anomaly based intrusion detection system for IoT networks through deep learning model. Comput. Electr. Eng. **99**, 107810 (2022)
6. Nazir, A., et al.: A deep learning-based novel hybrid CNN-LSTM architecture for efficient detection of threats in the IoT ecosystem (2024)
7. Albouni, A., Gunawan, T.S., Habaebi, M.H., Halbouni, M., Kartiwi, M., Ahmad, R.: CNN-LSTM: hybrid deep neural network for network intrusion detection system. IEEE Access **10**, 99837–99849 (2022). https://doi.org/10.1109/ACCESS.2022.3202354
8. Zhao, F., Zhao, J., Niu, X., Luo, S., Xin, Y.: A filter feature selection algorithm based on mutual information for intrusion detection. Appl. Sci. **8**(9), 1535 (2018). https://doi.org/10.3390/app8091535
9. McMahan, H.B., Moore, E., Ramage, D., Hampson, S., Aguera y Arcas, B.: Communication-efficient learning of deep networks from decentralized data. Google, Inc., Seattle (2017)
10. Deng, J., Guo, R., Jin, Z.: An intrusion detection scheme based on federated learning and self-attention fusion convolutional neural network for IoT (2023)
11. Saadat, H., Aboumadi, H., Mohamed, A., Erbad, A., Guizani, M.: Hierarchical federated learning for collaborative IDS in IoT applications. In: 2021 10th Mediterranean Conference on Embedded Computing, MECO, pp. 1–6. IEEE (2021)

12. Belarbi, O., Spyridopoulos, T., Anthi, E., Khan, A., Authors, A.: Federated intrusion detection system based on deep belief networks. [License: CC BY 4.0] (2023)
13. Jin, Z., Zhou, J., Li, B., Wu, X., Duan, C.: FL-IIDS: a novel federated learning-based incremental intrusion detection system (2023)
14. Javeed, D., Saeed, M.S., Adil, M., Kumar, P., Jolfaei, A.: A federated learning-based zero trust intrusion detection system for internet of things (2024)
15. Chen, Z., Lv, N., Liu, P., Fang, Y., Chen, K., Pan, W.: Intrusion detection for wireless edge networks based on federated learning. IEEE Access **8**, 217463–217472 (2020)
16. Alazab, A., Khraisat, A., Singh, S., Jan, T.: Enhancing privacy-preserving intrusion detection through federated learning. Electronics **12**(16), 3382 (2023). https://doi.org/10.3390/electronics12163382
17. Qin, T., Cheng, G., Wei, Y., Yao, Z.: Hier-SFL: client-edge-cloud collaborative traffic classification framework based on hierarchical federated split learning (2023)
18. Panigrahi, R., Borah, S.: A detailed analysis of CICIDS2017 dataset for designing intrusion detection systems. Department of Computer Applications, SMIT, Sikkim Manipal University, Majitar, India (2020)
19. Liu, Y., Kumar, N., Xiong, Z., Lim, W.Y.B., Kang, J., Niyato, D.: Communication efficient federated learning for anomaly detection in industrial internet of things. In: GLOBECOM 2020-2020 IEEE Global Communications Conference, pp. 1–6. IEEE (2020)
20. Arafeh, M., Hammoud, A., Otrok, H., Mourad, A., Talhi, C., Dziong, Z.: Independent and identically distributed (IID) data assessment in federated learning. In 2021 IEEE 2nd International Conference on Artificial Intelligence and Computational Intelligence (ICAICI), pp. 161–166. IEEE (2021). https://doi.org/10.1109/ICAICI52102.2021.9417543
21. Morafah, M., Vahidian, S., Chen, C., Shah, M., Lin, B.: Rethinking data heterogeneity in federated learning: introducing a new notion and standard benchmarks. In: FL-NeurIPS 2022 Workshop, 36th Conference on Neural Information Processing Systems (2022)
22. Zhou, Y., Qing, Y., Lv, J.: Communication-efficient federated learning with compensated overlap-FedAvg. arXiv preprint arXiv:2012.06706 (2020)
23. Drissi, M.J., et al.: Fed-ANIDS: federated learning for anomaly-based network intrusion detection systems (2023)
24. Zhang, Z., Zhang, Y., Guo, D., Zhao, S., Zhu, X.: Communication-efficient federated continual learning for distributed learning systems with Non-IID data (2022)
25. Belarbi, O., Khan, A., Carnelli, P., Spyridopoulos, T.: An intrusion detection system based on deep belief networks. Toshiba Europe Ltd., Bristol Research & Innovation Laboratory, Bristol, UK, and Cardiff University, School of Computer Science & Informatics, Cardiff, UK (2022)
26. Yin, Y., Jang-Jaccard, J., Sabrina, F., Kwak, J.: Improving multilayer-perceptron (MLP)-based network anomaly detection with Birch clustering on CICIDS-2017 dataset (2022)
27. Al-zubidi, A.F., Farhan, A.K., Towfek, S.M.: Predicting DoS and DDoS attacks in network security scenarios using a hybrid deep learning model. J. Intell. Syst. (2023). https://doi.org/10.1515/jisys-2023-0195

Unveiling Human Activity Patterns in Smart Cities Through a CNN-LSTM Approach

Imane Ameur[1](✉), Mohamed El Amine Ameur[1], Mazene Ameur[1], and Houssem Eddine Dagha[2]

[1] Laghouat University, 37G Ghardaia Road, 03000 Laghouat, Algeria
i.ameur@lagh-univ.dz
[2] Ghardaia University, 455 Ghardaia, 47000 Ghardaia, Algeria

Abstract. Over the past two decades, significant technological advancements have taken place, giving rise to novel applications of intelligent digital sensors. The proliferation of Internet of Things (IoT) services in various settings such as smart homes, buildings, cities, factories, and other intelligent environments has brought substantial benefits to individuals, industries, and public institutions. A typical smart home has undergone enhancements and is equipped with an array of sensors and actuators to provide services to its residents. One of the key aspects in many smart home applications is the identification of residents' daily routines. Automating the Human Activity Recognition system based on human behavior patterns is a complex task due to the intricate nature of human life within a home environment, where multiple inhabitants and residents coexist. To address this complexity, various deep learning algorithms, known for their effectiveness in diverse domains, have been explored to enhance the accuracy of unveiling the human activity pattern. This research aims to investigate and develop a Human Activity Recognition system for smart cities through the implementation of a hybrid deep learning approach known as CNN-LSTM.

Keywords: Smart cities · Smart home · HAR · Deep learning

1 Introduction

In the contemporary landscape, various technologies have introduced activity recognition to assist individuals in monitoring their daily movements. The primary challenge is to devise an algorithm that can label the activities carried out by residents based on sensor data collected in intelligent environments. Recently, there has been a surge of interest in human activity identification and discovery, driven by their substantial potential in context-aware computing systems, particularly within smart home environments. Researchers have frequently employed machine-learning techniques to address this challenge [1, 2]. In contrast, deep learning models consistently deliver superior accuracy [3]. The primary objective of employing smart devices in this study is to gather user data and apply Deep Neural Network algorithms for data classification. There exist numerous types of bodily activities, but our primary focus lies on classifying mobile activities, such as walking, sitting, and standing. We categorize human activities based on these specific types.

S. Chikhi et al. (Eds.): MISC 2024, LNNS 1267, pp. 43–52, 2025.
https://doi.org/10.1007/978-3-031-82112-7_4

To address the challenge of human activity recognition, we have developed and put into practice a hybrid classification model. This model incorporates both Convolutional Neural Networks (CNN) and Long-Short Term Memory Networks (LSTM). We tested the performance of our activity identification model using a database sourced from existing literature. The key findings of this paper can be summarized as follows:

- First, the exploration of the problem of Recognition of Human Activity Patterns in cities,
- Then, the development of a Human Activity Recognition system for smart cities through the deep learning approach.
- Finally, the implementation of a hybrid deep learning approach known as CNN-LSTM and comparing it with the existing approaches.

This paper is structured into five sections, each serving a distinct purpose within the study. The second section synthesizes existing literature on the Recognition of Human Activity Patterns problem. Furthermore, the third section outlines the research design and methodology employed for system development. Accordingly, the fourth section presents the evaluation process and the obtained results. Finally, this paper concludes with a general summary of findings and outlines potential avenues for future research.

2 Recognition of Human Activity Patterns Based on Deep Learning: Related Work

Currently, Deep Learning (DL) has seen increasing success in modeling high-level abstractions from complex raw data across various domains. DL models possess the ability to autonomously extract high-level features from raw signals without human intervention, a concept known as "end-to-end learning" [4].

Furthermore, in recent years, numerous studies have been conducted and published on the application of deep learning methods to HAR using sensor-based systems. It is important to note that our research specifically focuses on methods based on sensors. When it comes to time series classification in the context of HAR, two primary neural network techniques have proven to be effective: Convolutional Neural Network (CNN) and Recurrent Neural Network (RNN) models. In fact, this section delves into an exploration of notable research endeavors that introduce HAR models constructed through the application of DL methodologies, which include CNN, LSTM, and ultimately hybrid models.

2.1 CNN

In recent years, with the increasing computational capabilities, CNN have demonstrated remarkable performance in sensor-based HAR and have outperformed other state-of-the-art methods that necessitate advanced preprocessing or labor-intensive manual feature extraction [25]. For instance, in [5], one of the early works employing CNN for HAR, a simple CNN model was developed for accelerometer data. This model extracted scale-invariant properties and local dependencies from the acceleration time series, achieving classification accuracies of 88.19% for Skoda, 76.83% for Opportunity, and 96.88%

for WISDM datasets, respectively. However, it utilized a shallow model with just one accelerometer. [6] introduced a multi-sensor identification framework that incorporated a CNN model for dual accelerometers. The architecture proposed in [7] was based on temporal convolutions of inputs, specifically designed for inertial signals, with the aim of enabling real-time and accurate classification for low-power wearable devices.

CNN have gained widespread popularity and have been highly successful in image recognition tasks [8]. In the context of HAR, two distinct variations of CNN are commonly employed: 2D CNN for image processing and 1D CNN for sequence processing [4]. In the case of 2D CNN, studies like [9] and [10] have introduced an approach that transforms raw sensor data into a 2D image signal, subsequently employing a two-layer CNN to classify this signal image for activity identification. Their experiments demonstrate the adaptability of 2D-based structures for HAR. On the other hand, for the 1D CNN variety, [11] and [12] have favored a 1D-based architecture due to its superior feature extraction capabilities. Among these, [13] stands out as a notable work. The effectiveness of CNN in HAR stems from their capacity to capture discriminative and robust features. Given that the collected data is inherently time-dependent, it becomes essential to incorporate time information into the input data. To address this challenge, LSTM has been introduced [14].

2.2 LSTM

It is important to highlight that HAR within smart home environments involves the challenge of pattern identification in time series data with irregular sampling. In this context, RNN have demonstrated a strong capability to represent time series data or sequential multi-dimensional data. However, as mentioned in [15], RNNs encounter difficulties with long-term dependencies. To mitigate this issue, a variation known as LSTM has been introduced [24]. LSTMs are designed to address the long-term dependency problem encountered by traditional RNNs, making them more suitable for tasks like HAR in smart home environments.

In [16], a stacked LSTM network was proposed for the detection of six human behaviors using smartphone sensor data. This network consists of five LSTM cells that are trained end-to-end on the sensor data. Prior to the stacked LSTM network, a single-layer neural network is employed to preprocess the data. Remarkably, without the need for manual feature engineering, this proposed network achieves a significant improvement in average accuracy, enhancing it by 93%.

In a similar vein, BiLSTM techniques have gained popularity in HAR due to their success in feature extraction and prediction. The proposed BiLSTM model is built upon the foundation of residual blocks and bidirectional Long Short-Term Memory (BiLSTM). This model operates in three main steps: First, it autonomously extracts spatial attributes from multidimensional sensor data using the residual block. Second, it employs BiLSTM to capture bidirectional temporal feature relationships. Finally, the collected features are passed through a Softmax layer for the identification of human actions. Remarkably, this HAR approach achieves high accuracy rates of 96.95%, 97.32%, and 97.15% on the handmade, WISDM, and PAMAP2 datasets, respectively. Consequently, it offers superior performance with fewer parameters compared to some existing models [17].

Alawneh et al. conducted a comparative study on HAR accuracy between two types of LSTM models: unidirectional LSTM and bidirectional BiLSTM. They evaluated these two LSTM models on two distinct datasets, each featuring different classes of movements. The first dataset, UniMiB SHAR, encompasses 17 various activities and fall states, while the second dataset, WISDM, comprises six types of human motions recorded under controlled laboratory conditions. The results of their study indicate that BiLSTM outperforms LSTM in terms of identification accuracy. However, it is worth noting that the BiLSTM technique requires more time for training, which could limit its applicability to large datasets [18].

Nonetheless, Sequential data analysis systems are currently limited to handling primitive and simple actions, and they struggle with complex activities. Basic activities like walking, running, or turning on a light involve single actions or movements. In contrast, complex activities such as baking or typing consist of a series of actions that may include various interactions with objects, equipment, or other individuals [4]. According to existing literature, the only extension for handling complex actions in HAR using deep learning is the use of hierarchical LSTM (LTSM), incorporating two layers to address the intricacies of actions in video-based HAR [19].

2.3 Hybrid Models

Recent research in HAR has placed significant emphasis on hybrid models that blend CNN and RNN components. For instance, [20] developed a model that utilizes CNN with variations in kernel dimensions along with Bidirectional LSTM (BiLSTM) to capture features at multiple resolutions. This technique achieved a higher accuracy of 98.53% on the WISDM dataset when compared to the UCI dataset (97.05%).

In the same line, Nan et al. conducted research on a multichannel-based HAR system for elderly individuals using smartphones. Their study evaluated various models including 1D CNN, CNN-LSTM, multichannel CNN, and multichannel CNN-LSTM. The findings demonstrated that the CNN-LSTM with a multichannel model was the most effective among these configurations [21]. Another example of hybrid models like these is found in the work of [22], where they introduce a DeepConvLSTM model that combines convolutional layers with Recurrent Neural Networks (RNN). Their results show that their technique outperforms competing non-recurrent deep networks in HAR applications.

In the quest to recognize complex behaviors, a study investigated four hybrid deep learning models that combine CNN and RNN components, including LSTMs, GRUs, BiLSTMs, and BiGRUs [15]. Experiments conducted on the UTwente dataset revealed that the CNN-BiGRU model outperformed many other models in terms of identification accuracy. Additionally, another study [23] introduced a method designed to identify transitions from one activity to another. This method employs a deep CNN to extract features, followed by the utilization of GRU to capture long-term dependencies between different actions, enhancing the identification of complex activities. Consequently, the CNNGRU model achieves a detection accuracy of 96.79% [23].

In summary, prior research has demonstrated that deep learning algorithms, particularly CNNs, excel at feature extraction and exhibit rapid training phases, achieving accuracy levels comparable to LSTM models [12]. However, LSTM outperforms CNNs thanks to its capacity to capture long-term dependencies. Therefore, based on the existing literature, the hybrid combination of 1D CNN and LSTM structures is considered the most effective approach [4].

3 System Design

In this research, we present a Smart city Human Activity Recognition model utilizing smartphone-integrated sensors. This model represents a fusion of CNN and LSTM networks.

3.1 General System Architecture

Our aim is to develop an intelligent system with the capability to forecast and identify everyday human activities within smart surroundings, encompassing smart cities, structures, and residences.

Figure 1 depicts the structural design of HAR in a smart city setting, leveraging deep learning. Data is sourced from a variety of IoT sensors and intelligent devices. Utilizing deep learning, the dataset is processed to categorize unlabeled activity data in accordance with a HAR training model. This section offers an overall perspective on the components of the HAR framework, followed by an in-depth explanation of each element.

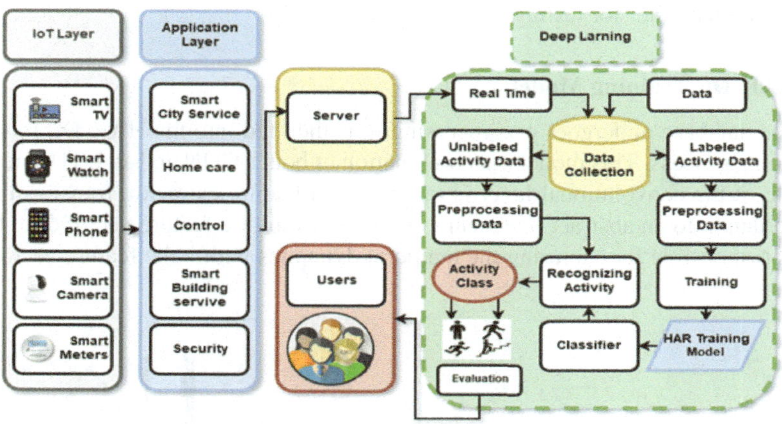

Fig. 1. Design Architecture of Deep learning for HAR in Smart cities

As indicated in Fig. 2, the framework we have developed consists of four key stages: data collection, preprocessing, the HAR-DL Training Model, and classification.

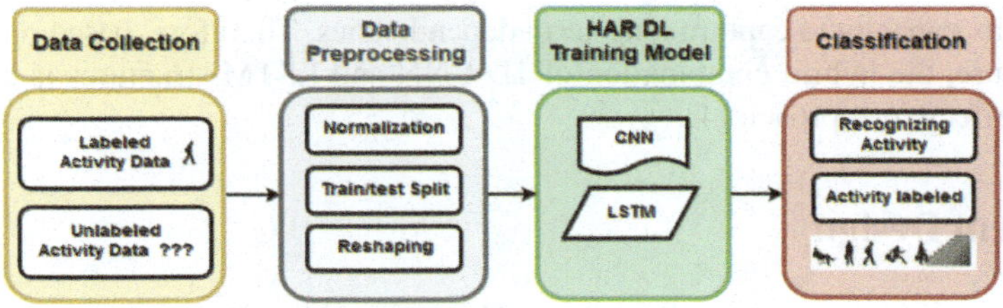

Fig. 2. System Stages Framework

3.2 Data Collection: WISDM Dataset Description

To evaluate the effectiveness of our system, we conducted experiments involving six distinct categories of everyday activities: sitting, standing, walking, going upstairs, going downstairs, and jogging. For this research, we opted to use activity-specific smart city HAR data from well-known activity datasets, notably the ACTi Tracker dataset, also referred to as the 'WISDM dataset' [26].

3.3 Data Preprocessing

In smart cities, the gathered data consists of both labeled and unlabeled data. Our aim is to develop a HAR model using this collected data to predict and identify activity labels that are not initially provided. To accomplish this objective, we begin by normalizing the acquired data and subsequently divide the normalized data into two sets: one for training and the other for testing.

3.4 HAR-DL Training Model

Once the database undergoes preprocessing, it is then fed into the HAR-DL model for the training process. This model is a combination of both a CNN and an LSTM network. The role of the convolutional layer is to serve as a feature extractor, transforming the original data into an abstract representation in the form of a feature vector. The LSTM layer is responsible for capturing the temporal dynamics within the feature vector.

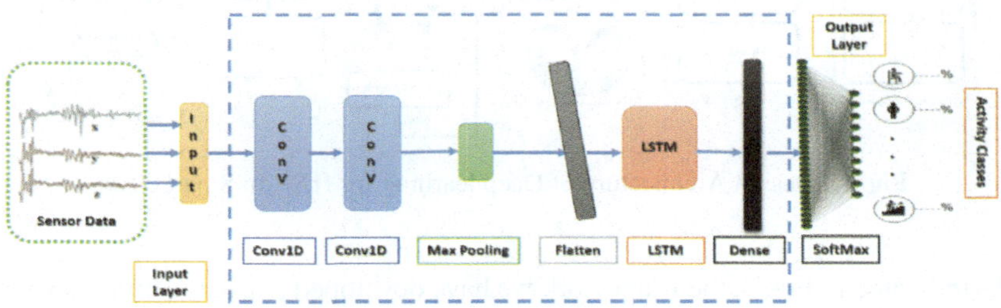

Fig. 3. HAR-DL Training Model

As depicted in Fig. 3, our model comprises two convolutional layers, one LSTM layer, and two dense layers.

3.5 Classification

The classification phase encompasses activity identification, which involves categorizing activities into specific activity classes. Activity recognition is carried out using the test data, following the completion of the HAR-DL training phase and model saving.

4 Implementation and Evaluation

4.1 Implementation

The CNN-LSTM architecture combines two components: firstly, CNN layers to extract pertinent features from raw data, and secondly, LSTM for sequence learning to detect and recognize human activities. In this setup, CNN analyzes the input data in blocks to extract features, while LSTM interprets the time-dependent features derived from each block.

4.2 Evaluation

To assess the effectiveness of our CNN-LSTM hybrid model, we conducted evaluations and validations using the WISDM database.

4.3 Simulation Results

Following the training of our HAR deep learning model, we proceeded to employ the test data for the purpose of recognizing and classifying activity data. Subsequently, we conducted a comparison of the results obtained with the actual data, using performance metrics such as accuracy and loss.

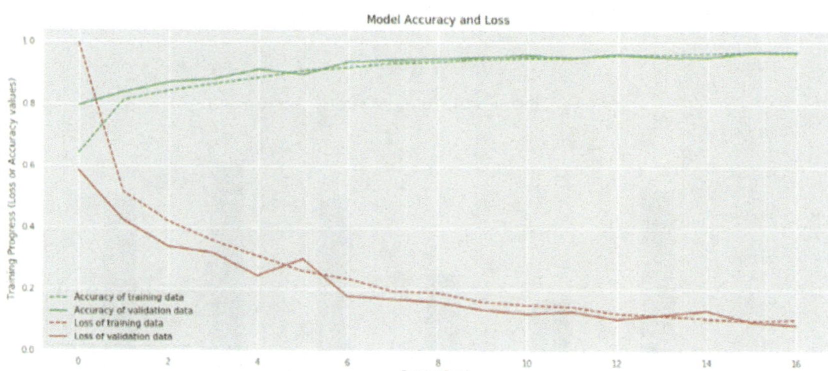

Fig. 4. Visual Accuracy and Loss Results

In Fig. 4, the training loss rate is represented as a dashed red line, while the validation loss rate is shown as a solid red line. The accuracy of the training data is depicted as a dashed green line, and the validation accuracy rate is presented as a solid green line. Notably, the training loss rate tends to be higher than the validation loss rate, while validation accuracy closely mirrors training accuracy. As the number of iterations increases, both training and validation accuracy rates progressively converge towards unity, accompanied by a gradual reduction in the loss rate. This indicates a gradual improvement in activity identification accuracy.

4.4 Comparison Between the Hybrid Model and Other Models

Likewise, we conducted comparisons between the hybrid model and several reference models from the literature, all using the same WISDM dataset.

Table 1. Comparison with other reference models.

Models		Accuracy
(Li and Wan, 2022)	**BiLSTM** [17]	**97.32%**
(Che et al., 2016)	**LSTM** [24]	92.1%
(Ign, 2018)	**CNN** [25]	93.32%
(Zen et al., 2014)	**CNN-Sharing** [5]	96.88%
(Naf et al., 2021)	**CNN-BiLSTM** [20]	**98.53%**
Hybrid	**CNN-LSTM**	**97%**

The accuracy results obtained by the hybrid models, CNN-LSTM and CNN-BiLSTM, outperformed the other models, as indicated in Table 1 and Fig. 5. Currently, the CNN-LSTM hybrid model achieved an accuracy of 97%, while the CNN and LSTM models achieved lower accuracies of 93.32% and 92.1%, respectively.

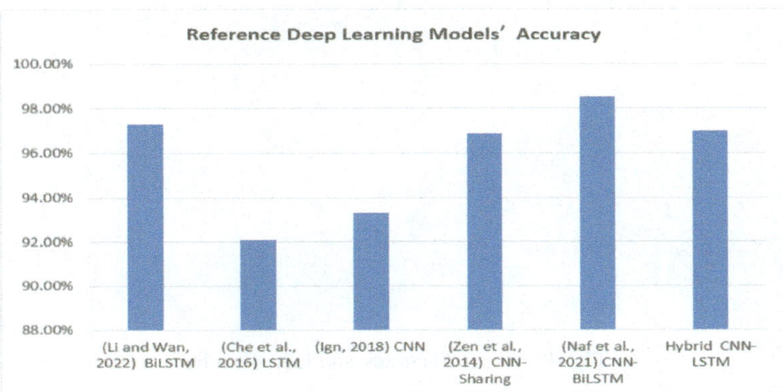

Fig. 5. Accuracy and Loss for Reference DL Models

It is worth noting that when these models incorporate internal enhancements, such as the CNN model with partial weight sharing, accuracy improves from 93.32% in CNN to 96.88% in CNN-Sharing. Similarly, the LSTM model with bidirectional internal processing sees its accuracy rise from 92.1% in LSTM to 97.32% in BiLSTM. Consequently, when these improved models are combined into the enhanced hybrid model, CNN-BiLSTM, they achieve a remarkable accuracy result of 98.53%, surpassing our basic hybrid model's accuracy of 97%. This highlights that the hybrid models significantly enhance activity identification and prediction results.

5 Conclusion

Deep Learning for Human Activity Patterns Identification and Recognition, holds a pivotal role in the daily lives of individuals. This paper delves into a study of recent research on HAR within smart cities and the methodologies employed to achieve or enhance results in this domain. Following the synthesis of related work in the second section, we have identified several key aspects that facilitate the development of an efficient and effective identification model for HAR. Our approach involves designing a hybrid activity recognition model in smart cities, which combines CNN with one of the most widely employed recurrent neural network techniques, LSTM. CNN is chosen for its ability to extract features from input data, while LSTM is preferred for detecting and recognizing activities with inherent temporal sequences. As for perspectives on this work: First, it is of interest to generalize our system beyond the confines of a smart home environment to encompass human activity identification across various datasets within smart cities. Additionally, exploring the potential of Autoencoder and transformer methods in smart cities for the recognition of human activity patterns holds promise. Finally, the integration of context-aware ontologies with hybrid methods could lead to more efficient human activity patterns identification.

References

1. Shahid, Z.: Distributed Machine Learning for Anomalous Human Activity Recognition using IoT Systems. Lulea University of Technology, Luleå (2021)
2. Sarker, I.H.: Deep learning: a comprehensive overview on techniques, taxonomy, applica–tions and research directions. SN Computer Science **2**(6), 1–2 (2021)
3. Fereidoonian, F., Firouzi, F., Farahani, B.: Human activity recognition: From sensors to applications. In: International CONFERENCE, Omni-layer Intelligent Systems (COINS), pp. 1–8. IEEE (2020)
4. Bouchabou, D., et al.: A survey of human activity recognition in smart homes based on IoT sensors algorithms: taxonomies, challenges, and opportunities with deep learning. Sensors **21**(18), 6037 (2021)
5. Zeng, M., et al.: Convolutional neural networks for human activity recognition using mobile sensors. In: International CONFERENCE, mobile computing, applications and services, pp. 6–8 (2014)
6. Hessen, H.O., Author, S., Tessem, A.J.: Human activity recognition with two body-worn accelerometer sensors. Dept. Comp. Eng., NTNU Univ., Trondheim, Norway (2016)

7. Ravi, D., et al.: Deep learning for human activity recognition: A resource efficient implementation on low-power devices. In: International CONFERENCE, wearable and implantable body sensor networks, pp. 71–76 (2016)
8. Mutegeki, R., Han, D.S.: A CNN-LSTM approach to human activity recognition. In: International CONFERENCE, artificial intelligence in information and communication (ICAIIC), pp. 362–366. IEEE (2020)
9. Jiang, W., Yin, Z.: Human activity recognition using wearable sensors by deep convolutional neural networks. In: 23rd Proceedings on Multimedia, pp. 1307–1310. ACM international conference, New York (2015)
10. Gochoo, M., et al.: Unobtrusive activity recognition of elderly people living alone using anonymous binary sensors and DCNN. IEEE J. Biomedical and Health Informatics **23**(2), 693–702 (2019)
11. Wang, F., et al.: Joint activity recognition and indoor localization with WiFi fingerprints. IEEE Access **7**, 80058–80068 (2019)
12. Singh, D., et al.: Human activity recognition using recurrent neural networks. In: International CONFERENCE, machine learning and knowledge extraction, pp. 267–274 (2017)
13. Cho, H., Yoon, S.M.: Divide and conquer-based 1D CNN human activity recognition using test data sharpening. Sensors **18**(4), 1055 (2018)
14. Sun, J., et al..: Sequential human activity recognition based on deep convolutional network and extreme learning machine using wearable sensors. Sensors **2018**(8580959), (2018)
15. Mekruksavanich, S., Jitpattanakul, A.: Deep convolutional neural network with RNNs for complex activity recognition using wrist-worn wearable sensor data. Electronics **10**(14), 1685 (2021)
16. Ullah, M., et al.: Stacked LSTM network for human activity recognition using smartphone data. In: 8th European Workshop on Visual Information Processing (EUVIP), pp. 175–180. IEEE, Roma (2019)
17. Li, Y., Wang, L.: Human activity recognition based on residual network and BiLSTM. Sensors **22**(2), 635 (2022)
18. Alawneh, L.: A comparison of unidirectional and bidirectional LSTM networks for human activity recognition. In: International CONFERENCE, Pervasive Computing and Communications Workshops, pp. 1–6. IEEE (2020)
19. Devanne, M., et al.: Recognition of activities of daily living via hierarchical long-short term memory networks. In: International CONFERENCE, Man and Cybernetics, pp. 3318–3324. IEEE (2019)
20. Nafea, O., et al.: Sensor-based human activity recognition with spatiotemporal deep learning. Sensors **21**(6), 2141 (2021)
21. Nan, Y., et al.: Deep learning for activity recognition in older people using a pocket-worn smartphone. Sensors **20**(24), 7195 (2020)
22. Ordóñez, F.J., Roggen. D.: Deep convolutional and LSTM recurrent neural networks for multimodal wearable activity recognition. Sensors **16**(1), 115 (2016)
23. Pushpalatha, S., Math. S.: Hybrid deep learning framework for human activity recognition. Int. Journal **13**(1), 1225–1237 (2022)
24. Chen, Y., et al.: LSTM networks for mobile human activity recognition. In: International CONFERENCE, Artificial Intelligence, pp. 50–53. Atlantis Press (2016)
25. Ignatov, A.: Real-time human activity recognition from accelerometer data using Convolutional Neural Networks. Int. Journal **62**, 915–922 (2018)
26. WISDM Homepage: https://archive.ics.uci.edu/dataset/507/wisdm+smartphone+and+sma%20rtwatch+activity+and+biometrics+dataset. Last accessed 16 December 2023

Emerging Systems and Technologies a Comparative Study

Amel Houha[1](✉) and Abdelkamel Tari[2]

[1] Laboratoire d'Informatique Médicale et des Environnements Dynamiques et intelligents (LIMED), Université de Bejaia, Bejaia 06000, Algérie
amel.houha@univ-bejaia.dz

[2] Laboratoire LITAN, École Supérieure en Sciences et Technologies de L'Informatique et du Numérique, RN 75, Amizour, Bejaia 06300, Algérie
tari@estin.dz

Abstract. Recently, in collaborative systems, several terms have been frequently used, such as "System-of-systems", "Mega-systems", "Holonic-systems" and "Autonomic systems", although, despite their frequent usage, there is a remarkable lack of consensus on their precise definitions. This work aims to clear up any confusion about the fundamental characteristics of these systems, how to distinguish them from classical systems, and how their evolution differs from other systems, even if there is less agreement on these points. We provide definitions, a simplified taxonomy, and an essential set of design problems and principles. This paper also discusses the necessity of recognizing categories in system design, as well as some of the potential issues raised by misclassification.

Keywords: System-of-systems · Holonic-Systems · Autonomic Systems · Mega-systems

1 Introduction

As the world becomes increasingly interconnected, "information technology" (IT) systems are exposed to the rapid evolution of communication technologies and processing capabilities [1]. According to "Standard for Systems and Software Engineering - recommended practice for architectural description of software-intensive systems," ISO/IEC 42010 IEEE Std 1471–2000: A single system consists of being made up of a variety of components arranged to do specific tasks [1]. However, in certain cases, more than one system must work together to accomplish certain tasks since a single system is unable to perform them. The creation of a group of interconnected systems working together to accomplish global objectives, is necessary in this circumstance [1–3].

In many different fields, such as aerospace, military, and environmental monitoring, complex systems—which are defined by complex structures and heterogeneous constituents—have become more relevant, other examples of applications are presented in [3–8]. When integrating different aspects for collective aims, these systems give

S. Chikhi et al. (Eds.): MISC 2024, LNNS 1267, pp. 53–68, 2025.
https://doi.org/10.1007/978-3-031-82112-7_5

importance to optimizing performance, robustness, and reliability. Examples like satellite constellations and air traffic control networks highlight how important coordination and interoperability are to the resilience and efficiency of systems. These systems are complex and require an interdisciplinary approach based on network science, control theory, and systems engineering [1]. Furthermore, the emergent properties arising from component interactions present both challenges and opportunities in the field of complex systems engineering, highlighting the need for continued research and innovation in this domain,

Recently, in components-based systems (CBS) field, especially the two terms "System-of-Systems" (SoS) and "Mega-systems" (MS), have gained considerable popularity, These concepts involve the integration of diverse, autonomous systems, often of vast dimensions, under specific rules of interaction to achieve overarching objectives. Even though the term SoS is the most frequently used term to describe the integration of several systems, there are other new or old related terms, Despite extensive literature on these related concepts, several fundamental concerns remain unanswered, and there is no comprehensive comparative study to date.

This work represents an initial step toward bridging the gap between the fundamental concepts of CBS by a discussion and a comparative study between relevant systems. As we go through this topic in further details, it becomes evident that these composed-systems present many research challenges and open-ended problems that require groundbreaking concepts and rigorous verifications, such as: What is the right term to describe every situation where we integrate multiple systems? Where and when did those concepts emerge? And when has their development trend reached its peak? What are the design challenges? What are the used modelling tools?

In order to provide answers to the above cited issues, the reminder of the paper is organized as follows, in the first section, it highlights the differences between traditional systems engineering and CBS components-based systems engineering. Also, it presents some in-depth definitions (out of many possible) of every system. In the second section, we provide a general overview of modelling methods, the most famous architectures, and the research-challenging topics for the cited systems. In the last section we identify gaps in CBS modelling through gap analysis and stakeholder interaction. Prioritized research themes were established based on these outcomes.

2 Related Surveys

Several recent surveys provide valuable insights into the research surrounding System-of-Systems (SoS). Notably, the works in [9–14], offer comprehensive overviews of current approaches to SoS architecture selection. These surveys provide a holistic understanding of the challenges and methodologies associated with SoS design, highlighting key research areas and suggesting directions for future studies. However, when it comes to "Mega Systems," the available literature is notably scarce. Few studies use the term "Mega Systems," with the most relevant resource being [3], which discusses related concepts and applications, but there is still a significant gap in dedicated research on this topic. For "Holonic Systems," several literature reviews and studies have been conducted, such as those in [15–20], which explore the application of holonic principles across various fields. These works address the hierarchical and modular nature of holonic systems, offering insights into their design and operational behavior. Lastly, for "Autonomic Systems," key resources include the papers in [14, 21–23], which investigate the self-managing and adaptive capabilities of autonomic systems in complex environments. Diverse research groups have used these concepts individually, with little effort put into creating useful comparison between contexts. Furthermore, there is a significant lack of integration between these ideas and robust approaches, tools, and modelling frameworks based in Systems Theory. This ambiguity highlights the need for a more systematic, comparative approach that connects these imprecisely defined concepts with formalized modelling methodologies and engineering systems principles. Such an approach would not only enhance conceptual clarity but also facilitate more effective cross-disciplinary collaboration and knowledge transfer in complex systems research and development.

3 Definitions, History and Concepts

The definitions provided here are not intended to be exhaustive; rather, they are intended to serve as a starting point for further in-depth exploration. A system is an integrated collection of components working together to achieve a specific goal, including products, people, procedures, information, methods, facilities, services, and other supporting components, to fulfil a mission or purpose [24, 25]. In traditional engineering, the whole is only understood after it has been broken down into its most basic constituent pieces. After those parts have been studied, the whole is reconstructed and explained in terms of how each part functions. This approach is methodologically reductionist, but in CBS, the system is a whole and its development, behaviour, and structure differ due to the assembly of relationships and components.. In Fig. 1. We cite upcoming studied systems with key questions about relations and differences (see Fig. 1), for further details about components and system thinking we cite [26] and Comparative tables between a system and CBS are in [27].

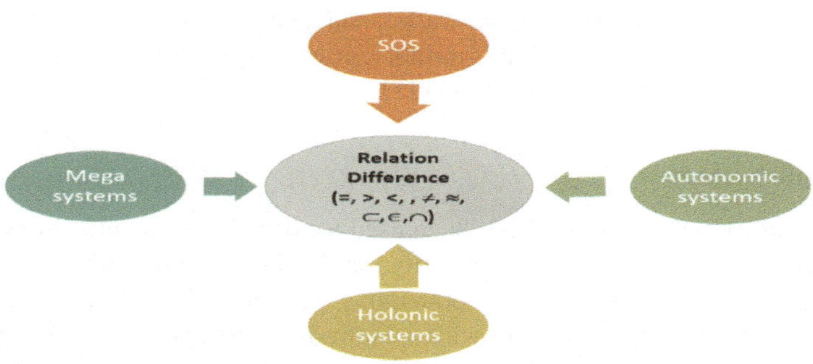

Fig. 1. Systems and key questions.

We will adopt this abbreviation for upcoming formal discussion, Mega-systems (MS), System-of-systems (SoS), holonic- systems (HS) and (AS) for autonomic systems.

3.1 System(s)-of-Systems

The concept of a System-of-Systems is the most important term to discuss, with no standard definition, despite that the SoS has been used since 1950, it is recently (after the mid-2000s) getting more attention and represents an active research area since it is used to describe complex systems made up of different, interdependent systems that work together in order to offer services that cannot be obtained separately [3]. Jaradat's histogram analysis traces its history from 1926 to 2011 for more history details refer to [28], we enumerate only three of many potential definitions:

Definition 1. Systems of systems exist when there is a presence of a majority of the following five characteristics: operational and managerial independence, geographic distribution, emergent behavior, and evolutionary development [29, 30].

Definition 2. Systems of systems are large-scale concurrent and distributed systems that are comprised of complex systems [31, 32].

Definition 3. In relation to joint war-fighting, System-of-Systems is concerned with interoperability and synergism of Command, Control, Computers, Communications, and Information, and Intelligence, monitoring, and Reconnaissance Systems [33].For further details Jamshidi [32] and Dagli [34] provided a detailed literature overview and discussion of these concepts.

3.2 Mega-Systems

Historically, the term Mega-system, originates from systems engineering, which emerged in the early 20th century and gained popularity during World War II due to complex military and defense requirements [3, 35]. MS are the large, complex systems that cross traditional boundaries to provide a level of functionality not achieved by their component elements, and to address complicated, large-scale challenges. For example MS are, large-scale projects like air traffic management, satellite constellations, and integrated industrial processes, today's military, government, and business domains, and new ones are being defined and developed [3].

3.3 Holonic Systems

The expression "holonic systems" has been used for decades in manufacturing and systems engineering, a holon represent an entity that is both a whole and a part of a larger system, combining autonomous and cooperative activities [36]. Holonic systems had initially gained popularity in industry with the introduction of Holonic industry Systems (HMS). HMS was introduced in the 1990s with the goal of increasing production flexibility and efficiency by arranging processes into holons that could run independently or collaboratively as required [16]. A distributed system, represented as a holon, is a hierarchical composition of other systems, with sub-holons below each holon. The hierarchy ends at the smallest possible systems, known as leaf holons, which are degenerate distributed systems running on individual physical nodes [15].

3.4 Autonomic Systems

The goal of Autonomic Systems were announced by IBM in 2001. Since then a research agenda beyond IBM has been formed around this idea of Autonomic Computing Systems [21]. Autonomic computing and autonomous systems promotes self-managing systems that adapt to their operating environment and user needs. It is defined by self-attributes, which include self-configuration, self-healing, self-protection, and self-optimization and self-management, noted as self.* capabilities. These attributes are not orthogonal, allowing systems to adapt to dynamic environments without manual administration. For example, a network to heal a component requires reconfiguration and resource optimization [23, 37], for further definitions we cite [21, 37–41].

Other related terms can be found in [3], for example: A federation of systems is a type of system-of-systems where components are managed by different organizations with their own methods, technologies, and schedules. It is more autonomous and widely distributed, focusing on managerial independence. A network of systems is geographically distributed components that interact through communication networks. A family of systems consists of independent systems that can be arranged or tailored to meet specific needs. An enterprise system is a social system implementing automated business processes and technical infrastructure.

4 Characteristics and Classifications of Systems

In this section we analyse briefly some of the characteristics and the most cited classifications of CBS.

4.1 System-of-Systems

SoS exist when there is a majority of the following five characteristics [13]:

Operational Independence of the Elements. If the system-of-systems is disassembled into its component systems, the components must be able to usefully operate independently. The system-of-systems is composed of systems that are independent and useful in their own right.

Managerial Independence of the Elements. The component systems not only can operate independently, they do operate independently. The component systems are separately acquired and integrated but maintain a continuing operational existence independent of the SoS.

Evolutionary Development. The system-of-systems does not appear fully formed. Its development and existence is evolutionary with functions and purposes added, removed, and modified with experience.

Emergent Behaviour. The system performs functions and carries out purposes that do not reside in any component system. These behaviours are emergent properties of the entire system-of-systems and cannot be localized to any component system. The principal purposes of the SoS are fulfilled by these behaviours.

Geographical Distribution. The geographical extent of the component systems is large. Large is a nebulous and relative concept as communications capabilities increase, but at a minimum it means that the components can readily exchange only information and not substantial quantities of mass or energy [13].

The most cited classification is proposed by Maier [42], who categorizes systems into directed systems, collaborative systems, and virtual systems.

Directed Systems-of-Systems. Are built and centrally managed to fulfil specific purposes. Maier cites an integrated air defence system as an example of this class of systems-of-systems because, while its component systems may operate independently, the entire system is usually centrally managed to defend a region against enemy air attack.

Collaborative Systems-of-Systems. Have a central management authority but lack coercive or directive power. As in the Internet, components of such systems voluntarily collaborate to fulfill agreed-upon central purposes.

Virtual Systems-of-Systems. Lack both a central management authority and, indeed, a centrally agreed-upon purpose. As in the example of the World Wide Web, large-scale behavior emerges but it does so through relatively invisible mechanisms [42].

4.2 Mega-Systems

Mega-systems are characterised by crossing traditional boundaries and formalizing interactions between inside and outside. They can be functional boundaries like military intelligence or commercial engineering. MS are formed through integration rather than monolithic creation, with components integrated at different stages in their life cycles and using different standards and design tenets. They can emerge from existing systems, be intentionally initiated as formal efforts, or be assembled in response to an immediate need or opportunity. Mega-systems can evolve in various ways, including from existing systems, intentional formal efforts, or assembled in response to an immediate need or opportunity [3], they are divided by Stevens [3] as follow:

Composed Mega-systems. Often referred to as "system-of-systems," are large-scale, complex systems formed from the integration of previously developed systems. These systems were initially developed separately to meet local requirements, with no expectation of operating as part of a larger whole. The mega-system is rarely formally structured as a single acquisition program, requiring interfaces with external systems.

Designed Mega-systems. Are also large, complex systems designed to develop new capabilities from a clean sheet. These systems can be massive undertakings, managed as single programs or families of cooperating programs. They can be managed as formal programs with a designated program manager, funding authority, and centralized systems engineering authority. Mega-systems are distinct from large monolithic systems due to their components being systems themselves.

Dynamically Assembled Mega-Systems. Are designed to meet immediate operational needs or opportunities, often incorporating existing systems used in unconventional and unanticipated ways. Development typically occurs through "glue-ware," software or gateways used to integrate systems not previously linked. These systems often consist of regular systems, commercial or government off-the-shelf components, and operational prototypes.

4.3 Holonic Systems

Key characteristics of HS are cited by [43], the holon comprises of both information and physical processing parts. The main features of HS are as follows: (1) recursivity, in which holons can be parts of larger holons; (2) autonomy in decision-making and execution; (3) cooperation among holons to achieve system goals; and (4) self-organization in response to environmental changes. These holons function inside a holarchy that establishes the bounds and parameters of cooperation the system is capable of acting in an environment, and can adapt can adapt and reorganize in response to changes in the environment by dynamic reconfiguration. According to [43–45] holonic systems can be classified based on various criteria,

Functional Classification. Manufacturing Holonic Systems: focuses on manufacturing processes and flexible automation. Robotics Holonic Systems: used in the coordination and control of robotic systems, and service-oriented Holonic Systems: implemented in service industries for adaptive and customer-oriented solutions.

Structural Classification. Simple Holons: basic units with specific functions. Composite Holons: aggregations of simple holons forming more complex structures.

Behavioral Classification. Autonomous Holons: operate independently and make decisions based on local information. Cooperative Holons: interact and cooperate with other holons to achieve common goals.

Hierarchical Classification. Flat Holonic Systems: holons operate at the same hierarchical level. Hierarchical Holonic Systems: holons are structured in multiple hierarchical levels, with higher-level holons coordinating lower-level ones [43–45].

4.4 Autonomic Systems

The most important and major characteristics were cited by Mazeiar [46], we cite:

Self-Configuring is the capability of adapting automatically and dynamically to environmental changes. Self-Healing is the capability of discovering, diagnosing, and reacting to disruptions. Such a system must be able to recover by detecting a failed component, taking it off-line to be fixed, and replacing the fixed component into the system without any apparent disruption. Self-Optimizing is the capability to efficiently maximize resource allocation and utilization for satisfying requirements of different users. Self-Protecting is the capability of reliably establishing trust, and anticipating, detecting and recovering from the effects of attack, self-aware, and other minor characteristics are detailed in [46]. Based on this characteristics we can cite two relevant classification of autonomic systems; *Based on autonomous capabilities.* We have Basic

Autonomic Systems: Offer the basic major characteristics self-management such as self-monitoring and basic self-configuration, and the Advanced Autonomic Systems: including more advanced functionalities, the minor desired characteristics such as self-healing, optimization, and protection.

The second criteria for classification is based *on the autonomic element composition* Single-Level Autonomic Systems, have autonomic aspects that operate at a single level. Multi-Level Autonomic Systems: Autonomic elements act at numerous levels, resulting in more hierarchical and sophisticated autonomic behavior [47–49].

5 Architectures and Modelling Tools

Engineering and management processes are essential for optimizing system performance and achieving project goals. Projects should be evaluated across multiple contexts. Various modelling tools address challenges in complex systems, such as emergent behavior and interoperability. In Table 1, we cite the most relevant modelling tools for each system.

Table 1. Modelling Tools/Frameworks

System	Modelling Tools/Frameworks	Description/Use Case
SoS	- DoDAF (Department of Defense Architecture Framework) - TOGAF (The Open Group Architecture Framework) - Agent-Based Modeling (ABM) - NetLogo - MBSE (Model-Based Systems Engineering) - SysML (Systems Modeling Language) - Hypergraphs	- DoDAF: Models defense systems and ensures interoperability [55–57] - TOGAF: Enterprise architecture framework for designing SoS [32, 35] - ABM: Captures emergent behaviors of SoS - NetLogo: Agent-based modeling of complex [5], - MBSE: Supports system requirements, design, and analysis [5] - SysML: Used for specifying, analyzing, and verifying SoS [36] - Hypergraphs: Models various systems, such as databases and enterprise relationships [50]
MS	- Enterprise Systems Engineering Profiler - Agent-Based Modeling (ABM) - Discrete Event Simulation (DES) - Unified Modeling Language (UML) - Systems Modeling Language (SysML)	- Profiler: Characterizes the environment and context of MS [3] - ABM: Represents system interactions [51] - DES: Models logistics and manufacturing processes [52] - UML: Graphical notation for specifying MS [3] - SysML: Supports systems engineering applications

(*continued*)

Table 1. *(continued)*

System	Modelling Tools/Frameworks	Description/Use Case
HS	- Agent-Based Modeling (ABM) - Unified Modeling Language (UML) - Multi-Agent Systems (MAS) - Petri Nets - Holonic Control Architecture (HCA) - Distributed Control Systems (DCS)	- ABM: Models holonic system behaviors and interactions [16] - MAS: Represents holons as agents with specific goals [50] - Petri Nets: Analyzes concurrent and asynchronous processes - HCA: Control architecture for holonic systems in manufacturing and robotics [53]. - DCS: Architecture for large-scale holonic systems [54]
AS	- Autonomic Computing Blueprint - Model-Driven Architecture (MDA) - Reinforcement Learning (RL) - Control Theory - Self-Organizing Maps (SOM)	- Blueprint: Framework for autonomic system elements [47] - MDA: Defines structure and behavior of autonomic systems [55] - RL: Optimizes performance and adapts to environments - Control Theory: Develops feedback loops [56] - SOM: Recognizes patterns for adaptation [57]

The common used modelling technologies include Multi-Agent Systems (MAS), Petri Nets, Unified Modelling Language (UML), Reinforcement Learning (RL), Model-Driven Architecture (MDA), and Event-Driven Architecture (EDA). MAS is used to represent autonomous entities in SoS, HS, and AS. Petri Nets are useful for analyzing concurrent processes in SOS and HS. UML graphically illustrates the architecture and behaviors of all systems. RL improves performance in both autonomic and HS, whereas MDA abstracts and automates design processes, especially in AS. EDA controls real-time reactions in AS. The following sections will compare similarities and differences between systems.

6 Discussion and Comparison

The all cited terms in this paper address a common concept that is well recognized and, in many cases is the achievement of a fundamentally new and desirable capability through the cooperative interaction of what may otherwise have been separate systems. The existence of multiple terms indicates that different authors have each focused on particular dimensions of the same problem and suggests that no one-dimensional view will be sufficient. After this study, we noticed that there are some common points in challenges and modelling tools and some key differences, so complex systems, including SoS, MS, HS, and AS, have common fundamental issues in complexity management, scalability, adaptability, and interoperability. SoS focuses on the integration and interoperability of separate systems, whereas MS deal with large-scale and complex interdependences. AS emphasize self-management skills, whereas HS involve balancing holons' autonomy and collaboration. Scalability also relies on controlling component and interaction growth. To achieve successful interoperability, these systems must adapt to changes in the environment or requirements, as well as guarantee good and persistent communication and coordination across different systems.

The first difference lies in the nature of the systems and their scale. A MS is a single, large, and highly integrated system, whereas a SoS is a network of independent systems brought together to function cohesively as a larger, more capable entity. The choice between these concepts depends on the specific requirements and challenges of a given application or problem domain. HS, SoS, and AS are all concepts used to describe different aspects of complex and self-organizing systems. MS are even more complex systems than system systems. They are usually made up of hundreds or thousands of smaller systems, and they can be very difficult to manage and control. MS are often used for large-scale applications, such as defence, aerospace, or critical infrastructure. HS emphasize hierarchical structures and interrelated components within a single system..Autonomic systems emphasize self-management and adaptation, reducing the need for human intervention in system operation. These concepts can be applied in different domains, and their use depends on the specific requirements and goals of the system or organization. HS are a specific type of SoS with a focus on autonomous and cooperative components. And the potential application for SoS are: Supply chain management: Integrating inventory systems, transportation networks, and production facilities to optimize product flow. Disaster response: as the coordinating of emergency services, communication networks, and resource allocation during a crisis. Smart cities: Connecting and managing traffic lights, power grids, and public transportation systems for efficient urban operations [6]. HS are well-suited for situations requiring distributed decision-making, adaptation, and self-organization. Some potential applications including Autonomous robots, Traffic management systems, Manufacturing systems [15]. AS in scenarios where self-management is crucial for continuous operation and performance optimization. Self-driving cars: Continuously adapting to road conditions, traffic patterns, and potential hazards. Data center management: Automatically scaling resources based on real-time workload demands and ensuring system uptime. Self-healing networks: Identifying and resolving network issues without human intervention [21], finally MS are used in defense sysetems and entreprise systems [3]. Table gives a summary of discussed features.

For the Common challenges and issues Interoperability is a critical concern in complex systems, as they often have diverse components and protocols, making them difficult to manage and control. Scalability becomes increasingly challenging as these systems expand, and adaptability is crucial as they must evolve to dynamic environments. The interconnected nature of these systems introduces vulnerabilities in trust and security, necessitating robust mechanisms to safeguard system integrity. Holistic approaches are needed to address emergent properties and behaviors arising from complex interactions. Future research should focus on developing integrated solutions that maintain system flexibility and resilience (Table 2).

Table 2. Comparison recapitulation.

Feature	System-of-Systems (SoS)	Holonic Systems	Autonomic Systems	Mega-Systems
Components	Independent systems working together to achieve a goal	Self-regulating entities called holons, exhibiting autonomy and dependence on other holons	Self-managed systems capable of self-configuration, self-healing, self-optimization, and self-protection	Large, complex systems composed of numerous interconnected parts
Coordination	Requires explicit mechanisms for communication and collaboration between independent systems. (e.g., protocols, standards)	Coordination emerges from interactions between holons, driven by their autonomy and goals	Self-adaptation through internal mechanisms that monitor and adjust system behavior. (e.g., control loops, feedback mechanisms)	May require coordination mechanisms to manage interactions between different parts, especially if composed of heterogeneous systems
Adaptability	May require manual reconfiguration in response to changes in the environment or system requirements	self-reconfiguration due to holon autonomy. Holons can adapt their behavior and interactions with other holons to respond to changes	Self-adaptation to changing environments without significant external intervention. Autonomic systems strive for continuous optimal performance	May require complex adaptation strategies due to the scale and interconnectedness of the system. Mega-systems might adapt through a combination of autonomic behavior in sub-systems and centralized interventions
Examples	Supply chain network, transportation system, power grid (can be viewed as a SoS)	Nervous system, immune system, ecosystem (considered natural holonic systems)	Self-driving car, self-healing network, autonomic computing platform	Power grid (as a mega-system),satellites global economy, internet of things (IoT) ecosystem

As technologies, Cyber-Physical Systems (CPS) and the Internet of Things (IoT) are intricately linked to above complex system paradigms. CPS, which fuse computational and physical processes, often use IoT technologies for enhanced connectivity and data exchange. These systems frequently display SoS characteristics [58], such as the operational and managerial autonomy of their constituent components [46]. In contexts like smart cities, the scale and complexity of CPS and IoT implementations can categorize them as Mega Systems [34]. Furthermore, the demand for self-management in extensive CPS and IoT applications aligns with Autonomic Systems principles [51]. The Holonic Systems concept, which emphasizes both autonomy and cooperation among entities, has been leveraged to improve the flexibility and scalability of CPS and IoT architectures. Despite their unique focuses, these paradigms face shared challenges related to managing complexity, ensuring interoperability, and adapting to dynamic environments [29]. After the discussion we present a Venn diagramm to represent the relation between systems also a formal representation (Fig. 2 and Table 3):

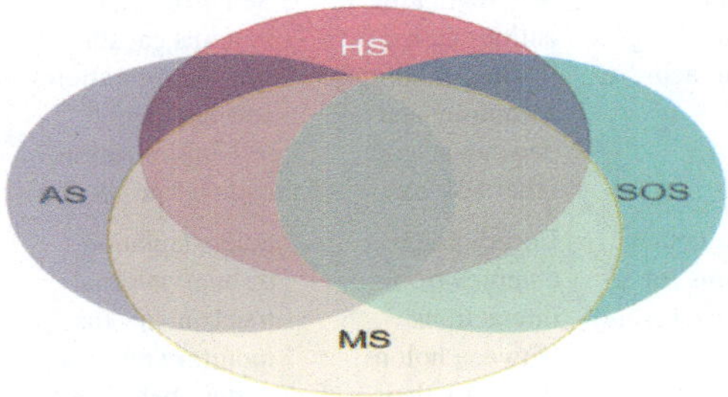

Fig. 2. Venn diagram of systems relations.

Table 3. Formal representation of relations between systems

MS and SoS	Some MS are SoS: MS $\cap$ SoS $\neq \emptyset$ Not all MS are SoS: MS $\subseteq$ SoS Not all SoS are MS, SoS $\subseteq$ MS
HS and SoS	Some HS are SoS: HS $\cap$ SoS $\neq \emptyset$ Not all HS are SoS: HS $\nsubseteq$ Sos Not all SoS are HS: SoS $\nsubseteq$ HS
AS and SoS	Some AS are SoS: AS $\cap$ SoS $\neq \emptyset$ Not all AS are SoS: AS $\nsubseteq$ SoS Not all SoS are AS: SoS $\nsubseteq$ AS
MS and HS	Some MS are HS: M $\cap$ H $\neq \emptyset$ Not all MS are HS: MS $\nsubseteq$ HS Not all HS are MS: HS $\nsubseteq$ MS
MS and AS	Some MS are AS: MS $\cap$ AS $\neq \emptyset$ Not all MS are AS: MS $\nsubseteq$ AS Not all AS are MS: AS $\nsubseteq$ MS
HS and AS	Some HS are AS: HS $\cap$ AS $\neq \emptyset$ Not all HS are AS: HS $\nsubseteq$ AS Not all AS are HS: AS $\nsubseteq$ HS
Intersection of All Four	There exists a subset that is part of all four sets: SoS $\cap$ MS $\cap$ HS $\cap$ AS $\neq \emptyset$

As an example for the intersection between systems, International Space Station (ISS) Combines elements of SoS (international collaboration), Mega Systems (complex interdependent systems), Holonic Systems (modular units like laboratories), and Autonomic Systems (self-maintaining life support and robotic systems).

7 Conclusion

After reviewing the literature on Component-Based Systems CBS and complex systems, we identified several contact points among the frequently used terms: System-of-Systems (SoS), Mega-Systems, Autonomic Systems, and Holonic Systems. This work aims to reduce confusion related to these terms by presenting definitions and frameworks, conducting a state-of-the-art review, and performing a comparative study to address their characteristics and challenges. Complex systems, including SoS, MS, HS, and AS, face common challenges such as managing complexity, scalability, adaptability, integration, and interoperability. Despite their differences, these systems share issues related to scale, uncertainty, variability, interdependencies, and emergent behavior. Common modelling tools like Multi-Agent Systems (MAS), Petri Nets, Unified modelling Language (UML), Reinforcement Learning (RL), Model-Driven Architecture (MDA), and Event-Driven Architecture (EDA) provide frameworks to address these challenges, facilitating the development of robust and efficient systems capable of self-management and seamless integration. The term "Mega-Systems" serves as an umbrella term encompassing notions of systems-of-systems, federations of systems, networks of systems, enterprise systems, and other closely related terms. This term reflects both the scale of these systems and their ability to cross traditional boundaries.

References

1. Antal, M., Pop, C., Cioara, T., Anghel, I., Salomie, I., Pop, F.: A system of systems approach for data centers optimization and integration into smart energy grids. Futur. Gener. Comput. Syst. **105**, 948–963 (2020). https://doi.org/10.1016/j.future.2017.05.021
2. Teixeira, P.G., et al.: Constituent system design: a software architecture approach. Proc. - 2020 IEEE Int. Conf. Softw. Archit. Companion, ICSA-C 2020. Pp. 218–225 (2020). https://doi.org/10.1109/ICSA-C50368.2020.00045
3. Stevens, R.: Engineering mega-systems: The challenge of systems engineering in the information age. CRC Press (2016)
4. Adler, C.O., Dagli, C.H.: Study of the use of a genetic algorithm to improve networked system-of- study of the use of a genetic algorithm to improve networked system-of-systems resilience. Procedia - Procedia Comput. Sci. **36**, 49–56 (2014). https://doi.org/10.1016/j.procs.2014.09.036
5. Acheson, P., Dagli, C., Kilicay-Ergin, N.: Model based systems engineering for system of systems using agent-based modeling. In: Procedia Computer Science, pp. 11–19. Elsevier B.V. (2013). https://doi.org/10.1016/j.procs.2013.01.002
6. Horvath, L.: Model system for the representation of smart system of systems in engineering. SOSE 2020 - IEEE 15th Int. Conf. Syst. Syst. Eng. Proc., pp. 603–607 (2020). https://doi.org/10.1109/SoSE50414.2020.9130485
7. Farhangi, H., Konur, D., Dagli, C.H.: Multiobjective system of systems architecting with performance improvement funds. Procedia - Procedia Comput. Sci. **95**, 119–125 (2016). https://doi.org/10.1016/j.procs.2016.09.301
8. Konur, D., Farhangi, H., Dagli, C.H.: A multi-objective military system of systems architecting problem with inflexible and flexible systems: formulation and solution methods. OR Spectr. **38**, 967–1006 (2016). https://doi.org/10.1007/s00291-016-0434-2

9. Fang, Z.: System-of-systems architecture selection: a survey of issues, methods, and opportunities. IEEE Syst. J. **16**, 4768–4779 (2022). https://doi.org/10.1109/JSYST.2021.3119294
10. Vargas, I.G., Braga, R.T.V.: Understanding system of systems management: a systematic review and key concepts. IEEE Syst. J. **16**, 510–519 (2022). https://doi.org/10.1109/JSYST.2020.3018068
11. Axelsson, J.: Systems-of-systems design patterns: a systematic literature review and synthesis. In: 2022 17th Annual System of Systems Engineering Conference (SOSE), pp. 171–176 (2022). https://doi.org/10.1109/SOSE55472.2022.9812681
12. Dridi, C.E., Benzadri, Z., Belala, F.: System of systems modelling: recent work review and a path forward. In: 2020 International Conference on Advanced Aspects of Software Engineering (ICAASE), pp. 1–8 (2020). https://doi.org/10.1109/ICAASE51408.2020.9380125
13. Karcanias, N., Hessami, A.G.: System of systems and emergence Part 1: Principles and framework. Int. Conf. Emerg. Trends Eng. Technol. ICETET. 27–32 (2011). https://doi.org/10.1109/ICETET.2011.51
14. Torkjazi, M., Raz, A.K.: A Taxonomy for System of Autonomous Systems. 198–203 (2022)
15. Blair, G., et al.: Holons: towards a systematic approach to composing systems of systems. Proc. 14th Int. Work. Adapt. Reflective Middlew. **5**(1–5), 6 (2015). https://doi.org/10.1145/2834965.2834970
16. Tchappi, I.H., Galland, S., Kamla, V.C., Kamgang, J.C., Nono, C.M.S., Zhao, H.: Holonification model for a multilevel agent-based system: application to road traffic. Pers. Ubiquitous Comput. **23**, 633–651 (2019). https://doi.org/10.1007/s00779-018-1181-y
17. Vetorazzi, C.N., Telles, G.N.: The holonic system paradigm [for manufacturing]. **1**, 88 (2001). https://doi.org/10.1016/S0079-6611(02)00064-2
18. Pascal, C., Panescu, D.: On increasing adaptability of holonic systems. Service Orientation in Holonic and Multi-Agent Manufacturing: Proceedings of SOHOMA2017, pp. 229–241 (2018)
19. Babiceanu, R.F., Chen, F.F.: Development and applications of holonic manufacturing systems: a survey. J. Intell. Manuf. **17**, 111–131 (2006)
20. Jlassi, N., Moyaux, T., Bouleux, G., Ouzrout, Y.: Holonic systems for decision-making in hospital mergers. In: Borangiu, T., Trentesaux, D., Leitão, P., Berrah, L., Jimenez, J.-F. (eds.) Service Oriented, Holonic and Multi-Agent Manufacturing Systems for Industry of the Future, pp. 113–123. Springer Nature Switzerland, Cham (2024)
21. Palmer, K.D.: Autonomic Computing and Special Systems Theory (2007)
22. Lesch, V., Krupitzer, C., Tomforde, S.: Emerging self-integration through coordination of autonomous adaptive systems. Proc. - 2019 IEEE 4th Int. Work. Found. Appl. Self* Syst. FAS*W 2019. 6–9 (2019). https://doi.org/10.1109/FAS-W.2019.00016
23. Lalanda, P., McCann, J.A., Diaconescu, A.: Autonomic Computing. (2013). https://doi.org/10.1007/978-1-4471-5007-7
24. Simpson, J.J., Dagli, C.H.: System of systems architecture evaluation using evolutionary computation. Incose. 1–9 (2008). https://doi.org/10.1002/j.2334-5837.2008.tb00911.x
25. Dimario, M., Verma, D.: System of Systems Engineering Management Framework (2008)
26. Arnold, R.D., Wade, J.P.: A definition of systems thinking: a systems approach. Procedia Comput. Sci. **44**, 669–678 (2015). https://doi.org/10.1016/j.procs.2015.03.050
27. Boardman, J., Sauser, B.: System of systems-the meaning o. In: 2006IEEE/SMC international conference on system of systems engineering, pp. 6–pp. IEEE (2006)
28. Jaradat, R.M., Keating, C.B., Bradley, J.M.: A histogram analysis for system of systems. Int. J. Syst. Syst. Eng. **5**, 193–227 (2014). https://doi.org/10.1504/IJSSE.2014.065750
29. Jamshidi, M.: System of systems engineering: innovations for the 21st century. Deakin University (2008)

30. Henshaw, M., et al.: Systems of systems engineering: a research imperative. ICSSE 2013 - IEEE Int. Conf. Syst. Sci. Eng. Proc. 389–394 (2013). https://doi.org/10.1109/ICSSE.2013.6614697
31. Dagli, C.H., Miller, A.K., Gideon, J.M., Dagli, C.H., Miller, A.: Taxonomy of Systems-of-Systems Taxonomy of Systems-of-Systems (2005)
32. Jamshidi, M.: System of systems engineering—an overview. Engineering Emergence, pp. 11–42 (2018)
33. Manthorpe, W.H.J.: The emerging joint system of systems: a systems engineering challenge and opportunity for APL. Johns Hopkins APL Tech. Dig. **17**, 305–313 (1996)
34. Kilicay, N., Dagli, C.H.: Methodologies for Understanding Behavior of System of Systems (2004)
35. INCOSE: INCOSE systems engineering handbook. John Wiley \& Sons (2023)
36. Rodriguez, S., Hilaire, V., Gaud, N., Galland, S., Koukam, A.: Holonic Multi-Agent Systems. In: Di Marzo Serugendo, G., Gleizes, M.-P., Karageorgos, A. (eds.) Self-organising Software: From Natural to Artificial Adaptation, pp. 251–279. Springer Berlin Heidelberg, Berlin, Heidelberg (2011). https://doi.org/10.1007/978-3-642-17348-6_11
37. Macdiarmid, A., Lindsay, P.: Can System of Systems Be Given Self-X Requirement Engineering Capabilities?
38. Bandh, T.: Coordination of autonomic function execution in Self-Organizing Networks Self-Organizing Networks (2013)
39. Karamolegkos, P.N., Patrikakis, C., Protonotarios, E.: Autonomic networking. Encycl. Internet Technol. Appl. 72–78 (2007). https://doi.org/10.4018/978-1-59140-993-9.ch011
40. Bourdenas, T., Tei, K., Honiden, S., Sloman, M.: Autonomic role and mission allocation framework for wireless sensor networks. Proc. - 2011 5th IEEE Int. Conf. Self-Adaptive Self-Organizing Syst. SASO 2011, pp. 61–70 (2011). https://doi.org/10.1109/SASO.2011.17
41. Siqueira, M.A., Verdi, F.L., Pasquini, R., Magalhães, M.F.: An architecture for autonomic management of Ambient Networks. Lect. Notes Comput. Sci. (including Subser. Lect. Notes Artif. Intell. Lect. Notes Bioinformatics). 4195 LNCS, pp. 255–267 (2006). https://doi.org/10.1007/11880905_21
42. Maier, M.W.: Architecting principles for. Aerosp. Corp. John Wiley Sons, Inc. Syst Eng. **1**, 267–284 (1999). https://doi.org/10.1002/(SICI)1520-6858(1998)
43. Giret, A., Botti, V.: Analysis and Design of Holonic Manufacturing Systems. 18th Int. Conf. Prod. Res. 2–6 (2005)
44. Van Brussel, H., Wyns, J., Valckenaers, P., Bongaerts, L., Peeters, P.: Reference architecture for holonic manufacturing systems: PROSA. Comput. Ind. **37**, 255–274 (1998)
45. Leitão, P.: Agent-based distributed manufacturing control: a state-of-the-art survey. Eng. Appl. Artif. Intell. **22**, 979–991 (2009)
46. Salehie, M., Tahvildari, L.: Self-Adaptive Software: Landscape and Research Challenges. V, 1–40 (2009)
47. Kephart, J.O., Chess, D.M.: The vision of autonomic computing. Computer (Long. Beach. Calif). **36**, 41–50 (2003). https://doi.org/10.1109/MC.2003.1160055
48. Dobson, S., et al.: A survey of autonomic communications. ACM Trans. Auton. Adapt. Syst. **1**, 223–259 (2006)
49. Huebscher, M.C., McCann, J.A.: A survey of autonomic computing—degrees, models, and applications. ACM Comput. Surv. **40**, 1–28 (2008)
50. Khalil, W., Merzouki, R., Ould-Bouamama, B., Haffaf, H.: Hypergraph models for system of systems supervision design. IEEE Trans. Syst. Man, Cybern. Part ASystems Humans. **42**, 1005–1012 (2012). https://doi.org/10.1109/TSMCA.2012.2183350
51. Bonabeau, E.: Agent-based modeling: Methods and techniques for simulating human systems. Proc. Natl. Acad. Sci. **99**, 7280–7287 (2002)

52. Alzraiee, H., Moselhi, O., Zayed, T.: A comparison of des and SD for simulating construction operation. Proceedings, Annu. Conf. - Can. Soc. Civ. Eng. **2**, 1263–1272 (2013)
53. Valckenaers, P., Van Brussel, H.: Holonic manufacturing execution systems. CIRP Ann. - Manuf. Technol. **54**, 427–432 (2005). https://doi.org/10.1016/S0007-8506(07)60137-1
54. Leitão, P.: Agent-based distributed manufacturing control: A state-of-the-art survey. Eng. Appl. Artif. Intell. **22**, 979–991 (2009). https://doi.org/10.1016/j.engappai.2008.09.005
55. Kleppe, A.G., Warmer, J.B., Bast, W.: MDA explained: the model driven architecture: practice and promise. Addison-Wesley Professional (2003)
56. Brun, Y., et al.: Engineering Self-Adaptive Systems through Feedback Loops. 48–70 (2009)
57. Usama, M., et al.: Unsupervised machine learning for networking: techniques. Applications and Research Challenges. IEEE Acess. **7**, 6557965615 (2019). https://doi.org/10.1109/ACCESS.2019.2916648
58. Fortino, G., Savaglio, C., Spezzano, G., Zhou, M.: Internet of things as system of systems: a review of methodologies, frameworks, platforms, and tools. IEEE Trans. Syst. Man, Cybern. Syst. **51**, 223–236 (2021). https://doi.org/10.1109/TSMC.2020.3042898

An Approach for UML Sequence Diagram Models/MetaModels Co-evolution Using Transformation Rules

Asma Chouikh[1](✉), Khaled Khalfaoui[2], ELHillali Kerkouche[2], and Allaoua Chaoui[3]

[1] LaRIA Laboratory, University of Jijel, 18000 Jijel, Algeria
asma.chouikh@univ-jijel.dz
[2] Department of Computer Science, University of Jijel, 18000 Jijel, Algeria
[3] MISC Laboratory, Abdelhamid Mehri University, Constantine 2, Constantine, Algeria

Abstract. Like all software artifacts, metamodels frequently evolve in response to various factors, such as technological advances or changing user requirements. When a metamodel evolves, a simple change can affect all associated artifacts, including models, code generators, transformation rules and consistency constraints. As a result, these artifacts may no longer be compliant. The main challenge is how to adapt and maintain the consistency and conformity of the initial artifacts to the updated metamodel. Several approaches have been proposed, focusing exclusively on model co-evolution during metamodel evolution. In this paper, we propose an approach to address the evolution of the UML Sequence Diagram metamodel. Our approach consists in calculating and analyzing the differences between the initial metamodel and the evolved metamodel. These differences are used to define transformation rules that enable the models to co-evolve. We illustrate our approach with a practical example.

Keywords: Model-Driven Engineering (MDE) · Metamodeling · Model Transformation · Model/metamodel Co-evolution · UML Sequence Diagram · Eclipse Modeling Frameworks

1 Introduction

The software industry is becoming increasingly complex, requiring continuous improvement in methods and development practices. Model-Driven Engineering (MDE) offers a higher level of abstraction to manage complexity. MDE generates applications from models, allowing developers to address abstract concerns compared to traditional programming [1].

MDE comprises three main artifacts: models, metamodels and transformation models. Models simplify system understanding, metamodels define modeling language syntax and transformation models generate target models from source models using predefined rules [2].

S. Chikhi et al. (Eds.): MISC 2024, LNNS 1267, pp. 69–84, 2025.
https://doi.org/10.1007/978-3-031-82112-7_6

As their modeling languages evolve, metamodels must undergo several modifications to remain aligned with the evolving requirements of the modeling languages.

The same applies to both General-Purpose Languages (GPL) and Domain-Specific Modeling Languages (DSMLs) [1]. For example, the Unified Modeling Language (UML) is undergoing a process of continual evolution with the objective of restructuring and improving its usability, implementation and adaptation. This evolution serves to highlight the significance of model co-evolution, particularly in artifacts such as the UML SD, which enables the reuse of older models and facilitates transitions to more advanced modeling practices. In general, modeling languages evolve for various reasons, including technological advances or changes in user needs [3].

Co-evolution is crucial to maintain compatibility with evolving metamodels, saving time and resources on model recreation. Over the past decade, several methods have been proposed to address co-evolution, including operator-based approaches and state-based approaches [4]. Operator-based approaches record changes during modifications, requiring metamodel developer intervention, while state-based approaches document changes declaratively, facilitating model transformation management [5].

This paper presents a state-based model transformation approach that use metamodel matching to adapt UML SD to the evolution of their metamodel. The approach involves defining the initial and new metamodels, calculating the delta metamodel to identify differences and extracting a correspondence table. Transformation rules are established based on the extracted correspondence table. The ATL transformation language is utilized to apply these rules and transform an input model that conforms to the initial metamodel into output model, ensuring compatibility with the new metamodel version.

The document is organized as follows: Sect. 2 reviews related work. Section 3 discusses the basic concepts of MDE and model/metamodel coevolution. Section 4 deals with UML sequence diagrams, while Sect. 5 focuses on Eclipse modeling frameworks. Section 6 details the proposed co-evolution approach and Sect. 7 presents the case study. The final section presents conclusion.

2 Related Word

The problem of metamodel evolution and model co-evolution in MDE has been extensively studied. Approaches vary from manual updates as discussed by Rose et al., who introduced the Epsilon Flock language for model-to-model transformation [6], to operator-based methods like those of Vermolen and Visser, which have provided a transformation language for coupled evolution [7]. Narayanan et al. developed a Model Change Language (MCL) for migrating models between metamodels [8], while Sprinkle and Karsai used a domain-specific visual language (DSVL) for model co-evolution [9]. Automated approaches, such as those proposed by Kebaili et al., trace tests affected by metamodel changes [10], while Gruschko and Kolovos emphasized computing differences between metamodel versions [11]. Anguel et al., presented a hybrid approach, starting with metamodel matching, followed by migration operations [4]. Ruscio et al., focused on correlating metamodel and ATL transformation rule evolution [12].

Tools like GrammarTransformer by Zhang et al. [13] and frameworks such as Edelta by Bettini et al. support the co-evolution process [14]. Lastly, Khelladi et al. explored syntactic reasons for different resolutions in metamodel evolution, achieving high syntactic and semantic co-evolution [15].

3 Model-Driven Engineering and Models/MetaModels Co-evolution

After the object-oriented approach and "everything is an object" in 1980, software engineering now focuses on MDE and "everything is a model" [16]. MDE is a software engineering approach that uses high-level abstractions and automation to reduce the complexity of software development and maintenance and makes it easier to develop software systems by utilizing modeling techniques such as domain-specific modeling languages and generative techniques [5] (Fig. 1).

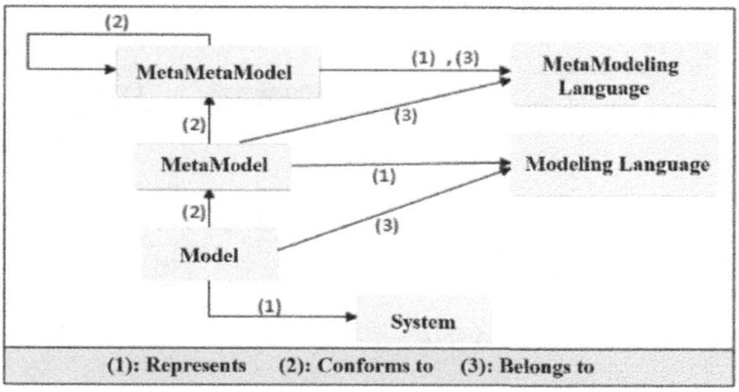

Fig. 1. Meta-Level Architecture in Model-Driven Development [4]

A metamodel defines the structure, semantics and constraints of models in a specific domain, ensuring consistency and interoperability between different tools and platforms [18]. Metamodels are crucial in MDE, defining entities, relationships, and rules for models. The Object Management Group (OMG) introduced the Meta-Object Facility (MOF) as a standard for defining metamodels, which serves as the foundation for standards like the Unified Modeling Language (UML) [19]. The model is a simplified view of a system based on the rules and structures defined by metamodels, established to represent various aspects of a system, such as its architecture, behavior, data and processes [15]. They are used to abstract complex systems from the real world, making them easier to manage and understand [18].

Model transformation is important in MDE approach to maintain model integrity and structure. In general, model transformations can be implemented in different ways. They can be declarative or imperative, focusing on the end result and defining necessary steps [1]. These rules can range from simple mappings between source and target models to complex ones requiring direct mappings between similar elements.

In this paper, we focus on transformations expressed in a specialized model transformation language.

Metamodels provide the necessary structure and guidelines for reducing errors in the development process of complex software systems [21]. Metamodels must adapt to software system evolution due to new requirements, improved performance and functional integration often necessitating changes in their design [6]. In addition, technological advances and the adoption of new platforms or tools may require updating the metamodel to remain compatible with these innovations [7].

Metamodel evolution management involves updating metamodels, Ensuring the non-conformity and facilitating co-evolution to maintain coherence and correctness. Model co-evolution aligns models with metamodel structure and semantics, preserving relationships, constraints and features [8]. This method improves productivity and reduces human error risk, allowing reuse in different metamodel versions [11]. Consequently, efficient model co-evolution strategies are crucial to ensure the long-term maintainability and scalability of MDE projects [22] (Fig. 2).

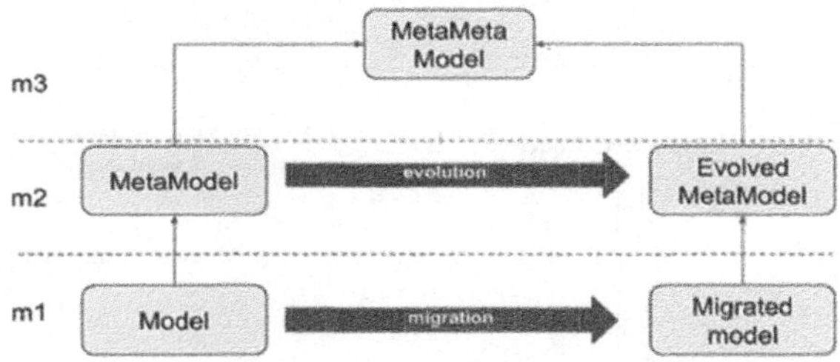

Fig. 2. Evolving metamodels and Co-evolving Models

4 UML Sequence Diagrams

A sequence diagram is a type of interaction diagram in the UML that shows how objects interact in a specific sequence [3]. UML SD diagrams are crucial in object-oriented software development for understanding the order and manner of object collaboration. They provide sequential representations of object interactions, including conditions and iterations, which are essential for managing system behavioral aspects during the project requirements phase. The UML SD, like any other artifact, has evolved from UML 1.5 to UML 2.5. The transition from UML 1.x to UML 2.0 introduced combined fragments, such as “alt” (alternatives), “loop” (repetitions), “opt” (optional interactions), “par” (parallel interactions), “ref” (references to other diagrams) and “region” (critical regions) to model complex interaction scenarios more sophisticatedly [23].

This allows for clear depiction of conditions, repetitions and concurrent interactions, crucial for accurately representing real-world systems. UML 2.5 SD have been enhanced, enhancing their power and versatility in system modeling, dynamic behavior depiction and stakeholder communication [23] (Fig. 3).

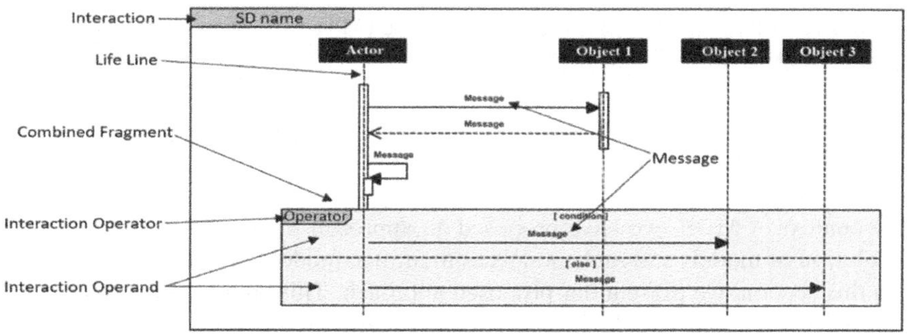

Fig. 3. UML SD diagram

5 Eclipse Modeling Frameworks

Eclipse Modeling Frameworks for MDE, are essential for creating and managing structured models, supporting a model-centric approach in software development. These tools streamline model transformation and analysis, enhancing the efficiency and consistency of development processes [24].

Eclipse Modeling Framwork (EMF) Is a modeling and code generation tool that enables the creation of tools and applications using a structured model. EMF (Core) comprises three essential elements: a metamodel (Ecore) to provide a description of models and runtime, EMF.Edit: reusable generic classes and XEMF. Codegen: to generate everything required to build a complete editor for an EMF model [25].

EMF Compare Is an Eclipse project that provides tools for comparing and merging models to the EMF framework, supporting both structural and semantic comparisons. The objectives of this component are to enable a stable and efficient generic implementation of model comparison, and to offer an extensible framework for specific requirements [26].

ATL – ATL is a model transformation technology that enables users to convert source models into target models. It is a hybrid language that combines declarative and imperative constructs, focusing on establishing relations between source and target models [20]. The ATL Integrated Environment on the Eclipse platform includes standard development tools for easy ATL transformation development. ATL is designed to be modular and maintainable, similar to traditional programming languages [27].

Papyrus Eclipse Papyrus provides an integrated, user-consumable environment for editing any type of EMF model, with particular support for UML and related modeling languages such as SysML and MARTE. Papyrus provides diagram editors for EMF-based modeling languages such as UML 2 and SysML, as well as advanced support for UML profiles, enabling users to define editors for DSLs based on the UML 2 standard and its extension mechanisms [28].

Acceleo Is an Eclipse-based tool developed by French company Obeo that follows the Model-Driven Architecture (MDA) approach, transforming EMF models into code or files using templates written in a specific language (MTL). It offers high customization, interoperability, and easy kick-off [29].

6 Proposed Approach

In the context of MDE, we have proposed an approach to resolve the problem of the co-evolution of models and metamodels as their metamodels evolve.

In this section, we present our proposed approach. This approach ensures a smooth transition and maintains the high accuracy of model-driven development.

The envisioned essential steps are:

- Defining metamodels: Define the source and evolved metamodels MM1 and MM2.
- Detection of changes: Calculate the delta metamodel by using the matched technique and defining the differences between both metamodels.
- Extracting the Table of Correspondences T: The detected changes are used to identify correspondences between metamodels.
- Defining transformation rules (R): to define the transformation rule, we use the Table of Correspondences (T).
- Mapping: Load the source model M1 that complies with MM1, apply the transformation rules and update the models to reflect the changes. Perform a validation to verify the conformity of the obtained model M2 with the evolved version of the metamodel (MM2) (Fig. 4).

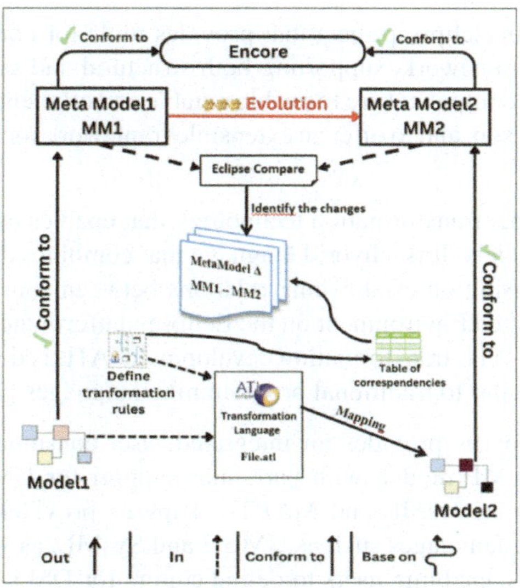

Fig. 4. The proposed Approach

6.1 Defining Metamodels

Defining metamodels is essential for managing model evolution. The source metamodel, $\mathbf{MM_1}$, contains the elements $\{\mathbf{m_1}, \mathbf{m_2}, \ldots, \mathbf{m_x}\}$ where each element $\mathbf{m} \in \mathbf{MM_1}$. The evolved metamodel $\mathbf{MM_2}$ comprises elements $\{\mathbf{n_1}, \mathbf{n_2}, \ldots, \mathbf{n_y}\}$ with each element $n \in MM_2$ where **m** and **n** can be a class **C**, an attribute **A** or a reference **R**. Formally, they are expressed repectively as:

$$\mathbf{MM_1} = \{\mathbf{m_1}, \mathbf{m_2}, \ldots, \mathbf{m_x}\} \ \mathbf{where} \ \forall \mathbf{m_i} \in \mathbf{MM_1}; \ \mathbf{mi} \in (\mathbf{C_1} \cup \mathbf{A_1} \cup \mathbf{R_1}),$$

$$\mathbf{MM_2} = \left\{\mathbf{n_1}, \mathbf{n_2}, \ldots, \mathbf{n_y}\right\} \ \mathbf{where} \ \forall \ \mathbf{n_i} \in \mathbf{MM_2}; \ \mathbf{ni} \in (\mathbf{C_2} \cup \mathbf{A_2} \cup \mathbf{R_2}).$$

In our approach, we use the EMF framework, based on a meta-model (Ecore) which is a powerful tool for describing metamodels and models. The following figures represent metamodels, providing a detailed view of the components and their relationships.

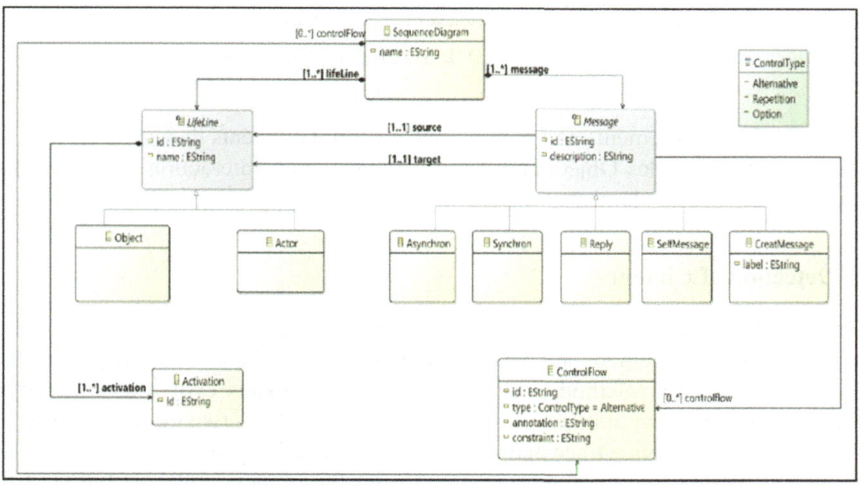

Fig. 5. UML SD Source metamodel

The MM1 in Fig. 5, defines a UML SD Diagram metamodel containing multiple LifeLine elements, each lifeline associated with one or more Activation instances. Within the UML SD, there are various types of Messages, including Asynchron, Synchron, Reply, SelfMessage and CreatMessage which relate with LifeLine elements serving as the source and target for these messages. Additionally, the metamodel includes Object and Actor elements and defines ControlFlow elements that are connected to the UML SD. The ControlType enumeration specifies the possible types for ControlFlow.

The MM2, shown in Fig. 6, similarly describes a UML SD but includes Interaction elements connected to LifeLine elements. These LifeLine elements are associated with multiple ActivationBar instances. The Messages can be Asynchron, Synchron, Reply, SelfMessage or CreatMessage, linking the source and target LifeLine elements. Furthermore, the metamodel introduces CombinedFragment elements, which group multiple

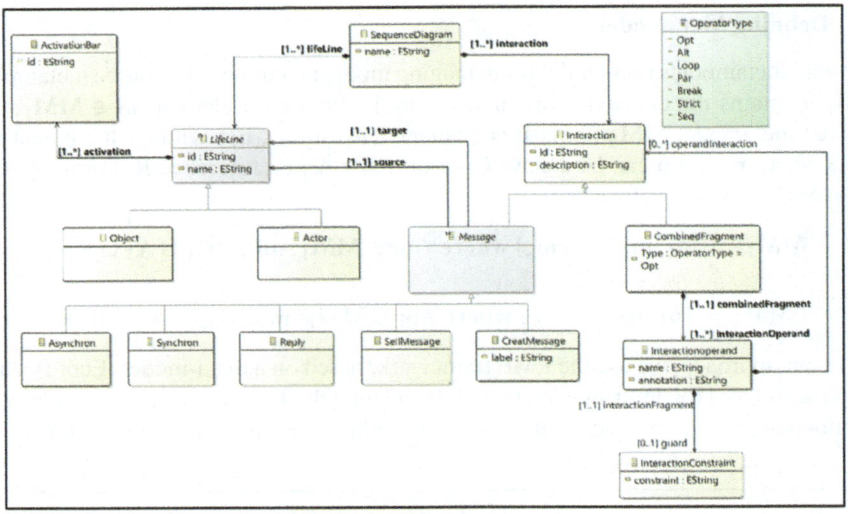

Fig. 6. Evolved UML SD diagram metamodel

InteractionOperand elements and InteractionConstraint elements that act as guards. The metamodel also includes Object and Actor elements, with Interaction elements linked to OperandInteraction elements.

6.2 Detection of Changes

Understanding the evolution of metamodels and the changes made to them is crucial for effective model management and development practices and it can be achieved through two methods: the first method understands how metamodels evolve and identifying differences [19]. Direct comparison is a method that compares two metamodel versions without requiring a change trace and works with any modeling tool. The second method uses change operations recording to capture traces during metamodel evolution [30].

In our approach, to defining the metamodel of differences, we use the formula $\Delta_{MM} = \{ d | d \in MM_1 \cup MM_2 \perp d \notin MM_1 \cap MM_2 \}$, whose **d** represents an element within the metamodels. The union (∪) of MM_1 and MM_2 encapsulates all elements in both versions, while the intersection (∩) identifies elements common to both versions, representing the unchanged core. In this manner, Δ_{MM} isolates elements that are in either MM_1 or MM_2 but not in both. This helps identify differences or changes, categorizing them as additions, deletions, moves and updates. We use the Eclipse Compare tool for calculate Δ_{MM}. This tool is well integrated with EMF [45] (Fig. 7).

The table below details the changes made to the metamodel, classified by type of change: addition, deletion, update, move or equivalence. Elements are also classified by type (Table 1).

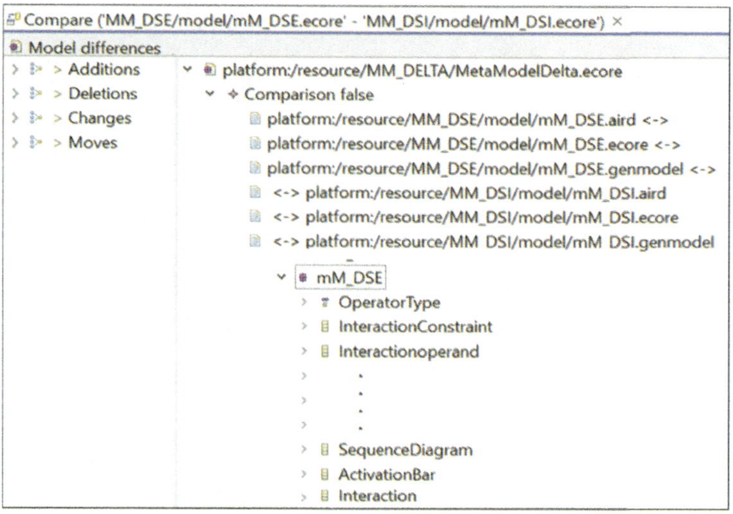

Fig. 7. Ecore Model Comparison: MetaModel Delta

6.3 Extracting the Table of Correspondences

The process of evolving from $UML_{1.5}$ to $UML_{2.5}$ involves several critical steps, one of which is extracting the table of correspondences. This step is crucial as it identifies the relationships between elements in the source metamodel $\mathbf{MM_1}$ and the target metamodel $\mathbf{MM_2}$. The formal representation of this process defined as:

$$\mathbf{T} = (\mathbf{m}, \mathbf{n})|\ \mathbf{m} \in \mathbf{MM1}, \mathbf{n} \in \mathbf{MM2} \wedge \mathbf{m} \leftrightarrow \mathbf{n}.$$

This formula indicates that for each element **m** in the source metamodel $\mathbf{MM_1}$, there is a corresponding element n in the target metamodel $\mathbf{MM_2}$, such that **m** and **n** are equivalent or related. The set of correspondences denoted as **T**. The relation indicates that there is a correspondence or equivalence between **m** and **n**.The transformation process involves a table of correspondences, which establish relationships between elements in the metamodel. This table maps each element in the $UML_{1.5}$ metamodel to its corresponding element in the $UML_{2.5}$ metamodel, providing a clear blueprint for transformation. The transformation of elements like UML SD between versions is straightforward, with elements like LifeLine, Actor and Message remaining unchanged. However, ActivationBar requires specific renaming rules. The introduction of CombinedFragment in $UML_{2.5}$ replaces ControlFlow, requiring a more complex transformation of attributes. The table below represents correspondences (Table 2).

6.4 Defining Transformation Rules

The process of defining transformation rules is essential in the evolution of UML metamodels from $version_{1.5}$ to $version_{2.5}$. Transformation rules dictate how each element in the source metamodel $\mathbf{MM_1}$ should be transformed into an element in the target metamodel $\mathbf{MM_2}$. These rules are defined using the Table of Correspondences **T** that maps

Table 1. Changes to UML SD Metamodel Elements

Type	Element	Addition	Deletion	Update	Move	Equivalence
Classes	SequenceDiagram					✓
	LifeLine					✓
	Activation			✓		
	Object					✓
	Actor					✓
	Message				✓	
	Asynchron				✓	
	Synchron				✓	
	Reply				✓	
	SelfMessage				✓	
	CreatMessage				✓	
	ControlFlow		✓			
	Interaction	✓				
	CombinedFragment	✓				
	InteractionOperand	✓				
	InterctionConstraint	✓				
Attributes	name (SequenceDiagram)					✓
	id (LifeLine)					✓
	name (LifeLine)					✓
	id(Activation)					✓
	id(Message)				✓	
	description(message)					✓
	label(CreatMessage)					✓
	id(ControlFlow)		✓			
	type(ControlFlow)		✓			
	annotation(ControlFlow)		✓			
	constraint(ControlFlow)		✓			
	id(Interaction)				✓	
	type(CombinedFragment)	✓				
	annotation (Interactionoperand)	✓				
	constraint (InteractionConstraint)	✓				
Relations	lifeline(SequenceDiagram)					✓
	controlflow(SequenceDiagram)		✓			
	message(SequenceDiagram)				✓	
	source(LifeLine)					✓
	target(LifeLine)					✓
	controlflow(Message)		✓			
	activation(LifeLine)					✓
	interactions(SequenceDiagram)	✓				

(*continued*)

Table 1. (*continued*)

Type	Element	Addition	Deletion	Update	Move	Equivalence
	operandInteraction(Interaction)	✓				
	interactionOperand(CombinedFragment)	✓				
	Guard(InteractionConstraint)	✓				
Enumerations	ControlType		✓			
	OperatorType	✓				

Table 2. Correspondence Table

$UML_{1.5}$ Element	$UML_{2.5}$ Element
SequenceDiagram	SequenceDiagram
LifeLine	LifeLine
Actor	Actor
Object	Object
Message	Message
Synchron	Synchron
Asynchron	Asynchron
Reply	Reply
SelfMessage	SelfMessage
CreatMessage	CreatMessage
Activation	ActivationBar
ControlFlow	CombinedFragment
ControlType(ControlFlow)	OperatorType(CombinedFragment)
Annotation(ControlFlow)	Annotation(InteractionOperand)
Constraint(ControlFlow)	Constraint(InteractionConstraint)

elements from $\mathbf{MM_1}$ to $\mathbf{MM_2}$. The formal representation of defining transformation rules can be expressed as:

$$\forall(\mathbf{m}, \mathbf{n}) \in \mathbf{T}, \exists \mathbf{r_i} \in \mathbf{R}\ \textbf{such that}\ \mathbf{m} \rightarrow \mathbf{n} \mid \mathbf{R} = \{\mathbf{r_1}, \mathbf{r_2}, \ldots, \mathbf{r_i}\}.$$

The transformation of a pair (m, n) in a correspondence table is determined by a specific rule $\mathbf{r_i}$ in the set of rules **R**. This rule defines how **m** from $\mathbf{MM_1}$ is transformed into n in $\mathbf{MM_2}$ [27]. The transformation $\mathbf{m} \rightarrow \mathbf{n}$ represents the application of rule $\mathbf{r_i}$ and the set $\mathbf{R} = \{\mathbf{r_1}, \mathbf{r_2}, \ldots, \mathbf{r_i}\}$ encompasses all necessary transformation rules for the metamodel transformation.

The figures below illustrate the important helpers functions used in the ATL transformation. Each image captures the details of a specific helper function (Figs. 8 and 9).

```
-- Helper pour mapper ControlType vers OperatorType
helper def : mapControlTypeToOperator(cType : MMinit!ControlType) : MMevol!OperatorType =
    (if cType = #Alternative then
        #Alt
    else if cType = #Repetition then
        #Loop
    else if cType = #Option then
        #Opt
    else
        #Opt |
    endif endif endif);
```

Fig. 8. Helper for Mapping ControlType to OperatorType

```
helper def: mapMessageID(m : MMinit!Message) : String = m.id;
helper def : isRepetitiveMessage(m1 : MMinit!Message, messages : Sequence(MMinit!Message)) : Boolean =
    messages->exists(m2 | m1 <> m2 and m1.description = m2.description and m1.source = m2.source and m1.target = m2.target);
helper def : transformMessage(m : MMinit!Message) : MMevol!Message =
    (if m.oclIsTypeOf(MMinit!Synchron) then
        thisModule.transformSynchronousMessage(m.oclAsType(MMinit!Synchron))
    else if m.oclIsTypeOf(MMinit!Asynchron) then
        thisModule.transformAsynchronousMessage(m.oclAsType(MMinit!Asynchron))
    else if m.oclIsTypeOf(MMinit!SelfMessage) then
        thisModule.transformSelfMessage(m.oclAsType(MMinit!SelfMessage))
    else if m.oclIsTypeOf(MMinit!Reply) then
        thisModule.transformReplyMessage(m.oclAsType(MMinit!Reply))
    else if m.oclIsTypeOf(MMinit!CreatMessage) then
        thisModule.transformCreationMessage(m.oclAsType(MMinit!CreatMessage))
    else
        OclUndefined
    endif endif endif endif endif);
```

Fig. 9. Helper for Message Transformation in ATL

The essential transformation rules used in the ATL transformation are represented in the following figures. Each figure captures the details of a specific rule and its application in transforming elements from the initial metamodel to the evolved metamodel (Figs. 10 and 11).

```
rule SequenceDiagramToSequenceDiagram {
    from
        sd : MMinit!SequenceDiagram
    to
        newSD : MMevol!SequenceDiagram (
            name <- sd.name,
            lifeLine <- sd.lifeline -> collect(t | thisModule.transformLifeLine(t)),
            interaction <- sd.controlFlow -> collect(cf | thisModule.transformControlFlow(cf))
        )}
```

Fig. 10. Rule for Transforming SequenceDiagram to SequenceDiagram

```
rule MessageToLoopFragment {
    from
        cf : MMinit!ControlFlow (cf.Type = #Repetition)
    to
        newCf : MMevol!CombinedFragment (
            id <- cf.id,
            type <- #Loop,
            operand <- thisModule.createInteractionOperand(
                cf.annotation,
                cf.constraint,
                cf.message->select(m | not thisModule.isRepetitiveMessage(m, cf.message))
                ->collect(m | thisModule.transformMessage(m))->asSet())) }
```

Fig. 11. Rule for Transforming Message to Loop Fragment

6.5 Applying Transformation Rules

The mapping step involves loading the source model $\mathbf{M_1}$ which is conform to the source metamodel $\mathbf{MM_1}$, applying the transformation rules and updating the models to reflect the changes. This process ensures that each element in the source model is appropriately transformed and mapped to the target model, aligning with the target metamodel $\mathbf{MM_2}$. The formal representation of this step expressed as: $\forall \mathbf{e} \in \mathbf{M_1} \wedge \mathbf{M_1} \models \mathbf{MM_1}, \exists \mathbf{e'} \in \mathbf{M_2} \mid \mathbf{e'} = \mathbf{r_i}$. This formula indicates that for each element e in the source model M1 (which conforms to the metamodel $\mathbf{MM_1}$), there exists a corresponding element e′ in the target model $\mathbf{M_2}$ such that e′ is the result of applying the transformation rule $\mathbf{r_i}$.

The mapping step ensures accurate transformation of each element in the source model to the target model. The source model $\mathbf{M_1}$, conforming to the source metamodel $\mathbf{MM_1}$, is loaded and transformed using defined rules **R**. The rules dictate how each element in M1 should be mapped to the corresponding element in the target model $\mathbf{M_2}$. The target model $\mathbf{M_2}$ is updated based on the transformation rules, with each transformed element $\mathbf{e'}$ added to $\mathbf{M_2}$, ensuring it conforms to the target metamodel $\mathbf{MM_2}$. This systematic approach maintains the original model's integrity while adapting to new features and structures introduced in the evolved metamodel. The mapping step is crucial in the overall model transformation process, facilitating a seamless transition from the source to the target metamodel.

6.6 Validation

The final step in model transformation is to validate the conformity of the transformed model M2 with the evolved metamodel $\mathbf{MM_2}$. This ensures that the transformation rules have been correctly applied and $\mathbf{M_2}$ adheres to the structural and semantic constraints defined by $\mathbf{MM_2}$. The process is expressed as $\mathbf{M_2} \models \mathbf{MM_2}$, indicating $\mathbf{M_2}$ conforms to $\mathbf{MM_2}$. Automated tools perform this validation by comparing $\mathbf{M_2}$ with $\mathbf{MM_2}$, reporting any discrepancies that need correction.

7 Case Study

We have tested our approach using the use case of an ATM withdrawal, a common and relevant use case that showcases various modeling concepts, including synchronous and asynchronous messages, combined fragments like loops and alternatives and error handling.

The example involves a series of interactions between a customer, an ATM and the bank system, including card insertion, PIN verification, amount entry, balance checking, cash dispensing and optional receipt printing. If the balance is insufficient, an appropriate message is displayed and the sequence manages multiple PIN entry attempts (Figs. 12 and 13).

To visualize the example graphically, we used Eclipse Papyrus. The initial model is shown in Fig. 14 and the obtained model is illustrated in Fig. 15.

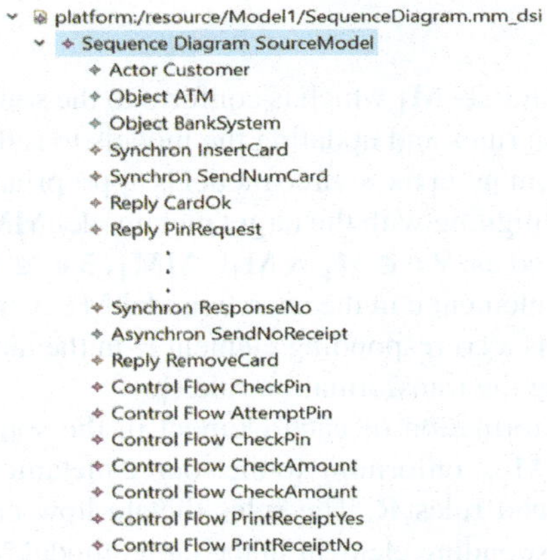

Fig. 12. SourceModel

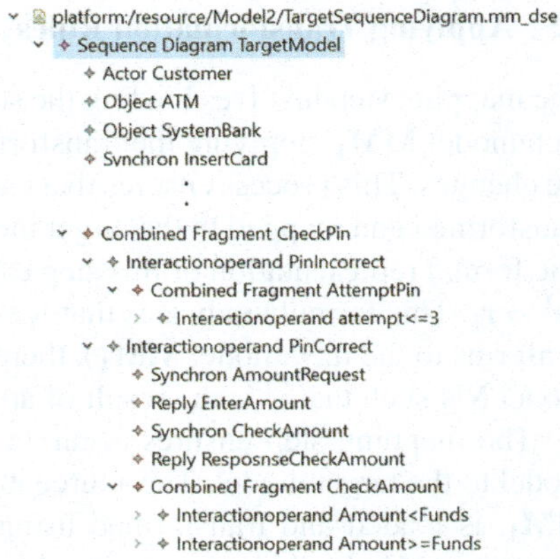

Fig. 13. TargetModel

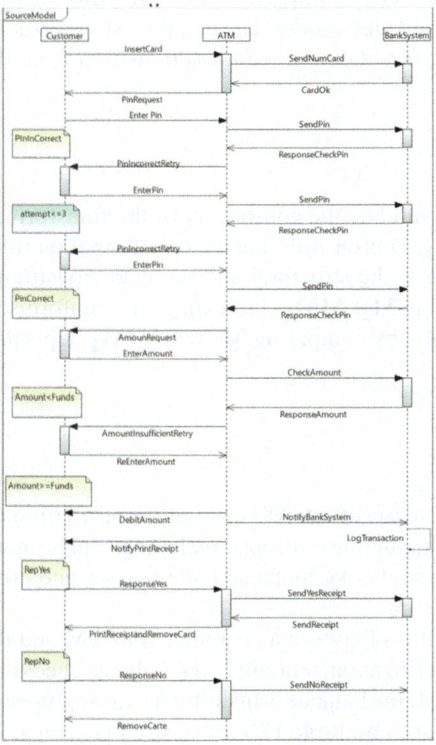

Fig. 14. SourceModel

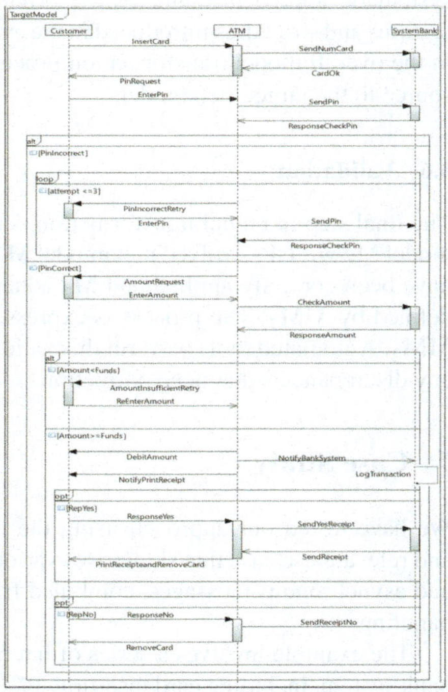

Fig. 15. TargetModel

8 Conclusion

The issue of metamodel evolution and model co-evolution in the field of MDE has been extensively studied over the past decade. In this paper, we address the challenge of model co-evolution and propose an approach using metamodel matching to minimize the effort required for model migration in response to changes in the metamodel. Our approach begins by defining both an initial metamodel and a new metamodel. A delta metamodel is then calculated to capture the differences between these two versions using EMF Compare. Next, the user interprets these differences to define transformation rules, which are implemented using the ATL Transformation Language (ATL). These rules are applied to an input model that conforms to the initial metamodel, adapting it to the new version of the metamodel. To validate our approach, we applied it to a UML SD, demonstrating a scenario of metamodel evolution and model co-evolution to align with the updated metamodel. The UML SD example involves withdrawing money from an ATM, showing the system interactions. Our approach ensures that this diagram remains consistent with the evolved metamodel, reflecting any changes in system functionality or structure. In future works, we aim to apply the proposed approach to other modeling languages and use other comparison techniques to enhance the robustness and applicability of our method.

References

1. Jézéquel, J.M., Combemale, B., Vojtisek, D.: Ingénierie Dirigée par les Modèles: des concepts à la pratique, p. 144. Ellipses (2012)
2. Saïd, B.M., Fleurquin, R., Barais, O.: Evaluation incrémentale des contraintes OCL (2009)
3. OMG: Object Management Group. Unified Modeling Language (UML) superstructure version 2.2. http://www.omg.org/spec/UML/2.2/ (2009)
4. Anguel, F., Amirat, A., Bounour, N.: Hybrid approach for metamodel and model co-evolution. In: Proceedings of Conference on Computer Science and Its Applications (CIIA), pp. 563–573 (2015)
5. Favre, J.-M., Estublier, J., Blay-Fornarino, M.: L'ingénierie dirigée par les modèles au-delà du MDA sous la direction de Hermes Science publications Lavoisier (2006)
6. Rose, L.M., Kolovos, D.S., Paige, R.F., Polack, F.A.C.: Model migration with epsilon flock. In: Proceedings of the International Conference on Theory and Practice of Model Transformations (ICMT), pp. 184–198 (2010)
7. Vermolen, S., Visser, E.: Heterogeneous coupled evolution of software languages. In: Proceedings of the International Conference on Model Driven Engineering Languages and Systems (MODELS), pp. 630–644 (2008)
8. Narayanan, A., Levendovszky, T., Balasubramanian, D., Karsai, G.: Automatic domain model migration to manage metamodel evolution. In: Proc. MODELS'09, LNCS Springer, vol. 5795, pp. 706–711 (2009)
9. Sprinkle, J., Karsai, G.: A domain-specific visual language for domain model evolution. J. Vis. Lang. Comput. **15**(3–4), 291–307 (2004)
10. Kebaili, Z.K., Khelladi, D.E., Acher, M., Barais, O.: Towards leveraging tests to identify impacts of metamodel and code co-evolution. In: International Conference on Advanced Information Systems Engineering, pp. 129–137. Cham: Springer International Publishing (2023)

11. Gruschko, B., Kolovos, D., Paige, R.: Towards synchronizing models with evolving metamodels. In: Proceedings of the Workshop on Model-Driven Software Evolution (MoDSE), pp. 1–9 (2007)
12. Di Ruscio, D., Iovino, L., Pierantonio, A.: What is needed for managing co-evolution in MDE? In: Proc. of the 2nd International Workshop on Model Comparison in Practice, IWMCP'11 (2011)
13. Zhang, W., Holtmann, J., Strüber, D., Hebig, R., Steghöfer, J.P.: Supporting meta-model-based language evolution and rapid prototyping with automated grammar transformation. J. Syst. Softw. **214**, 112069 (2024)
14. Bettini, L., Di Salle, A., Iovino, L., Pierantonio, A.: Supporting reusable model migration with Edelta. J. Syst. Softw. **212**, 112012 (2024)
15. Khelladi, D.E., Hebig, R., Bendraou, R.: Approaches to co-evolution of metamodels and models: a survey. IEEE Trans. Software Eng. **43**(5), 396–414 (2017)
16. Bézivin, J.: Sur les principes de base de l'ingénierie des modèles. RSTI-L'Objet **10**(4), 145–157 (2004)
17. Meta-Object Facility (MOF) Specification, Version. 1.4.1. OMG, formal/05–05–05 (2005)
18. Bézivin, J., Gerbé, O.: Towards a Precise Definition of the OMG/MDA Framework. In: Proceedings of the 16th IEEE international conference on Automated Software Engineering (ASE), p. 273. IEEE Computer Society Press, San Diego, USA (2001)
19. Seidewitz, E.d.l What models mean. IEEE Software **20**(5), 26–32 (2003)
20. Jouault, F., Allilaire, F., Bézivin, J., Kurtev, I.: ATL: A model transformation tool. Sci. Comput. Program. **72**(1–2), 31–39 (2008)
21. OMG: Meta Object Facility (MOF) Core Specification (2016). Retrieved from https://www.omg.org/spec/MOF/2.5.1/PDF
22. García, J., Díaz, O., Azanza, M.: Model transformation co-evolution: a semi-automatic approach. In: SLE, pp. 144–163 (2012)
23. Object Management Group: OMG Unified Modeling Language (UML) Specification (Version 2.5.1) (2017). Retrieved from https://www.omg.org/spec/UML/2.5.1/PDF
24. Bruneliere, H., Cabot, J., Dupe, G., Madiot, F.: MoDisco: a model-driven reverse engineering framework. Inf. Softw. Technol. **56**(8), 1012–1032 (2014)
25. Eclipse Foundation: Eclipse Modeling Framework (EMF) Overview: https://www.eclipse.org/modeling/emf/. Last accessed 15 Jul 2024
26. Eclipse Foundation: EMF Compare Overview: https://www.eclipse.org/emf/compare/. Last accessed 15 Jul 2024
27. Eclipse Foundation: ATL (Atlas Transformation Language) Overview.: https://www.eclipse.org/atl/. Last accessed 15 Jul 2024
28. Eclipse Foundation.: Papyrus Overview. Available at: https://projects.eclipse.org/projects/modeling.mdt.papyrus. Last accessed 15 Jul 2024
29. Eclipse Foundation.: Acceleo Overview. Available at: https://eclipse.dev/acceleo/overview.html. Last accessed 15 Jul 2024
30. Cicchetti, A., Ruscio, D.D., Eramo, R., Pierantonio, A.: Automating co evolution in modeldriven engineering. In: Proceedings of the International Enterprise Distributed Object Computing Conference (EDOC), pp. 222–231 (2008)

Artificial Intelligence and its Applications

Advancing Speech Emotion Recognition: A Comparative Study and Enhanced Performance in English and Algerian Dialect Using Uni and Multimodal Transformer

Sabrina Khemis(✉), Abdelouahab Moussaoui, and Skander Hamdi(✉)

Department of Computer Science, University of Setif 1, Setif, Algeria
sabrinakhemis1@gmail.com,
{abdelouahab.moussaoui,skander.hamdi}@univ-setif.dz

Abstract. In this study, we conduct a comprehensive comparative analysis of various deep learning techniques for the Speech Emotion Recognition (SER) task, ranging from LSTM to CNN-BiLSTM-Multihead-Attention. Our evaluation utilizes the CREMA-D dataset for English and a crowdsourced dataset for the Algerian dialect. We propose two novel transformer-based frameworks, including an unimodal Transformer model and a multimodal Transformer model integrating DziriBERT. The Transformer-based model achieved an accuracy of 95% on the CREMA-D dataset, while the multimodal Transformer model with DziriBERT attained an impressive 97.42% accuracy on the Algerian dialect dataset. The proposed frameworks, unimodal Transformer and multimodal Transformer+DziriBERT significantly outperformed previous studies that used the same employed datasets in English language and Algerian dialect, respectively.

Keywords: Speech Emotion Recognition · Deep Learning · Transformers · Attention Mechanisms · Pre-trained Language Model

1 Introduction

Emotions play a fundamental role in human interactions, conveying crucial information during daily exchanges [18]. The field of emotion recognition has gained significant attention, especially in human-computer interaction systems [17]. This interest has driven the development of various techniques for detecting and analyzing emotions across different modalities, including text, facial expressions, and speech. Among these, speech is particularly notable for its accessibility and naturalness in studying human emotions [11]. Speech Emotion Recognition (SER) focuses on identifying human emotions from voice signals, involving signal pre-processing, feature extraction, feature selection, and classification into emotional classes such as happy, sad, fearful, surprised, and angry.

S. Chikhi et al. (Eds.): MISC 2024, LNNS 1267, pp. 87–101, 2025.
https://doi.org/10.1007/978-3-031-82112-7_7

The applications of SER and audio signal classification are diverse and expanding across various domains. In healthcare, they have been used to detect the intensity of virus attacks through patients' voice signals, particularly relevant during the COVID-19 pandemic [4,8,9]. In customer service, real-time customer emotion recognition helps analyze customer sentiments during interactions, aiding companies in enhancing service quality and customer experience [10]. Educational technology has integrated SER into online learning platforms to analyze students' emotional states, informing personalized learning strategies, improving engagement, and fostering a supportive learning environment [2,19].

In this study, we propose and evaluate several advanced Deep Learning (DL) architectures for SER, including Long Short-Term Memory (LSTM) networks, Convolutional Neural Networks (CNN), hybrid CNN-LSTM models, Attention-based mechanisms, and Transformer-based approaches to English and Algerian Dialect. Our investigation focuses on both unimodal and multimodal techniques, considering the interplay between spectral features and linguistic content in emotional expression by injecting text embeddings as potential features besides the audio. The key contributions of this paper advocate the following:

- A thorough evaluation and comparative study between various DL architectures for the SER task, providing insights into their relative performance and suitability for different languages and dialects.
- The development and training of distinct DL models tailored for both the English language and Algerian Dialect, demonstrating the versatility and adaptability of these models across different linguistic contexts.
- Achieving a classification accuracy exceeding 94% for six emotion classes in the English language using a Transformer model on the CREMA-D dataset, highlighting the efficacy of advanced DL architectures in SER tasks.
- Achieving a remarkable classification accuracy of over 97% for four emotion classes in the Algerian Dialect by integrating a Transformer model with DziriBERT, significantly outperforming previous studies and setting a new benchmark for SER in this dialect.

2 Background and Related Work

The advent of DL has revolutionized the performance of SER systems. By incorporating diverse models such as neural networks, and hybrid approaches, researchers have achieved notable improvements in the accuracy and efficiency of SER tasks. This section provides a comprehensive review of the field, examining the methodologies, features utilized, datasets employed, and the results reported in various studies.

Kenan Donuk [5] focused on enhancing SER by extracting informative features from MFCC coefficient matrices from CREMA-D dataset samples using a CNN-based feature extraction method, followed by feature selection using Particle Swarm Optimization (PSO) to improve classification performance. The researchers reported 66.01% of accuracy as the best achieved performance.

In another work, F. A. Dal Ri et al. [16], employed CNN-based architectures along with convolutional attention block on standard datasets including *CREMA-D*. These works demonstrated variety in the reported performance, with an accuracy of 63% for the CREMA-D dataset.

D. Issa and M. [12] explored a comprehensive feature set comprising MFCCs, Chromagram, Mel-scale Spectrogram, Tonnetz representation, and Spectral Contrast features extracted directly from sound files. Their 1D-CNN model achieves high accuracy on RAVDESS, EMO-DB, and IEMOCAP datasets, surpassing existing benchmarks with results of 71.61%, 86.1%, and 95.71% of accuracy, respectively, in speaker-independent audio classification tasks.

Shifting the focus to Recurrent Neural Networks (RNNs), Wang et al. [20] proposed a dual-level model combining MFCC features and Mel-spectrograms. Each utterance is preprocessed into MFCC features and two mel-spectrograms at different time-frequency resolutions. A standard LSTM processes the MFCC features, while a novel LSTM architecture, denoted as Dual-Sequence LSTM (DS-LSTM), processes the two mel-spectrograms simultaneously. The outputs are averaged to produce a final classification of the utterance, achieving 72.7% weighted and 73.3% unweighted accuracy on IEMOCAP data.

In the context of model hybridization, Etienne et al. [6] computed spectrograms, which were used as input to their proposed combined CNN and Bi-LSTM architecture. This model achieved a weighted accuracy of 64.5% and an unweighted accuracy of 61.7% for four emotions on the IEMOCAP dataset.

On the other hand, Zhao et al. [22] introduced two CNN-LSTM networks (1D and 2D) for learning local and global emotion-related features. Their 2D-CNN-LSTM network achieved accuracies of 95.89% (speaker-independent) on EMO-DB and 52.14% (speaker-independent) on IEMOCAP.

By including the attention mechanism, Yu and Kim [21] combined IS09 features, commonly used in SER, with mel-spectrograms as input features for their proposed Attention-LSTM-Attention model. This model achieved a weighted accuracy of 68% on the original IEMOCAP dataset and 73% on a reconstructed version.

In another work that investigated attention mechanism, Aouani and Ayed [14] implemented six different architectures, focusing on models with attention mechanisms. They used 40 MFCC features and trained the models on SAVEE and RAVDESS datasets. The results showed that networks with attention mechanisms outperformed the others, with the CNN-ATN model achieving the best performance: 74% accuracy on SAVEE and 77% on RAVDESS, surpassing existing state-of-the-art systems for these datasets.

For the Algerian Dialect, Raoudha Y. C. et al. [15] compiled a dataset of 1202 audio recordings for four classes (Sad, Angry, Happy, and Neutral) from TV shows, then, explored multiple classification models, including ML and DL architectures such as CNNs, LSTM networks, and a hybrid LSTM-CNN model to train a DL model on MFCC features. The proposed hybrid model achieved a notable classification accuracy of 93.34%. Table 1 summarizes the mentioned related works and highlights the dataset, language or dialect, the employed methodology, and the reported accuracy (some of the studies reported weighted and unweighted accuracy).

Table 1. Summary of the aforementioned related works on SER

Work	Dataset	Language or Dialect	Classifier	Accuracy
Kenan Donuk [5]	CREMA-D	English	CNN+PSO+SVM	66%
F. A. Ri et al. [16]	CREMA-D	English	Att-CNN	63%
D. Issa and M. [12]	RAVDESS EMO-DB IEMOCAP	English German English	1D-CNN	71.61% 86.1% 95.71%
Wang et al. [20]	IEMOCAP	English	DS-LSTM	73.3%
Etienne et al. [6]	IEMOCAP	English	CNN-BiLSTM	64.5%
Zhao et al. [22]	EMO-DB	English	2D-CNN-LSTM	95.89%
Yu and Kim [21]	IEMOCAP	English	Att-LSTM-Att	73%
Aouani and Ayed [14]	SAVEE + RAVDESS	English	CNN-ATN	77%
Raoudha Y. C. *et al.* [15]	Crowdsourced	Algerian	LSTM-CNN	93.34%

3 Methodology

3.1 Employed Datasets

The performance of the proposed models was evaluated on two datasets: a crowd-sourced Algerian Dialect Dataset and the CREMA-D dataset. The CREMA-D dataset was chosen due to its reliability and usage in SER research. This widely used dataset efficiently compares our model's performance with other studies.

Algerian Dialect Dataset: The Algerian dataset used in our study consists of 2235 WAV files organized into two parts: The first one, named **"A"**, composed of **1202** WAV files, reported in [15]. These recordings were collected from well-known Algerian TV shows such as *"Imarat lhadj Lakhdhar"*, *"Soug nssa"*, and *"Taht lmourakaba"*. It consists of natural speech from diverse groups, including 21 men, 12 women, and 4 children, each expressing one of four basic emotions: *Neutral*, *Angry*, *Happy*, and *Sad*. This portion of data has been manually and carefully annotated by three native Algerian listeners where each one was asked to write down what was heard in the audio recording. The second part of the dataset, named **"B"**, consists of 1033 WAV files collected from the public society. This addition aims to include the variability in emotional expression across different speakers and contexts, thereby contributing to the generalizability of emotion recognition models. Both parts **"A"** and **"B"** are used for all experiments, while only **"A"** was used to train the multimodal Transformer model to ensure a fair comparison with the previous study.

CREMA-D Dataset: *CRowd-sourced Emotional Multimodal Actors* Dataset[1] is composed of 7442 original audio-visual clips from 91 actors (48 male and 43 female) aged between 20 and 74, from a variety of races and ethnicities. The clips were recorded using 12 sentences, with each sentence expressed through one of six different emotions: *Happiness*, *Sadness*, *Anger*, *Fear*, *Disgust*, and *Neutrality*.

3.2 Audio Feature Extraction and Data Augmentation

To effectively analyze and train DL models on audio features in the frequency domain, we employed MFCCs as spectral representation, due to their proven efficiency as reported in related works. MFCCs were selected for their effectiveness in capturing distinct audio properties, their robustness to noise, and their ability to provide a compact feature representation. The logarithmic frequency scaling in MFCCs aligns well with human auditory perception, facilitating meaningful spectral analysis. Furthermore, the widespread use of MFCCs in audio processing ensures access to established methodologies and tools. MFCCs can be formulated as follows:

$$\mathrm{MFCC}_n = \sum_{m=1}^{M} \log E_m \cos\left(\frac{\pi n(m-0.5)}{M}\right), n = 1, 2, \ldots, L \qquad (1)$$

where E_m represents the energies of the Mel-filtered spectrum. We chose to extract 40 MFCC features after performing an extensive hyperparameter tuning process using GridSearch.

Given the limited number of audio recordings, particularly in the Algerian Dialect dataset, data augmentation techniques are essential to enrich the dataset, introduce more variability, and avoid the risk of overfitting. Data augmentation techniques were implemented to enhance the model's robustness: **Noise injection**, **Time stretching**, and **Pitch shifting**. These methods were chosen to simulate diverse recording conditions and account for variations in speech tempo and intonation.

3.3 Text Feature Extraction and Data Augmentation

For models that involve textual data, the process involved tokenization followed by applying a pre-trained language model, resulting in 768-dimensional text embeddings, these embeddings capture both linguistic nuances and emotional content inherent in the text. To enhance the dataset's diversity and model robustness, four text augmentation techniques were employed: **Synonym replacement**, **Random insertion**, **Random swap**, and **Random deletion**. For the DziriBERT model [1], a specific **dictionary of synonyms** for the Algerian dialect was developed to facilitate Synonym replacement and Random insertion techniques. In contrast, the AraBERT model [3] utilized the Arabic WordNet (AWN) [7] resource from the (Natural Language Toolkit) NLTK[2].

---
1 https://www.kaggle.com/datasets/ejlok1/cremad.
2 https://www.nltk.org.

3.4 Baselines and Proposed Classifier

In this section, we present a range of architectures for the SER task, encompassing both baseline models and our proposed classifier. Our baseline models include various configurations such as LSTM, CNN, CNN-LSTM, CNN-BiLSTM, CNN-Multihead-Attention, and CNN-BiLSTM-Multihead-Attention, each designed to leverage different strengths in handling temporal and spatial features of audio data. Additionally, we explore the use of a Transformer model that processes only audio data to further benchmark performance. Our main contribution, however, is the development of a Multimodal Transformer that combines audio features extracted from MFCCs with text embeddings generated using AraBERT at the first stage and DziriBERT at the final stage. By gradually adding more components to the basic architecture, we aim to achieve a more refined extraction of patterns, enhancing the robustness and accuracy of SER. This innovative approach integrates complementary information from both audio and text modalities.

LSTM: Given the sequential nature of speech data, we employed LSTM architecture. It consists of two LSTM layers with 64 and 128 units, followed by a Dense layer (64 units) with ReLU activation. L2 regularization (0.01) and dropout (rates 0.2, 0.3, and 0.5) were applied across layers to avoid overfitting.

CNN: The model comprises two 1D convolution layers with 64 and 128 filters to process MFCC sequences, each followed by Batch Normalization and ReLU activation. Dropout (0.1) and Max Pooling with a pooling window size of 4 are applied for regularization and feature extraction. The flattened output feeds into two Dense layers (256 and 128 units) with an intermediate Dropout (0.3). The final layer uses softmax activation for classification.

CNN-LSTM: This hybrid model incorporates two LSTM layers with 64 and 128 units following the second convolution layer. Dropout was used to prevent overfitting. LSTM output is then processed by a fully connected network for final classification, similar to the previous approach.

CNN-BiLSTM: In this hybrid model, we kept the same structure as the CNN-LSTM architecture 3.4, however, we replaced the two LSTM layers with a BiLSTM layer containing 128 units. The advantage of using a BiLSTM lies in its ability to capture dependencies in both forward and backward directions, ensuring a more comprehensive understanding of the temporal dynamics in the audio data.

CNN Multi-head Attention: This model integrates multi-head attention with CNN to enhance the focus on relevant features.

The architecture utilizes 8 parallel attention heads, each performing scaled dot-product attention. The attended outputs are then concatenated and linearly transformed. A GlobalAveragePooling1D layer generates a fixed-length vector, which is subsequently fed into fully connected layers for classification. The convolutional and fully connected layers, along with their unit parameters, remain identical to those in the CNN architecture detailed in Sect. 3.4. The key addition here is the attention mechanism, which aims to improve the model's ability to discern and emphasize important features within the audio data.

CNN BiLSTM Multi-head Attention: This architecture extends the CNN Multi-head Attention model by incorporating a BiLSTM layer between the CNN and attention components. The model processes MFCC features through the following sequence: it starts with two convolutional layers, followed by a BiLSTM layer with 256 units and L2 kernel regularization. The output then passes through a multi-head attention mechanism with 8 heads, similar to the setup in the CNN Multi-head Attention model (Sect. 3.4). After the attention mechanism, a Global Average Pooling 1D layer produces a fixed-length vector, which is then fed into fully connected layers for classification. This hybrid approach combines the spatial feature extraction capabilities of CNNs, the sequential learning strengths of BiLSTMs, and the focused feature weighting provided by attention mechanisms. The convolutional and fully connected layers retain the specifications from the original CNN architecture described in Sect. 3.4.

Transformer-Based Model: A novel transformer architecture was applied for classifying emotions by using MFCC features extracted from audio signals. As we are dealing with a classification task, the model employs only the encoder part. Figure 1 showcases the proposed Transformer-based model.

The Transformer-based model comprises four key elements:

- **Embedding Layer:** Initially, the MFCC features transformed into a higher-dimensional vector of size 512. This step enhances the model's capacity to capture complex patterns within the acoustic data.
- **Positional Encoding:** To preserve sequential information, positional encoding is then applied to these embedded vectors.
- **Transformer Encoder Layers:** The embedded and position-encoded vectors are subsequently processed through two transformer encoder layers. Each encoder layer employs 12 attention heads, combining self-attention mechanisms to capture salient sequence information.
- **Classification Layer:** A final fully connected layer maps the processed data to emotion class predictions.

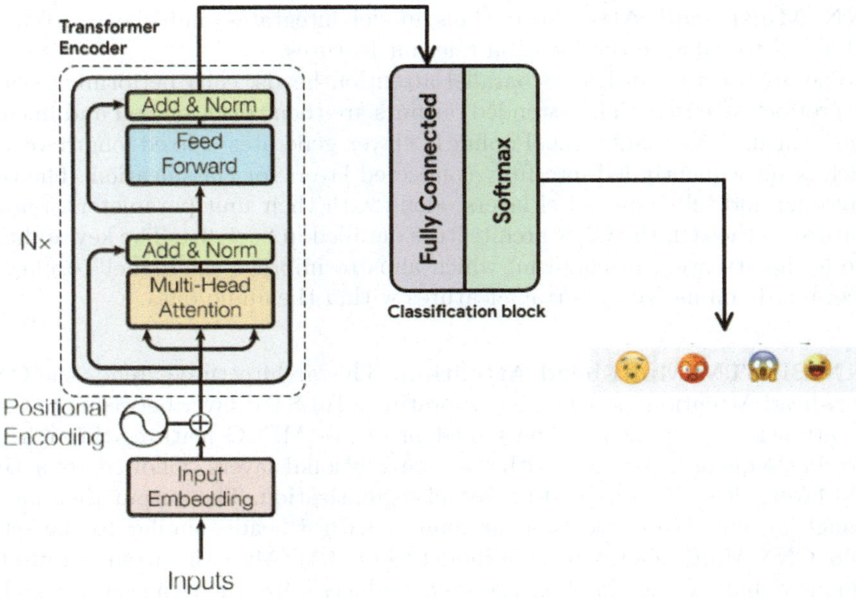

Fig. 1. Block diagram of the proposed Transformer-based model

Multimodal Transformer: This section presents our novel approach to SER, which utilizes a multimodal transformer architecture that receives both audio and textual data. This methodology aims to combine the complementary strengths of spectral and linguistic features for more comprehensive emotion analysis. A block diagram of the full framework that uses the proposed Multimodal Transformer is illustrated in Fig. 2. Multimodal Transformer processes dual input channels:

- **Audio channel:** MFCCs extracted from audio recordings are fed into a transformer encoder.
- **Text channel:** Transcripts are processed through a pre-trained language model to generate *embeddings* that capture linguistic features and emotional content.

After computing MFCC features, and extracting text embeddings, the model combines both input channels, enabling parallel processing of *spectral* and *linguistic* features. This merged representation is further processed through fully connected layers for final emotion classification.

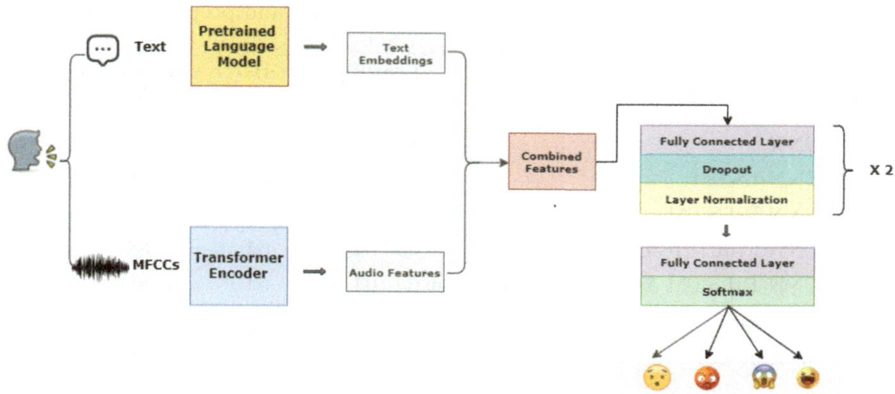

Fig. 2. Proposed framework that utilizes a multimodal transformer model where spectral features and text embeddings are combined and forwarded to the learning and classification phase

4 Experimental Setup

4.1 Evaluation Metrics

To comprehensively evaluate the performance of the proposed architectures, we used various evaluation metrics, including **Accuracy (%)**, **Precision (%)**, **Recall (%)**, and **F1-score (%)**. These metrics are essential for understanding the strengths and weaknesses of the models.

4.2 Training Phase and Hyperparameter Tuning

The models were trained on different hardware setups: some were trained on a local machine with an i5-1145G CPU and 16 GB of RAM, while others utilized Kaggle notebooks[3] equipped with 2 NVIDIA P4 GPUs and an NVIDIA T4 GPU. Each of the used datasets was split into two parts: 80% for training and 20% for test.

We employed GridSearch to determine the optimal set of hyperparameters for our models. The hyperparameters tuned include the number of computed MFCC features, the number of fully connected layers, their respective number of units, learning rate, batch size, and the number of epochs. Several hyperparameters have been mentioned in the previous sections, and Table 2 shows the optimal values found for the learning rate, batch size, and number of epochs.

All models were trained using the Adam optimizer [13] and categorical cross-entropy as the loss function. We employed several callbacks: *early stopping*, *reducing learning rate on plateau*, and *model checkpointing* to enhance model performance and prevent overfitting. Early stopping was configured with a patience of

[3] https://kaggle.com/code.

Table 2. Tuned hyperparameters of each of the baselines and proposed architectures

Model	Learning rate	Batch size	Epochs
LSTM	1e−03	32	300
CNN	1e−04	32	300
CNN-LSTM	1e−04	64	300
CNN-BiLSTM	5e−04	64	200
CNN Multi-Head Attention	6e−04	64	300
CNN-BiLSTM Multi-Head Attention	6e−04	64	200
Transformer-based model	1e−04	64	150
Multimodal Transformer	1e−04	32	50

10 epochs, monitoring the validation loss. Reducing the learning rate was set to monitor validation accuracy and reduce the learning rate by a factor of 0.5 after 5 epochs of no improvement, with a minimum learning rate of 1e−05.

5 Results and Analysis

The achieved performance is presented in this section for each dataset separately.

5.1 Performance on CREMA-D Dataset

Table 3 shows the performance metrics achieved by the proposed baseline models as well as the novel Transformer-based approach on the CREMA-D dataset. This comprehensive comparison highlights the strengths of each architecture. Figure 3 illustrates the training and validation accuracy and loss curves of our novel Transformer-based approach, demonstrating the learning dynamics and model convergence over epochs.

Table 3. Performance of the baselines and the proposed Transformer-based model

Model	Acc. (%)	Prec. (%)	Rec. (%)	F1 (%)
LSTM	55.77	57.80	55.75	55.62
CNN	67.30	67.88	67.19	67.29
CNN-LSTM	76.65	76.96	76.85	76.74
CNN-BiLSTM	83.31	87.61	80.73	84.33
CNN Multi-Head Attention	87.10	90.48	80.95	85.72
CNN BiLSTM Multi-Head Attention	94.54	94.91	93.66	94.04
Transformer Encoder	**95.00**	**95.35**	**95.13**	**95.34**

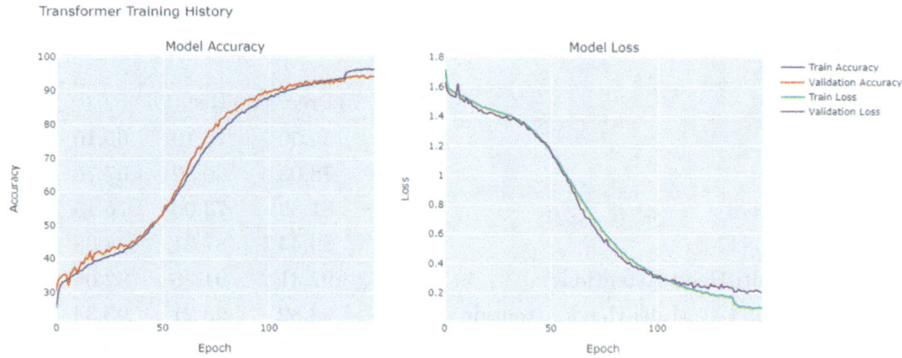

Fig. 3. Training history of Transformer-based model on CREMA-D dataset

From Table 3 LSTM model exhibits the lowest performance (55.77% of accuracy), indicating that only temporal dependencies are not enough to distinguish between emotions. The CNN model shows significant improvements (67.30% of accuracy), benefiting from spatial and spectral feature extraction. Furthermore, the CNN-LSTM model enhances performance by combining spatial and temporal modeling, highlighting the benefits of hybridization. Introducing bidirectional LSTM to CNN-LSTM improves context capture, yielding better model overall accuracy (83.31%). On the other hand, the multi-head attention, through several parallel attention heads for both CNN and BiLSTM multi-head attention, the model shows a significant improvement for all metrics (94.54% of accuracy) due to the particular focus on the most relevant features. Our Transformer-based architecture, outperforms all baselines, achieving the highest metrics with overall accuracy of 95%. Its ability to capture long-range dependencies and focus on relevant features through self-attention mechanisms proves highly effective for emotion classification. Figure 3 shows a highly consistent and smooth progression of the training process. There is no observable gap between the training and validation curves, indicating that the models generalize well to unseen data. The smoothness of the curves, devoid of perturbations or fluctuations, further highlights the stability and robustness of the training process. This consistent performance reflects the models' effectiveness in capturing and learning from the MFCC features without overfitting or underfitting.

5.2 Performance on Algerian Dialect Dataset

Table 4 summarizes the performance metrics of our novel Multimodal Transformer combined with DziriBERT on the Algerian dialect dataset compared to the proposed baselines. The table highlights the model's effectiveness in terms of accuracy, precision, recall, and F1-score. Following this, Fig. 4 illustrates the training and validation accuracy and loss curves, demonstrating the model's learning progression and stability.

Table 4. Performance of our Multimodal Transformer + DziriBERT compared to the baselines on the Algerian dialect dataset

Model	Acc. (%)	Prec. (%)	Rec. (%)	F1 (%)
LSTM	67.29	55.50	76.10	65.10
CNN	69.79	78.03	60.32	67.76
CNN-LSTM	77.12	81.30	73.06	76.15
CNN-BiLSTM	88.44	89.44	87.31	88.08
CNN Multi-Head Attention	92.19	92.41	91.36	92.05
CNN-BiLSTM Multi-Head Attention	93.38	93.82	93.21	93.34
Transformer Encoder	90.71	90.73	90.32	90.50
Multimodal Transformer + AraBERT	91.76	88.12	86.17	88.27
Multimodal Transformer + DziriBERT	**97.42**	**97.59**	**96.61**	**97.08**

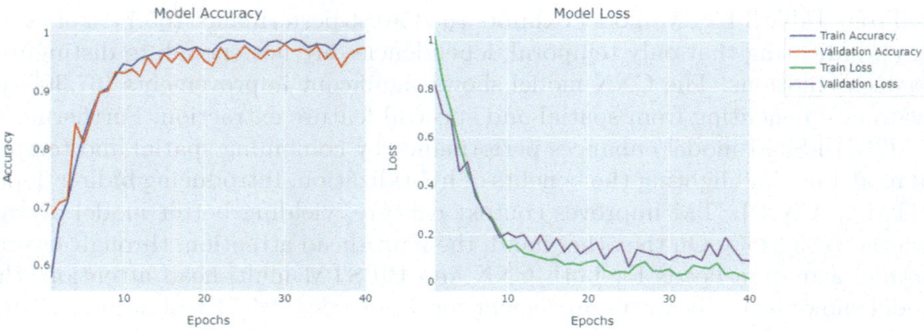

Fig. 4. Training history of Multimodal Transformer with DziriBERT model on Algerian dialect dataset

The results on the Algerian dialect dataset, shown in Table 4, demonstrate that more advanced models significantly outperform simpler ones. While LSTM and CNN models show moderate accuracy, hybrid models with BiLSTM and multi-head attention, like the CNN-BiLSTM Multi-Head Attention, achieve up to 93.38% accuracy. The Transformer Encoder alone performs well at 90.71%, but the Multimodal Transformer with DziriBERT excels at 97.42% accuracy. This improvement is due to DziriBERT being trained on a large-scale Algerian dialect dataset, capturing its linguistic characteristics better than AraBERT, which is trained on a large-scale Arabic dataset. This superiority mirrors trends seen in the CREMA-D dataset, emphasizing the efficacy of advanced architectures and attention mechanisms for speech emotion recognition. Regarding the training process, highlighted in Fig. 4, it shows a consistent, and smooth evolution. This demonstrates stable learning and effective generalization on unseen data.

5.3 Comparison with Prior Works

In this subsection, we present a comparative analysis of our proposed models against prior studies using the CREMA-D and the Algerian dialect datasets. This comparison aims to highlight the effectiveness and improvements brought by our approaches in the SER task for English language and Algerian dialect. Tables 5 and 6 summarize the performance metrics of related works and our models for CREMA-D and Algerian dialect datasets, respectively, providing a clear perspective on advancements and benchmarks in the field. Our Unimodal Transformer-based model significantly outperforms prior works on the CREMA-D dataset by achieving 95.00% accuracy, compared to 63.00% and 66.01%.

Table 5. Comparison of our Unimodal Transformer-based model on CREMA-D dataset with prior works (we shall note that the mentioned studies reported only the accuracy metric)

Model	Accuracy (%)	Precision (%)	Recall (%)	F1 score (%)
F. A. Dal Rì *et al.* [16]	68.00	–	–	–
Kenan Donuk [5]	66.01	–	–	–
Our work	**95.00**	**95.35**	**95.13**	**95.34**

On the other hand, our novel Multimodal Transformer + DziriBERT model shows an improvement of more than 4% compared to the prior work that proposed a hybrid LSTM-CNN.

Table 6. Comparison of our Multimodal Transformer + DziriBERT model on Algerian Dialect data with prior study that used the same dataset

Model	Acc. (%)	Prec. (%)	Rec. (%)	F1 (%)
Raoudha Y. C. *et al.* [15]	93.34	–	–	–
Our framework	**97.42**	**97.59**	**96.61**	**97.08**

6 Conclusion

In this study, we explored the effectiveness of different DL architectures for SER, demonstrating a significant progression in performance metrics as model complexity increased when data was more complex like the case in audio classification tasks. We started with basic LSTM models and progressed to more complex architectures, including CNNs, hybrid CNN-LSTM, CNN-BiLSTM, and CNN Multi-Head Attention models. The CNN-BiLSTM Multi-Head Attention model achieved notable accuracy levels of 93.83% on the Algerian dialect dataset and 94.54% on the CREMA-D dataset. The performance further improved with a Transformer-based architecture, reaching 95% accuracy on the CREMA-D dataset and 90.71% on the Algerian dialect dataset. Furthermore, our novel multimodal Transformer, combined with DziriBERT's text embeddings, attained remarkable results, achieving 97.42% accuracy on the Algerian dialect dataset, This superior performance underscores the potential of attention mechanisms and multi-modal-based methods in enhancing the robustness and accuracy of SER systems. However, the performance of the proposed approach for the Algerian dialect is still questionable due to the small size of the dataset (1202 audio recordings). This limitation should be addressed in future work. We are currently working on annotating the remaining 1303 recordings for the Algerian dialect and exploring the use of facial expressions as another data modality, which may further enhance the model's performance and generalizability.

Acknowledgments. We want to thank Raoudha Y. C. for providing the spoken Algerian dialect dataset, which was invaluable for conducting our research.

References

1. Abdaoui, A., Berrimi, M., Oussalah, M., Moussaoui, A.: DziriBERT: a pre-trained language model for the Algerian dialect. arXiv preprint arXiv:2109.12346 (2021)
2. Abdelhamid, A.: Speech emotions recognition for online education. Fusion Pract. Appl. **10**, 78–87 (2022). https://doi.org/10.54216/FPA.100104 https://doi.org/10.54216/FPA.100104 https://doi.org/10.54216/FPA.100104
3. Antoun, W., Baly, F., Hajj, H.: AraBERT: transformer-based model for Arabic language understanding. In: LREC 2020 Workshop Language Resources and Evaluation Conference, 11–16 May 2020, p. 9 (2020)
4. Despotovic, V., Ismael, M., Cornil, M., Call, R.M., Fagherazzi, G.: Detection of COVID-19 from voice, cough and breathing patterns: dataset and preliminary results. Comput. Biol. Med. **138**, 104944 (2021). https://doi.org/10.1016/j.compbiomed.2021.104944. epub 13 Oct 2021
5. Donuk, K.: CREMA-D: improving accuracy with BPSO-based feature selection for emotion recognition using speech. J. Soft Comput. Artif. Intell. **3**(2), 51-57 (2022). https://doi.org/10.55195/jscai.1214312 https://doi.org/10.55195/jscai.1214312 https://doi.org/10.55195/jscai.1214312
6. Etienne, C., Fidanza, G., Petrovskii, A., Devillers, L., Schmauch, B.: CNN+ LSTM architecture for speech emotion recognition with data augmentation. arXiv preprint arXiv:1802.05630 (2018)
7. Goodman, M.W.: Wn: a python library for working with wordnets. https://github.com/goodmami/wn. Accessed 21 June 2024
8. Hamdi, S., Moussaoui, A., Oussalah, M., Saidi, M.: Autoencoders and ensemble-based solution for COVID-19 diagnosis from cough sound. In: Chikhi, S., Diaz-Descalzo, G., Amine, A., Chaoui, A., Saidouni, D.E., Kholladi, M.K. (eds.) Modelling and Implementation of Complex Systems, pp. 279–291. Springer, Cham (2023)

9. Hamdi, S., Oussalah, M., Moussaoui, A., Saidi, M.: Attention-based hybrid CNN-LSTM and spectral data augmentation for COVID-19 diagnosis from cough sound. J. Intell. Inf. Syst. (2022). https://doi.org/10.1007/s10844-022-00707-7
10. Hashem, A., Arif, M., Alghamdi, M.: Speech emotion recognition approaches: a systematic review. Speech Commun. **154**, 102974 (2023). https://doi.org/10.1016/j.specom.2023.102974, https://www.sciencedirect.com/science/article/pii/S0167639323001085
11. Huang, L., Dong, J., Zhou, D., Zhang, Q.: Speech emotion recognition based on three-channel feature fusion of CNN and biLSTM. In: Proceedings of the 2020 the 4th International Conference on Innovation in Artificial Intelligence, ICIAI 2020, pp. 52–58. Association for Computing Machinery, New York (2020). https://doi.org/10.1145/3390557.3394317
12. Issa, D., Fatih Demirci, M., Yazici, A.: Speech emotion recognition with deep convolutional neural networks. Biomed. Signal Process. Control **59**, 101894 (2020). https://doi.org/10.1016/j.bspc.2020.101894, https://www.sciencedirect.com/science/article/pii/S1746809420300501
13. Kingma, D., Ba, J.: Adam: a method for stochastic optimization. In: International Conference on Learning Representations (2014)
14. Mountzouris, K., Perikos, I., Hatzilygeroudis, I.: Speech emotion recognition using convolutional neural networks with attention mechanism. Electronics **12**(20) (2023). https://doi.org/10.3390/electronics12204376, https://www.mdpi.com/2079-9292/12/20/4376
15. Raoudha, Y.C., Moussaoui, A., Frahta, N., Berrimi, M.: Effective speech emotion recognition using deep learning approaches for Algerian dialect. In: 2021 International Conference of Women in Data Science at Taif University (WiDSTaif), pp. 1–6 (2021). https://doi.org/10.1109/WiDSTaif52235.2021.9430224
16. Rí, F.A.D., Ciardi, F.C., Conci, N.: Speech emotion recognition and deep learning: an extensive validation using convolutional neural networks. IEEE Access **11**, 116638–116649 (2023). https://doi.org/10.1109/ACCESS.2023.3326071
17. Saxena, A., Khanna, A., Gupta, D.: Emotion recognition and detection methods: a comprehensive survey. J. Artif. Intell. Syst. **2**, 53–79 (2020). https://doi.org/10.33969/AIS.2020.21005
18. Selvaraj, M., Bhuvana, R., Karthik, S.P.: Human speech emotion recognition. Int. J. Eng. Technol. (IJET) **8**, 311–323 (2016)
19. Vistorte, A.O.R., Deroncele-Acosta, A., Ayala, J.L.M., Barrasa, A., López-Granero, C., Martí-González, M.: Integrating artificial intelligence to assess emotions in learning environments: a systematic literature review. Front. Psychol. **15**, 1387089 (2024). https://doi.org/10.3389/fpsyg.2024.1387089
20. Wang, J., Xue, M., Culhane, R., Diao, E., Ding, J., Tarokh, V.: Speech emotion recognition with dual-sequence LSTM architecture. In: ICASSP 2020-2020 IEEE International Conference on Acoustics, Speech and Signal Processing (ICASSP), pp. 6474–6478. IEEE (2020)
21. Yu, Y., Kim, Y.: Attention-LSTM-attention model for speech emotion recognition and analysis of iemocap database. Electronics **9**(5), 713 (2020)
22. Zhao, J., Mao, X., Chen, L.: Speech emotion recognition using deep 1D & 2D CNN LSTM networks. Biomed. Signal Process. Control **47**, 312–323 (2019)

Material Property Prediction with CNN-LSTM Hybrid Models and Periodic Table as Input Representation

Saad Boudabia[1(✉)], Youcef Mahdjoub[1], Ahmed Draoui[2], Fatima Zohra Bensalem[3], and Kheira Laiourate[3]

[1] Laboratoire de Mathematique et Sciences Appliquées, LMSA, University of Ghardaia, Ghardaia, Algeria
{boudabia.saad,youcef.mahdjoub}@univ-ghardaia.dz
[2] ENSL, Laghouat, Algeria
a.draoui@univ-lagh.dz
[3] University of Ghardaia, Ghardaia, Algeria

Abstract. The integration of machine learning technologies, especially convolutional neural networks (CNN) and Long Short Term Memory (LSTM) in the extraction of material properties from the periodic table represents a significant advance in materials science research. CNN are widely used in artificial intelligence and machine learning applications due to their ability to handle structured complex data, including text, images, and even multi-dimensional data. The aim of this study is to understand the relationship between the energy of formation of compounds with the chemical formula ABX_3 referred to as Perovskites and the arrangement in the periodic table of the elements that form them, using CNN-LSTM hybrid model. The model is trained on the Open Quantum Materials Database (OQMD) dataset and achieves a high accuracy of 0.98 and a mean prediction error of 0.055 compared to Density Functional Theory (DFT) calculations. These results demonstrate the ability of CNN and LSTM to integrate structural information from the periodic table to predict the properties of materials, paving the way for the development of new technologies to design materials and improve energy efficiency in the future.

Keywords: Convolutional neural networks (CNN) · Long Short Term Memory (LSTM) · Material properties · machine learning · OQMD dataset · Density Functional Theory (DFT) · Energy efficiency

1 Introduction

In the dynamic field of materials science, the task of comprehension and prediction of the properties of materials is fundamental to innovation and progress. Traditionally, this endeavor has relied on empirical experimentation and theoretical models to decrypt the complex interplay between chemical composition and material behavior [1]. However, with the emergence of machine learning technologies, and in particular Convolutional

S. Chikhi et al. (Eds.): MISC 2024, LNNS 1267, pp. 102–111, 2025.
https://doi.org/10.1007/978-3-031-82112-7_8

Neural Networks (CNNs) [2] and Long Short Term Memory (LSTM) [3], a new epoch of predictive modeling has opened up, offering unprecedented insights into the properties of materials. The integration of CNN into materials science research represents a significant advance, as it harnesses the power of artificial intelligence to extract valuable information from large and complex datasets. Renowned for their ability to process structured data in a variety of domains, including text, images and multi-dimensional structures, CNNs offer a promising avenue for exploring the complex relationships between chemical elements and material properties, while LSTM enhances the model's capacity to handle temporal tasks and improves its ability to capture sequential data structures.

The PTR (Periodic Table Representation) approach can effectively address the data scarcity challenge often encountered in materials research by using the inherent knowledge encoded in the periodic table to provide meaningful data for materials with limited experimental data. This representation can also facilitate transfer learning [4], allowing a model pre-trained on a large dataset to be refined on a smaller dataset with specific properties of materials.

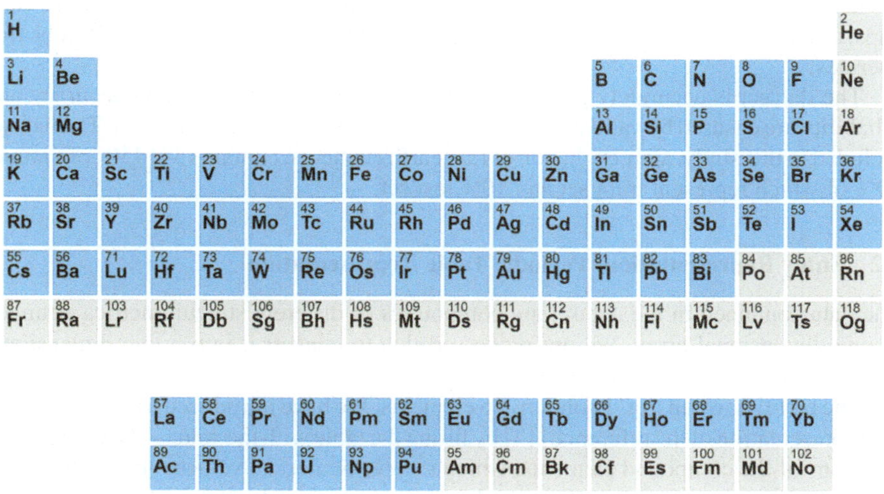

Fig. 1. The Periodic Table Representation

The combination of PTR and CNN offers immense potential to revolutionize materials discovery and property prediction. By enabling efficient and accurate prediction of material properties directly from the periodic table, this approach can accelerate the development of new materials with properties suited to a variety of applications. It could transform fields such as materials science, engineering and chemistry [5].

Perovskites compounds are chosen to demonstrate our CNN-LSTM and PTR approach. Perovskites compounds are used in sustainable technologies; they are used as electrode materials for solid oxide fuel cells where high ionic, electronic and mixed ionic/electronic conductivity materials are required [6]. Stability prediction of Perovskites compounds has been performed in the literature using machine learning methods

[7, 8], chemical models [9] and Theory of Functional Density (DFT) calculations [10, 11].

The rest of this article is organized as follows. In the Methods section, the architectures of the trained models are presented, as well as the representation of the model inputs, and the dataset used in this research. The results of the model training with a brief discussion are introduced in Sect. 3. Finally, in Sect. 4, we present our conclusions and a brief outlook on the future of CNN training for predicting the physical properties of materials.

2 Methods

2.1 Dataset

The dataset used was extracted from the Open Quantum Materials Database (OQMD) [12], which is a rich resource for materials science research, providing comprehensive data on the properties of various materials. It includes detailed structural information, formation energies, electronic properties such as band structures and electronic band gaps, as well as thermodynamic properties such as enthalpy of formation and Gibbs free energies.

The dataset is accurately generated using the DFT calculations, and accuracy and reliability are ensured by validation against experimental measurements [13]. The dataset included in this study, after exclusion of rare earth elements, contains 341450 compound, 90% of which represent train set and 10% test set.

2.2 Entry Representation: Periodic Table Representation

The data contained in the dataset are compounds of different stoichiometries arranged in one-dimensional array. We pre-processed this to convert it into a two-dimensional matrix to be used as input to our models.

The table shown in Fig. 1 contains 86 elements. We have introduced a modification to the representation of the reference [14] by filling the table with the value -1 except for the positions of the compound elements, which receive numerical values representing their fractions of the compound, and the total value is normalized to 83 or 82. For example, in the case of a ternary Perovskite ABX_3, 83 positions receive -1 (i.e. elements not in the compound) and 3 positions receive the values 16.6, 16.6 and 49.8 respectively.

2.3 CNN and LSTM Architectures

In the present work, to predict the enthalpy of formation of the Perovskites included in the dataset, we trained a CNN model and another called CNN-LSTM. CNN are used in many fields due to their ability to extract features from images and other structured data, which can be defined mathematically [15] as follow:

$$x(t) = \sum\nolimits_{\tau=-\infty}^{\infty} x(\tau)\delta(t-\tau) \tag{1}$$

where $\delta(t)$ is the sequence such that $\delta(0) = 1$ and for $t \neq 0$, $\delta(t) = 0$

- **CNN model:** The model as illustrated in Fig. 2, consists of three convolutional layers associated with activation functions, the first and third of which use 3×3 kernels, while the second uses a 5×5 kernel.

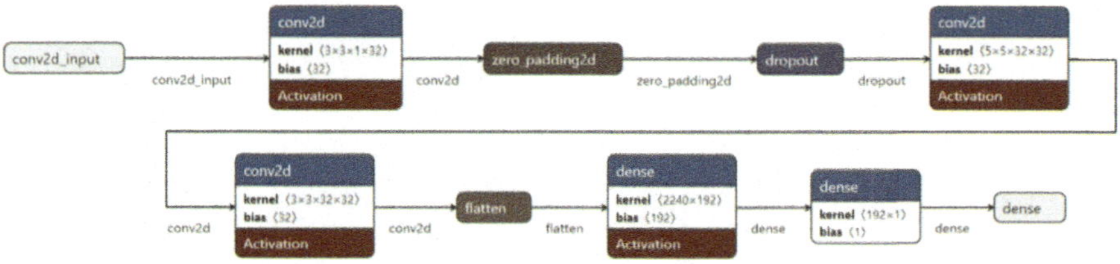

Fig. 2. CNN architecture

- **CNN-LSTM model:** LSTM stands for "Long Short-Term Memory". It is a Recurrent Neural Network (RNN) architecture designed to deal efficiently with problems where the sequence of data is important. Figure 3 shows the model, which is a combination of CNN and LSTM. Its architecture includes two LSTM-type layers after the three convolutional layers of the first model.

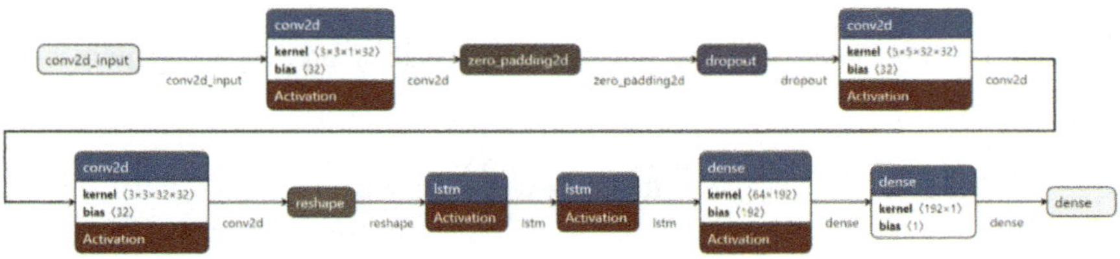

Fig. 3. CNN-LSTM architecture

During the training process, Mean Absolute Error (MAE) [16] and Huber loss [17] are used as loss functions.

The mathematical expression defined by Eq. 2 represents the MAE loss function, where N denotes the size of the data set. For an input vector x_n, the network calculates an output $y(x_n)$, which is compared with a target dn. In this way, the MAE evaluates the difference between the predicted output and the actual target.

$$E(\omega) = \frac{1}{N} \sum_{n} |d_n - y(x_n)| \tag{2}$$

Huber's loss is a loss function used in supervised learning, mainly in regression tasks. Its mathematical expression is given by Eq. 3.

$$f(d) = \begin{cases} 0.5d^2 & if\,|d| < 1 \\ |d| - 0.5 & otherwise \end{cases} \tag{3}$$

As an optimizer, we use SGD (Stochastic Gradient Descent) Nestarov with a learning rate equal to 0.005. The models run for 100 epochs with a batch-size of 32. Figure 4 shows the comparison between the two loss functions, MAE and Huber loss.

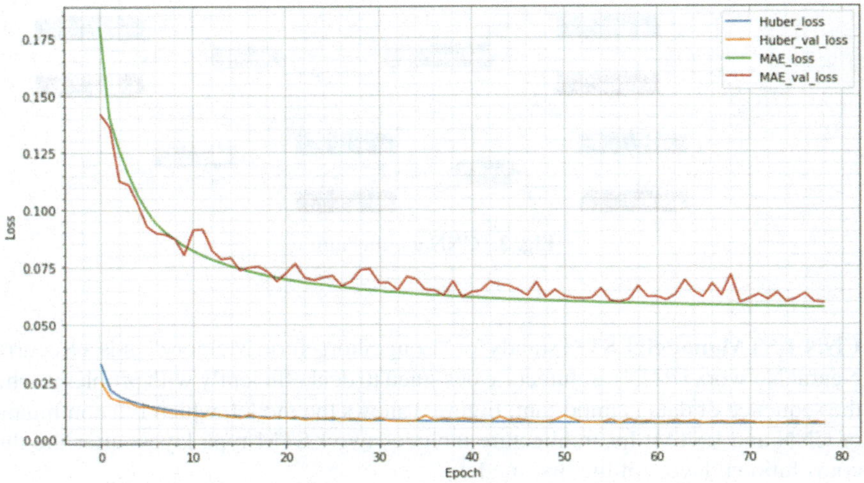

Fig. 4. Comparaison between two Loss, MAE and Huber loss

To measure model performance, the metric used is R^2 score, whose formula is given by Eq. 4.

$$R^2 = 1 - \frac{RSS}{TSS} \tag{4}$$

R^2 is the coefficient of determination.
RSS: Residual Square Sum.
TSS: Total Square Sum.

3 Results and Descussion

To demonstrate the applicability of CNN for predicting physical attributes of materials, we train two models to calculate the energy of formation, which is an important criterion for predicting perovskite stability. The learning curves of CNN and CNN-LSTM models in terms of loss and accuracy are presented in Figs. 5 and 6, respectively. The results show that both models perform well, with a slight improvement for CNN-LSTM. Indeed, the first achieves an accuracy of 0.9827 and the second 0.9807.

The trained models reveal formation energy values with an error of 0.0070 eV/atom and 0.0063 eV/atom for CNN and CNN-LSTM respectively. Finally, Figs. 7 and 8 show the histograms of the true values of the energies versus the predicted values for the two models respectively.

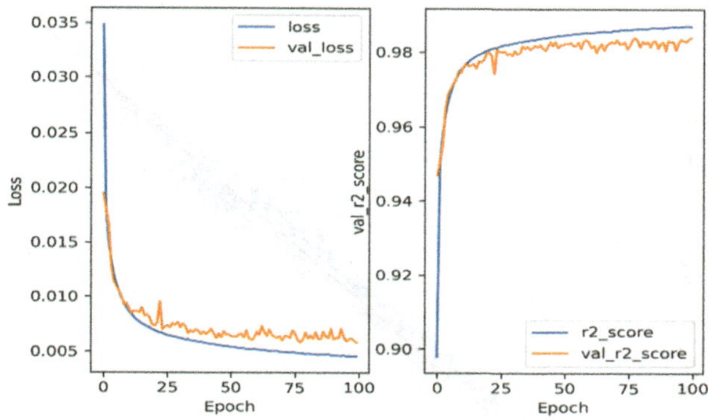

Fig. 5. Learning curves for CNN model

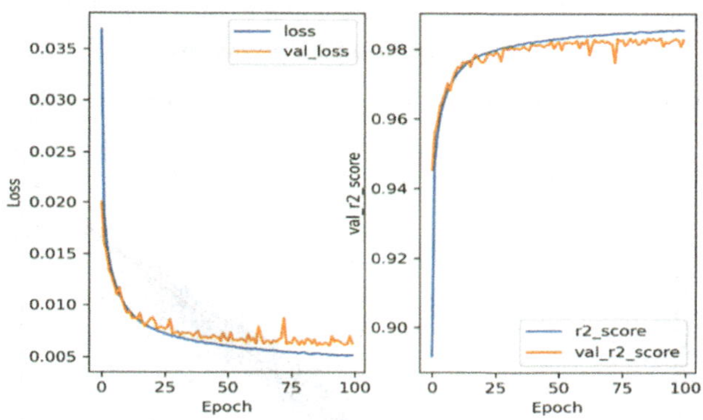

Fig. 6. Learning curves for CNN-LSTM model

The Fig. 9 shows trends in stable Perovskites. Li, Rb and Cs are most favorable in 1a, with stability decreasing for other elements. Non-metals, transition metals, and light elements form very few stable systems in 1a. In contrast to 1a, position 1b prefers light elements like B and W. Interestingly, Au and Ru can also stabilize perovskites in this position.

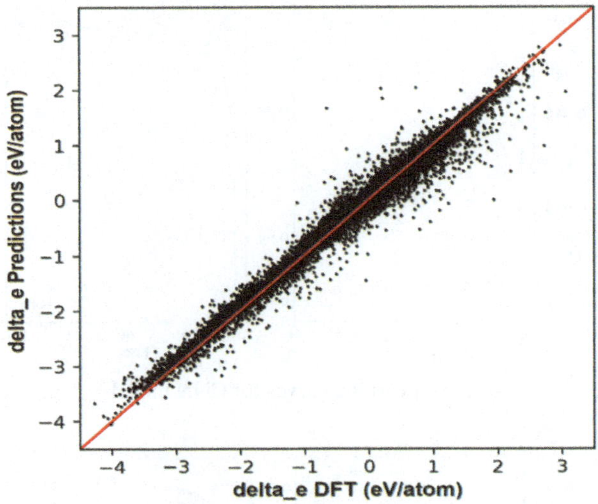

Fig. 7. Histogram of the true vs the predicted values of energies (CNN model)

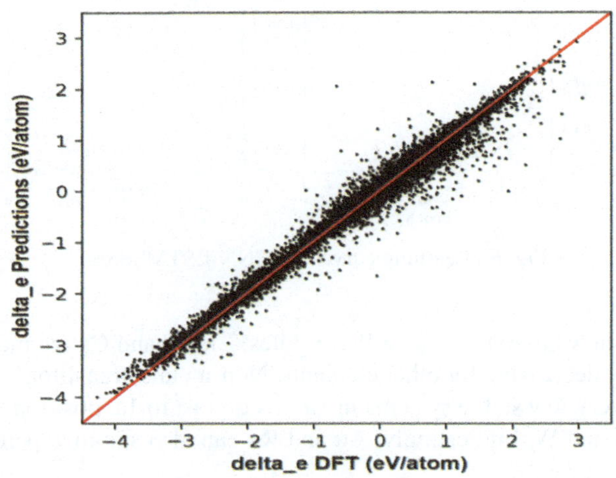

Fig. 8. Histogram of the true vs the predicted values of energies (CNN-LSTM model)

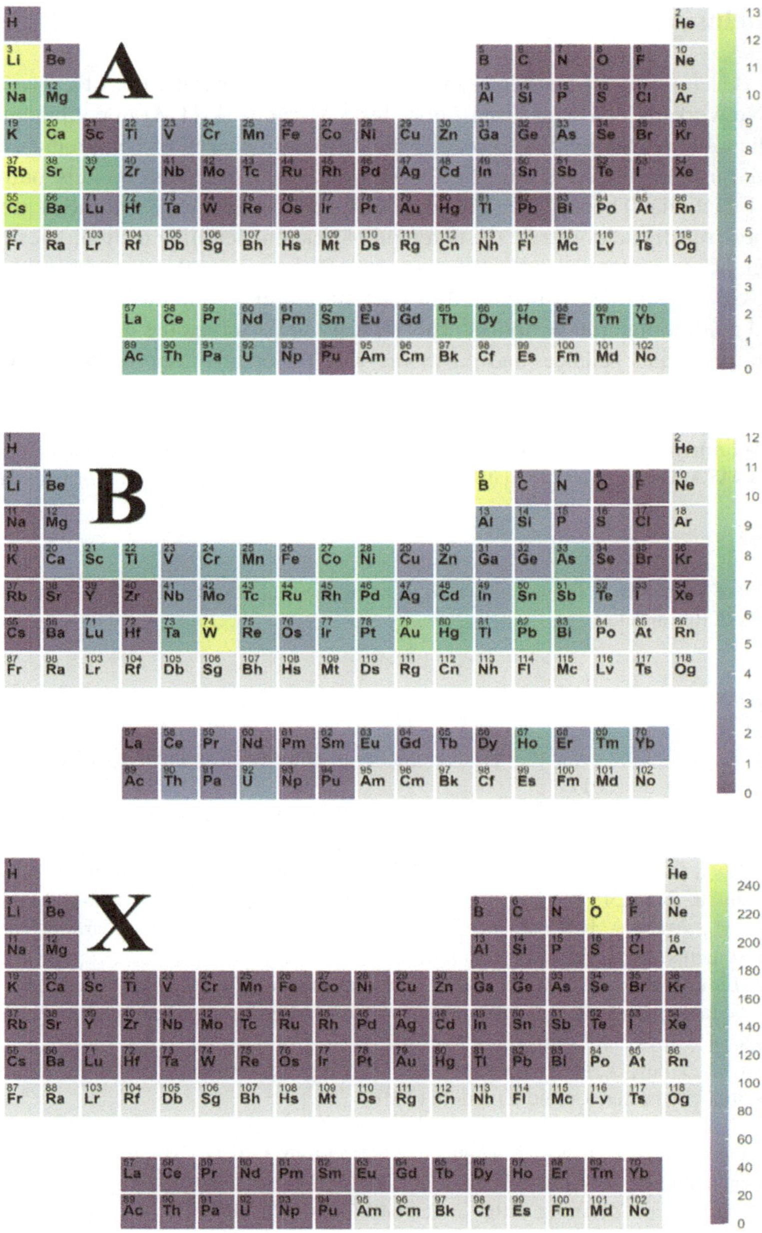

Fig. 9. Periodic Tables showing the stable structures for possible element A, B and X.

4 Conclusion

A major development in materials science research has been the use of Convolutional Neural Networks (CNN) and Long Short Term Memory (LSTM) in conjunction with a periodic table representation to predict the formation energy of compounds with the chemical formula $\mathbf{ABX_3}$. Using the OQMD dataset, the trained models outperform Density Functional Theory (DFT) estimates in terms of accuracy and mean prediction error. These results demonstrate the potential of machine learning technologies to guide the design of new materials and improve energy efficiency by validating the effectiveness of using CNN combined with LSTM to read structural information from the periodic table and predict material attributes. This strategy underscores the importance of AI in advancing materials science, and paves the way for creative approaches to materials design.

Acknowledgments. This work was carried out at the Laboratory of Mathematics and Applied Sciences LMSA. Departement of mathematics and informatics MI. University of Ghardaia.

Disclosure of Interests.. The authors have no competing interests to declare that are relevant to the content of this article.

References

1. Feng, R., et al.: First-principles prediction of high-entropy-alloy stability. npj Computational Materials **3**(1), 50 (2017)
2. Ji, S., Zhang, C., Xu, A., Shi, Y., Duan, Y.: 3D Convolutional neural networks for crop classification with multi-temporal remote sensing images. Remote Sens. **10**, 75 (2018)
3. Van Houdt, G., Mosquera, C., Nápoles, G.: A review on the long short-term memory model. Artif. Intell. Rev. **53**(8), 5929–5955 (2020). https://doi.org/10.1007/s10462-020-09838-1
4. Iman, M., Arabnia, H.R., Rasheed, K.: A review of deep transfer learning and recent advancements. Technologies **11**, 40 (2023)
5. Shuai Li a, D.L.A.J.Y.C.M.Z.B., Li b, S.: Hardness prediction of high entropy alloys with periodic table representation of composition, processing, structure and physical parameters. Journal of Alloys and Compounds **967** (2023)
6. Richard, J., Tilley, D.: Perovskites Structure–Property Relationships, pp.10–13. John Wiley & Sons, ltd., chichester: United Kingdom (2016)
7. Oliynyk, A.O., et al.: Chem. Mater. **28**, 7324–7331 (2016)
8. Legrain, F., Carrete, J., van Roekeghem, A., Madsen, G.K.H., Mingo, N.: The journal of physical chemistry. B (2017)
9. Kreiner, G., et al.: Z. Anorg. Allg. Chem. **640**, 738–752 (2014)
10. Sanvito, S., et al.: Sci. Adv. **3**, e1602241 (2017)
11. Zhang, X., Yu, L., Zakutayev, A., Zunger, A.: Adv. Func. Mater. **22**, 1425–1435 (2012)
12. Open quantum materials database. http://oqmd.org/
13. Kirklin, S., et al.: Te open quantum materials database (oqmd): assessing the accuracy of df formation energies. npj Computational Materials **1**, 15010 (2015)
14. Jha, D., et al.: ElemNet: deep learning the chemistry of materials from only elemental composition. Scientific Reports **8**(1) (2018)

15. Baudoin, G., Bercher, J.F.: Éléments de traitement du signal. Ecole Nationale d'Ingénieurs en Electrothechnique et Electronique, version 0.89 (1998)
16. Pettersson, L.: Thesis" Deep Learning Approach to Material", Department of Physics j Division of Mathematical Physics (2020)
17. Owen, A.B.: A robust hybrid of lasso and ridge regression. Contemp. Math. **443**, 59–72 (2007)

Attentional Seq2Seq Model for Arabic Opinion Question Generation

Leila Ouahrani(✉) and Djamal Bennouar

LIM Laboratory, Computer Science Department, Faculty of Applied Sciences, Bouira University, 10000 Bouira, Algeria
{l.ouahrani,djamal.bennouar}@univ-bouira.dz

Abstract. We tackle the task of generating opinion-based questions, focusing on providing users with samples of questions that express opinions relevant to their queries when utilizing search engines. The motivation is that Community Question Answering forums provide a suitable dataset for this purpose, as they reflect informal expressions of people's thoughts and facilitate users in posing questions and obtaining answers from others. Our approach involves developing a sequence-to-sequence model with an encoder-decoder architecture that incorporates an attention mechanism with automatically produced embedding which enhances the capability of the model to concentrate on relevant segments of the source text. To achieve this goal, we adapted Community Question Answering datasets and carried out practical experiments to create Arabic baselines, addressing the limited research in this area for the Arabic language. Following this, we performed experiments in English to facilitate a thorough comparative study, ensuring language-independent validation of the proposed approach.

Keywords: Automatic question generation · Natural language generation · Community Question Answering · Sequence-to-sequence model

1 Introduction

Question generation (QG) systems seek to formulate natural questions from a provided sentence or paragraph, converting statements or segments of text into questions. Due to their usefulness in conversational agents, machine reading comprehension, and educational tools, these systems have recently gained popularity [1]. In education, they assist in developing quizzes and assessments that foster active learning and boost student engagement [2, 3]. For intelligent assistants and chatbots, QG enhances interaction by enabling these systems to pose pertinent follow-up questions, thereby enriching the user experience [4]. Another application of QG is to improve search engine queries by offering users refined suggestions to obtain accurate results. Additionally, QG can support the question answering community by generating well-structured opinion questions, enhancing the effectiveness of their systems [5, 6]. In this paper, we utilize QG to improve search engine queries through Community Question Answering (CQA) systems. CQA systems are designed to handle inquiries that span multiple sentences that are generated by users.

S. Chikhi et al. (Eds.): MISC 2024, LNNS 1267, pp. 112–126, 2025.
https://doi.org/10.1007/978-3-031-82112-7_9

This results in a wide range of response quality, depending on the contributors' knowledge and effort. In our perspective, CQA forums provide a suitable dataset for generating opinion question as they capture informal and conversational speech, reflecting people's real thoughts. These forums allow users to pose questions and receive answers from the community. This method could enhance user experience and boost the relevance of search results. Our approach entails designing a sequence-to-sequence model featuring an encoder-decoder architecture that incorporates an attention mechanism. This mechanism, combined with automatically generated word embedding, improves the model's ability to focus on pertinent parts of the input sentence. We adapt CQA datasets for QG and conduct practical experiments to establish Arabic baselines and include comprehensive automatic and manual evaluation compared to English baselines. The structure of the rest of this paper is as follows: Sect. 2 covers the related work on Question Generation. In Sect. 3, we describe the method and data used. Section 4 discusses the system evaluation and the results. Finally, Sect. 5 presents the study's conclusions.

2 Related Work

Early methods for Question Generation (QG) predominantly utilized rule-based systems. These approaches depended on manually crafted rules and templates to create questions from text [7, 8]. While these methods were straightforward and interpretable, they often lacked flexibility and failed to generalize well to diverse text inputs. The advent of machine learning brought significant advancements to QG. Statistical methods, such as those based on n-grams and probabilistic models, allowed for more dynamic question generation [9]. However, the reliance on predefined features and inability to identify complex semantic relationships in the text still made these methods limited. The emergence of deep learning, particularly neural network models, revolutionized the QG task. Sequence-to-sequence (seq2seq) models became a foundational approach. These models (encoder/decoder), often using Long Short-Term Memory (LSTM) networks, are capable of generating questions by encoding the input text and then decoding it into a question [10, 11]. However, these models have limitations: some questions may still be syntactically incorrect or irrelevant. Additionally, the network may struggle to retain information from longer sentences, resulting in less effective question generation. The advent of Transformer models represented a major advancement in the field of NLP, including the area of Question Generation [12]. Transformers, with their self-attention mechanisms, outperformed traditional RNN-based models in both speed and accuracy but they present several challenges. Transformers are highly computationally intensive, making them difficult to load into memory. Additionally, they require the length of paragraphs to be limited to a certain number of tokens, which can constrain the input data. Furthermore, any changes made to the transformer architecture necessitate retraining the model from scratch, a process that is both computationally and resource-intensive. There have been limited efforts to tackle Question Generation (QG) in the Arabic language. Existing research predominantly relies on rule-based methods to manually create question formats [13]. These models are constrained by a lack of linguistic diversity, and the complexity of the tasks increases significantly with longer texts.

3 Method and Data

3.1 Problem Formulation

The formal definition of the question generation task is as follows: Given an answer, A that contains N words: A = (a_1, a_2, a_3… a_n), of which a_i is a token in the answer and N is the length of the sequence. The aim is to generate a Q question of M tokens, with Q = (q_1, q_2, q_3… q_m), and M the maximum length of the question. In order to generate the question, we will have to maximize the conditional probability P(Q|A), which is presented in formula (1):

$$p(Q|A) = \prod_{1}^{M} p(q_t|q_{1:t-1}, A) \quad (1)$$

This implies that in order to find the probability of q_m, we will have to take into account the entire previously generated sequence as well as the answer A. We represent p(Q|A) as a series of word predictions. A series of words (q_1, q_2, q_3,…, q_m) is predicted based on another series of words (a_1, a_2, a_3,…, a_n).

3.2 Proposed Question Generation Model

The proposed attentional Encoder-Decoder question generator is designed to enhance the quality of search engine results by generating natural-sounding, well-formed and accurate questions based on provided answers, simulating the conditional probability p(Q|A). As illustrated in Fig. 1, the model employs a recurrent neural encoder-decoder architecture [14] to transform input sequences into output sequences. Both the encoder and decoder utilize Gated Recurrent Unit (GRU) networks [15] to process sequences of different lengths. Utilizing the final hidden state, the encoder transforms the input text into a vector representation. The decoder initializes with this encoded vector as its initial hidden state and sequentially predicts words to form the output sequence. Integrating an attention mechanism is essential for improving the model's performance by enabling it to concentrate on relevant sections of the input text. Two key advantages of the model include the direct training on source and target phrases and its flexibility in managing input and output text sequences of varying lengths.

Encoder. Consists of an embedding layer and a hidden layer.

Embedding Layer. Rather than employing count-based discrete representations, we employ embedding for both the encoding and decoding processes to better capture nuanced semantic relationships [16] and avoid repeating words in the target sentence. Our model's objective is to forecast the vectors extracted from a source sentence during the encoding and decoding phases, taking into account any semantic repetitions [17]. The purpose of the embedding layer is to convert One-hot word representations into word embedding. After encoding the dataset, a one-hot vector is produced for each word, reflecting its specific position within the vocabulary. A one-hot vector represents each word before being passed through the embedding layer. This process generates embedding, which are compact representations that preserve the semantic relationships between words in the text. In this model, words will be embedded in a 256-dimensional vector space. The creation of these embedding leads to a distribution of words that captures their semantic similarities. This distribution allows the neural network to learn the relationships between words and their contexts, enabling it to make predictions that are more accurate. By representing words in a continuous vector space, the model can better comprehend the meaning of each term in the context of the words around it. The embedding serve as a bridge between the input text and the hidden layer, helping to improve the overall performance of the model.

Hidden Layer (GRU). We used Gated Recurrent Neural Networks (GRUs) that are a simplified variant of Long Short-Term Memory (LSTM) networks, incorporating two gates: the reset gate and the update gate. These gates control the resetting and updating of the hidden state. GRUs are more computationally efficient and have fewer training parameters than LSTMs, making them particularly effective for less complex sequence tasks sequences [15, 18]. We chose GRU cells for our model due to their faster convergence and iterative efficiency. The model consists of 1024 GRU units. The sparse Categorical Cross Entropy function (formula (2)) is employed to compute the loss throughout the training process.

$$-\frac{1}{N}\sum_{i=1}^{n}\sum_{c=1}^{n} 1_{y_i \in C_c} \log P_{model}\left[y_i \in C_c\right] \tag{2}$$

In this context, n denotes the total number of observations, C represents the number of distinct classes, which corresponds to the number of unique words in the vocabulary used for One-hot encoding, and P indicates that observation "i" is associated with class "c".

Attention mechanism. The encoder-decoder's effectiveness declines significantly when the input sentence lengthens [14]. To address this deficiency, we incorporate an attention mechanism within the encoder [19].This mechanism considers the entire set of hidden states across different time steps. At each time step, it generates a context vector by using all of the encoder's hidden states. The model directs its focus to specific segments of the input sequence by computing alignment scores between each hidden state of the encoder and the decoder's previous hidden state. For generating a word, q_i the attention mechanism assigns importance to different words in the input sequence, emphasizing the encoder's hidden states that are most relevant. These attention weights P_i are derived using the SoftMax function, applied to the alignment scores from the encoder. Once all attention weights are computed, a context vector is formed based on these weights (formula (3)). In the decoder, each node (GRU_i) receives input from the preceding node (q_{i-1}) and the context vector (C_t) generated by the attention mechanism at the current time step t. The initial node of the decoder (GRU_1) is fed with the encoder's final hidden state and the initial context vector (C_t) produced by the attention mechanism.

$$Context\ Vector = \sum\nolimits_{i=1}^{n} P_i h_i \tag{3}$$

h_i : The hidden state at step i,
n: The source sentence's total word count, and.

P_i: The h_i weight.

Question Generator. The contextual representation is used at each stage, alongside the pre-existing word embedding. In the vocabulary distribution produced, the word with the greatest probability is generated:

Embedding Layer. Converts each word's digital representation into its corresponding embedding vector.

The GRU Layer. The decoder initiates the process by taking the encoder's last hidden state, commonly known as the vector encoder, as its starting input. The hidden state captures essential information from every word in the input sentence. To produce a word q_i at each time step t, the decoder depends on the hidden state, the output from the previous time step, and the embedding of the word generated in the preceding step of the question.

The Output Layer. The SoftMax function acts as a classifier within the output layer of the decoder, taking the GRU layer's output and generating a probability distribution across integers, each representing a word in the vocabulary. The word with the greatest probability is picked, resulting in the formation of a single question. Although this approach is effective, there is potential for further improvement. To improve the decoding process, the Beam Search Decoding Algorithm [20] is utilized. This algorithm conducts an approximate search to generate multiple question candidates. The algorithm examines every possible subsequent step, retains the resulting sequences, and regulates the parallel searches at each phase of the probability sequence. The number of generated questions is determined by the beam size.

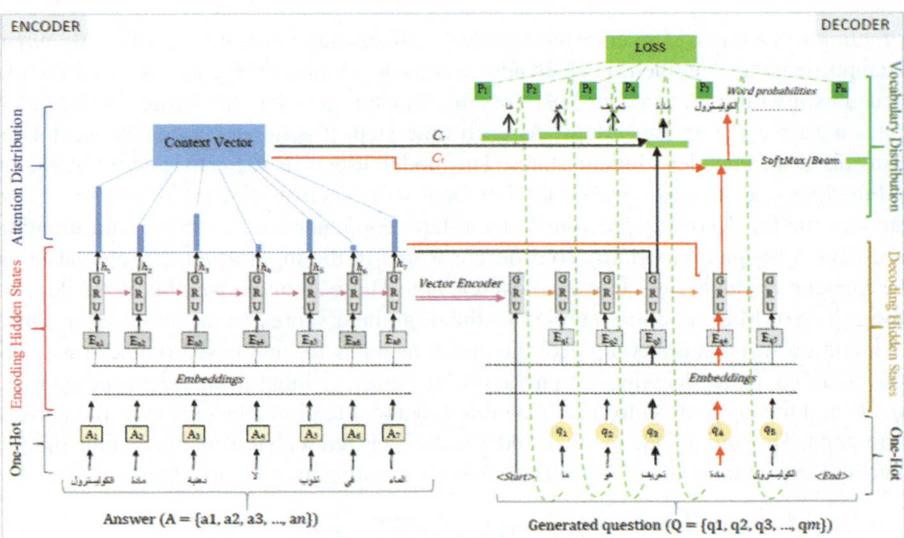

Fig. 1. The proposed question generator model employs an encoder-decoder architecture. It leverages key words from the source sentence to generate new words for formulating the question.

3.3 Data Collection and Adaptation

The proposed approach involves using a dataset from CQA systems to train a question generation model. This model can be utilized subsequently to generate automatically questions, thereby improving the quality of search engine query results. For this purpose, we adapted CQA datasets and performed empirical tests to assess the efficacy of the proposed question generation model. Initially, these experiments were carried out in Arabic to establish baselines due to the scarcity of research on Community Question Answering (QA) in this language. We subsequently conducted experiments in English for a comprehensive comparative study. Using English allows us to compare our model with previous research, providing a validation that is independent of language.

Arabic Dataset Acquisition. We obtained our Arabic dataset from the SemEval-2016 Task 3[1] [21]. We preprocessed and customized the bio-medical Arabic community question answering dataset, cQA-MD[2], which was featured in the SemEval 2016 subtask D Arabic, "*Rank the correct answers for a new question*" of task 3. The cQA-MD dataset originates from three websites: WebTeb[3], Al-Tibbi[4], and Islamweb[5]. Its primary objective is to reclassify the correct answers for new questions. Initially, this dataset included 1,531 contexts and 45,164 question-answer pairs. For the creation of the SemEval-2016 dataset, the 1,531 contexts were sourced from WebTeb, while the corresponding question-answer pairs were drawn from Al-Tibbi, contributing 69,582 pairs, and Islamweb, contributing 31,714 pairs. The data is sourced from three websites: original questions are taken from WebTeb, while related question-answer pairs are gathered from the other two sites. About 100,000 question-answer pairs were gathered and searched to locate answers for WebTeb questions. Each question-answer pair is categorized with one of the following labels depending on its relevance to the unanswered question: ("Direct" indicates a response that directly answers the query, "Relevant" suggests an answer that is connected to the question, and "Irrelevant" denotes a response that is not related to the query).

Arabic dataset adaptation. To train the model for the question generation task, we only need question-answer pairs, as our approach does not make use of class annotations. This adaptation involved utilizing a question-answering dataset to train a question generation model. We then customized the initial 45,164 question-answer pairs to suit our specific research requirements in question generation. We encountered several obstacles in preparing the dataset for automatic question generation, due to its community-driven nature that combines informal and formal language, various Arabic dialects, and metadata (URLs, source, etc.). These challenges included issues with grammatical accuracy and consistency, the presence of various dialects, spelling errors such as letter omissions and word merging, and handling question-answer pairs that were either excessively detailed or too brief. To address these issues, we undertook the following preprocessing steps: A challenge emerges when dealing with the lengthy introductions and pleasantries

1 https://alt.qcri.org/semeval2016/task3/.
2 https://alt.qcri.org/semeval2016/task3/index.php?id=data-and-tools.
3 http://www.webteb.com/.
4 https://altibbi.com/.
5 https://www.islamweb.net/ar/consult/.

common in Arabic, which often include religious salutations at both the beginning and end. According to [22], a significant portion (58%) of questions start with these types of introductions, and half (50%) of responses conclude with prayers or similar expressions. To maintain the core content of the questions and answers, we meticulously reviewed the dataset manually to exclude these introductory and concluding sections. Patients frequently offer extensive medical histories and treatment details, resulting in lengthy questions. On the other hand, doctors usually start by defining the medical condition before directly answering the question. We imposed specific length constraints on the entries (Questions: 4 to 35 tokens, Answers: 10 to 35 tokens) to deal with the constraints entries of the model. In a particular request, we divided all the included questions and their corresponding answers. An example of a single question that encompasses four interconnected questions.

Question: "ما هو الصرع ما هو علاجه ما هي أعراضه و كيف يتم التعايش معه?"(What is epilepsy, how is it treated, what are its symptoms, and how can one manage living with it?).

It transforms into four separate questions, with each one paired with its own specific answer:

- ما هو الصرع(What is epilepsy?)
- ما هو علاج الصرع(What is the treatment for epilepsy?)
- ما هي أعراض الصرع(What are the symptoms of epilepsy?)
- كيف يتم التعايش مع الصرع(How to live with epilepsy)

The resulting dataset went through preprocessing: punctuation marks were taken out, word formats were normalized, diacritic marks were removed (التشكيل)and elongation marks were taken out of some Arabic words (التطويل). After applying the necessary manual and automated preprocessing steps, we ultimately compiled a dataset of 20,112 question-answer pairs. The resulted dataset is referred to as Adapted-cQA-MD.

English Dataset. The English dataset is extracted from the Amazon QA dataset [23]. The Amazon QA dataset is a community-driven collection featuring over 1.4 million question-and-answer pairs, where customers pose inquiries about 191,000 products spanning 21 categories available on the Amazon website. Our selection of an English dataset was constrained because most English-language datasets for question generation are found in educational, rather than community contexts. The selected dataset includes the following categories of pairs of question-and-answer (QA) extracted using the Amazon dataset: baby products (13,126 QA), health products (35,580 QA), telephone accessories (40,583), office products (19,845 QA), home and kitchen products (84,922 QA), and household tools and habitat improvement (55,944 QA), totaling 225,000 QA pairs we used to train and test the proposed model. We enforced identical constraints on input length and carried out several preprocessing steps, including removing punctuation, converting text to lowercase, eliminating abbreviations, and deleting metadata (URLs, source, etc.).

4 Evaluation of the Generated Questions and Results

4.1 Experiment Setup

Conducting a comprehensive evaluation helps us establish baselines for Arabic and estimate the model's performance. In accordance with the information provided in Table 1, we assigned 85% of each dataset for training purposes, and the remaining 15% for testing (80%, 20% for the English dataset). Data preparation encompasses preprocessing data (as specified previously) and tokenization. In order to use a one-hot encoding of words, the datasets were prepared by extracting a vocabulary that includes all distinct terms found inside the dataset. The vocabulary comprised 101,000 distinct words in English and 52,000 unique words in Arabic. We used TensorFlow version 2.3 along with Keras 2.1 to develop the model. The training took place on Google Colab, utilizing an Nvidia Tesla K80 GPU equipped with 12 GB of DDR5 RAM [24]. The model is trained with batches of 64 sentence pairs from the dataset. The ADAM stochastic optimizer is employed, with a learning rate configured at 0.001 for the training process, which extends over 20 epochs with the sparse Categorical Cross Entropy function to compute the loss. A dropout rate of 0.3 (30%) was applied as a regularization technique. This helps the model make predictions by averaging over various parameter settings, thereby combining them into the final output. This approach enhances the model's generalization capabilities and mitigates the risk of over fitting. During inference, all units are utilized.

English baselines. Due to the absence of dedicated results for the Arabic language in this context, the performance of the model is evaluated using the English dataset for comparison. In our evaluation of the model for English, we draw comparisons with the work of Chali and Baghaee [6], who performed experiments using the Amazon question/answer dataset. Their approach employed a sequence-to-sequence attentional model with a 600-dimensional LSTM hidden layer. To avoid sentence repetition, they incorporated a coverage mechanism and used 300-dimensional Glove pre-trained word embedding [25] model for the encoder and decoder, which featured a vocabulary size of 50,000 words.

Table 1. Dataset segments for train and evaluate the question generation model

Dataset	Training	Test	Total
Adapted- cQA-MD (Arabic)	17,095 (85%)	3017 (15%)	20112
Amazon QA (English)	180,000 (80%)	45,000 (20%)	225000

4.2 Automatic Evaluation of Generated Questions

Metrics. We utilize various metrics for automatic evaluation, including BLEU 1, BLEU 2, BLEU 3, BLEU 4 [26] for assessing n-gram precision, METEOR [27] for evaluating text similarity, and ROUGE-L [28] for measuring n-gram recall. BLEU computes the average precision of n-grams by comparing the generated text with reference sentences, penalizing brevity in the generated sentences. BLEU-n refers to a metric that calculates the BLEU score by taking into account n-grams of a specified length to evaluate how often they appear together. METEOR is a recall-focused evaluation metric that measures how similar the generated text is to the original input sentences, considering synonyms, stemming variations, and generated content. ROUGE is employed frequently to measure the n-gram recall of summaries against established references, while ROUGE-L, relying on the longest common subsequence, is highlighted specifically in our findings.

Results and Discussion. Figure 2 illustrates examples of questions produced by the proposed model in the sample test for Arabic and English. In Table 2, we present the results of the proposed model, with the best-performing model highlighted in bold. Our model trained on the Amazon dataset outperforms the English baseline across Bleu1, Bleu2, Bleu3, and METEOR metrics. On the Rouge-L metric, however, it performs somewhat worse, with a difference of -1,45. Although comparing results across different datasets may be unfair, our model trained on the Arabic dataset shows slightly lower performance compared to the English one. However, it achieves high Bleu1 and the best Bleu4 scores. This indicates that the decoder accurately replicates the n-gram alignments of the answer within the generated question. When evaluated with METEOR, our model on Arabic performs comparably to the best model and surpasses the baseline in English, though it underperforms in ROUGE-L. The analysis of our model's performance on both Arabic and English datasets compared to the baseline shows that question generation is much better when embeddings are made within the model (the embedding layer) instead of using ones that have already been trained. Our model's use of an attention mechanism with automatically produced embeddings is a vital component that improves performance against the baseline. Taking into account the findings from the QA literature, it is evident that BLEU scores tend to be notably higher than ours are. This is because QA datasets, such as SQuAD [29] are often human-generated leading to well-structured, grammatically correct sentences. However, in the context of CQA systems, sentence structures are frequently inconsistent and informal and contain irrelevant details and symbols. According to Lavie [30]: "*BLEU scores above 30 typically signify translations that are understandable, while scores exceeding 50 indicate translations of good quality and fluency*". This suggests that the questions generated in Arabic and English are of good quality.

Producing various questions through the beam search algorithm. In Tables 3 and 4, we report the outcomes achieved with the beam search-decoding algorithm for generating multiple questions from both Arabic and English datasets. By varying the beam size to 1, 4, 7, and 10 questions per source answer, we evaluate the model's performance and calculate the average metrics accordingly. As an illustration, consider answer A1 (in Arabic) shown in Fig. 2, where using Beam 4 results in the generation of four questions:

- ما هو تاثير دواء بيرياكتين مضر ام لا("What is the effect of Periactin, harmful or not?")
- هل دواء له تاثير على المستوى البعيد("Does the drug have a long-term effect? ")
- ما هي تاثير دواء على المدى البعيد("What is the long-term effect of a drug")
- اريد ان استخدم دواء لعلاج حب الشباب("I want to use a medicine to treat acne")

We illustrate also answer A1 (in English) shown in Fig. 2, where using Beam 4 results in the generation of four questions:

- are these really truly stainless steel when not what finish? and will they rust ?
- do these rust ?
- does the product rust?
- i have read that they rust.

As observed, the score diminishes as the number of questions generated grows, which is expected since the average metric includes both the best and worst cases. Even though the metric scores have declined, they remain accurate and continue to outperform the baselines for both Bleu and Meteor. With a beam size of one, the algorithm selects the most probable question identical to those generated by the SoftMax function. Exploring the reasons behind the inclusion of certain irrelevant words in the question would be an area for further investigation as seen in the fourth generated question:
"اريد ان استخدم دواء لعلاج **حب الشباب**"("I want to use a medicine to **treat acne**").

4.3 Manual Evaluation of Generated Questions

We assessed manually the quality of the questions generated by the system, focusing on two key aspects: Syntactic correction and Relevance. Syntactic correction involves evaluating the naturalness, grammatical accuracy, and fluency of questions. Relevance measures how well a generated question is meaningful and appropriately aligned with the provided answer. For each sample test, we selected at random, 100 answer-question pairs and had three human experts evaluate them. The ratings ranged from one to 5, where a score of 5 indicated the highest level of quality in both syntactic accuracy and relevance. We conducted the same evaluation for question-answer pairs in both English and Arabic. The three human-generated scores were averaged, as shown in Table 5. Similar to [6]'s method, which achieved a relevance score of 3.37 out of 5 on the Amazon dataset, the English questions had a relevance score of 3.3 out of 5, maintaining 66% relevance to the answers. Overall, our model's queries exhibited a high level of relevance. In terms of syntactic correctness, our questions have an average score of 4.5 out of 5, slightly lower than the Chali and Baghaee (2018) model's [6] score of 4.52 out of 5. Nevertheless, this score highlights the effectiveness of our model in generating coherent and well-structured queries. While there are no comparative results from similar studies for the

Arabic (sample test)	
A1	يحتوي بيرياكتين على الريبافيرين وهو يستخدم كمضاد للفيروسات يستخدم لعلاج التهابات الجهاز التنفسي الفيروسيه في ويستخدم مع الفا انتر فيرون لعلاج الالتهابات المزمنه والتهاب الكبد الفيروسي ج Periactin contains ribavirin and is used as an antiviral used to treat viral respiratory infections in and is used with alpha interferon to treat chronic infections and viral hepatitis C
Q1	ما هي دواعي استخدام علاج بيرياكتين؟ What are the reasons for using Periactin treatment?
G-Q1	ما هو تاثير دواء بيرياكتين مضر ام لا؟ What is the effect of Periactin, harmful or not?
A2	يستخدم هذا المنتج لانقاص الوزن ولكن مثل هذه المنتجات غير مصرح بهت من قبل مؤسسه الغذاء والدواء الامريكيه لذلك يجب قبل اللجوء لمثل هذه المنتجات استشاره الطبيب او اخصائي التغذيه لاتباع نظام غذائي صحي وممارسه الرياضه بشكل منتظم This product is used to lose weight, but such products are not approved by the US Food and Drug Administration. Therefore, before resorting to such products, you must consult a doctor or nutritionist to follow a healthy diet and exercise regularly.
Q2	اود ان اسال على حبوب التنحيف سليمنج هل هي فعاله؟ I would like to ask about "Slimming" slimming pills are they effective?
G-Q2	ما رايكم ب حبوب التخسيس سليمنج وهل هو مضر؟ What do you think of "Slimming", slimming pills and is it harmful?
A3	لقد ثبت ان 40% من اسباب العقم تكون عند الزوج و40% اخرى تكون بسبب الزوجه وال 20% المتبقيه ناتجه عن اسباب مجتمعه بين الاثنين ليس هناك عقم وراثي It has been proven that 40% of the causes of infertility are caused by the husband, another 40% are caused by the wife, and the remaining 20% result from combined causes between the two. There is no hereditary infertility.
Q3	الاحظ حالات عقم كثيره في العائله الواحده فهل يمكن ان يكون العقم وراثيا؟ I notice many cases of infertility in one family. Can infertility be hereditary?
G-Q3	عندي مشكله منذ سنه تقريبا لم يحدث حمل لحد الان؟ I have had a problem for about a year, and I haven't gotten pregnant yet
English (sample test)	
A1	they do not rust but they are really really small
Q1	does it trust?
G-Q1	are these really truly stainless steel when not what finish? and will they rust?
A2	it may state that it is waterproof but i would not trust it to be so i would not recommend keeping this safe outdoors
Q2	is this safe water proof? we are considering it for an outdoor location so this is an important consideration
G-Q2	can this product be used outside in rain?
A3	there is no stamp on the kettle or base of country of manufacture assume china
Q3	where is it made?
G-Q3	can you confirm if it is made in china?

Fig. 2. Examples of generated questions produced by systems Arabic & English (Answer (Ai), ground truth questions (Qi), and Generated Question (G-Qi)).

Arabic dataset, the generated sample produced promising results overall (relevance = 2.79/5; syntactic correctness = 3.75/5). This indicates the model's potential to perform effectively in Arabic. Moreover, investigating methods to improve the relevance scores could further enhance the model's performance on Arabic datasets.

Table 2. Automatic evaluation for various systems using BLEU, ROUGE-L, and METEOR

Dataset	Bleu1	Bleu2	Bleu3	Bleu4	ROUGE-L	Meteor
Adapted- cQA-MD Arabic (ours)	41,14	11.60	10.59	**10.03**	15.08	11.61
Amazon QA (English) (ours)	**42,60**	**25,60**	**15,58**	9,48	23,65	**15,63**
Amazon QA (English) [6]	14,67	7,74	–	–	**25,10**	9,43

Table 3. The beam search decoding evaluation on the Arabic dataset (Average Metric).

BEAM	Bleu1	Bleu2	Bleu3	Bleu4	ROUGE-L	Meteor
1	**41,49**	**11,61**	**10,60**	**10,02**	**14,97**	**11,60**
4	37,51	8,40	7,45	6,93	11,42	08,33
7	36,87	7,91	6,98	6,47	10,76	07,76
10	36,33	7,52	6,62	6,12	10,31	07,39

Table 4. The beam search decoding evaluation on the English dataset (Average Metric).

BEAM	Bleu1	Bleu2	Bleu3	Bleu4	ROUGE-L	Meteor
1	**42,79**	**25,71**	**15,64**	**9,52**	**23,61**	**15,62**
4	41,81	24,83	14,68	8,85	19,73	13,29
7	40,62	24,06	14,65	8,68	18,54	12,44
10	39,90	23,33	13,33	8,40	17,47	12,02

Table 5. Human evaluation of generated questions (Average between three Annotators)

Dataset	Syntactic correction/5	Relevance/5
Adapted- cQA-MD Arabic (Ours)	3,75	2,79
Amazon QA (English) (Ours)	4,50	3,30
Amazon QA (English) [6]	4,52	3,37

5 Conclusion

In this study, we designed an encoder-decoder model with attention mechanisms to generate opinion-based questions. Our goal is to enhance the user experience by offering a variety of question samples relevant to their search queries. To train the model, we utilized community-based question-answer datasets that feature informal speech. These datasets include sentences that often deviate from strict grammatical rules, reflecting everyday speech and aligning closely with people's thoughts and opinions. Our experiments, performed in both Arabic and English, show an enhancement compared to the baseline system for English and tackle the challenging task of generating opinion-based questions in Arabic. This work also establishes baselines, as few studies have tackled this task in Arabic. Qualitative assessments by human evaluators indicate that the questions produced are adequately structured, grammatically accurate, and relevant to the provided answers. The proposed approach advances natural language processing and offers a wide range of practical uses. It can enhance search engine performance, create personalized educational assessments, and improve conversational AI in customer support systems. Furthermore, the model's capability to function in both Arabic and English enhances its flexibility, making it applicable across various linguistic contexts and improving user interactions in sectors such as education, customer service, and information retrieval. We identify two primary directions for future research. First, exploring how fine-tuning transformer models on the Arabic dataset can enhance the quality of the generated questions. Second, although we have adapted an Arabic dataset, it remains insufficient in terms of coverage and diversity for the model to generalize effectively. We could focus on employing various data augmentation techniques to create large scale, diverse and comprehensive datasets for training and evaluating the question generation model. To improve generalization, we may include domain-specific fine-tuning, adopting transformer-based architectures, and applying data augmentation techniques. To facilitate reproducibility, data and codes, including models and tests, for both Arabic and English are shared here: https://github.com/Arabic-OQG/Opinion-QG.

References

1. Mulla, N., Gharpure, P.: Automatic question generation: a review of methodologies, datasets, evaluation metrics, and applications. Prog. Artif. Intell. **12**(1), 1–32 (2023). https://doi.org/10.1007/s13748-023-00295-9
2. Steuer, T., Filighera, A., Tregel, T.: Investigating educational and noneducational answer selection for educational question generation. IEEE Access **10**, 63522–63531 (2022). https://doi.org/10.1109/ACCESS.2022.3180838
3. Srivastava, M., Goodman, N.: Question generation for adaptive education. In: ACL-IJCNLP 2021 - 59th annual meeting of the Association for Computational Linguistics and the 11th international joint conference on natural language processing, proceedings of the conference, Association for Computational Linguistics (ACL), pp. 692–701 (2021). https://doi.org/10.18653/v1/2021.acl-short.88
4. Pan, B., Li, H., Yao, Z., Cai, D., Sun, H.: Reinforced dynamic reasoning for conversational question generation. In: ACL 2019 - 57th annual meeting of the association for computational linguistics, proceedings of the conference, association for computational linguistics (ACL), pp. 2114–2124 (2020). https://doi.org/10.18653/v1/p19-1203

5. Yang, Z., Hu, J., Salakhutdinov, R., Cohen, W.W.: Semi-supervised QA with generative domain-adaptive nets. In: ACL 2017 - 55th annual meeting of the association for computational linguistics, proceedings of the conference (long papers), association for computational linguistics (ACL), pp. 1040–1050 (2017). https://doi.org/10.18653/v1/P17-1096
6. Chali, Y., Baghaee, T.: Automatic opinion question generation. In: INLG 2018 - 11th international natural language generation conference, proceedings of the conference, pp. 152–158 (2018). https://doi.org/10.18653/v1/w18-6518
7. Becker, L., Basu, S., Vanderwende, L.: Mind the gap: learning to choose gaps for question generation. In: NAACL HLT 2012 - 2012 conference of the North American chapter of the association for computational linguistics: human language technologies, proceedings of the conference, pp. 742–751 (2012). https://aclanthology.org/N12-1092
8. Labutov, I., Basu, S., Vanderwende, L.: Deep questions without deep understanding. In: ACL-IJCNLP 2015 - 53rd annual meeting of the association for computational linguistics and the 7th international joint conference on natural language processing of the asian federation of natural language processing, proceedings of the conference, association for computational linguistics (ACL), pp. 889–898 (2015). https://doi.org/10.3115/v1/p15-1086
9. Heilman, M., Smith, N.A.: Good question! Statistical ranking for question generation. In: NAACL HLT 2010 - human language technologies: the 2010 annual conference of the North American chapter of the association for computational linguistics, proceedings of the main conference, pp. 609–617 (2010). https://aclanthology.org/N10-1086
10. Du, X., Shao, J., Cardie, C.: Learning to ask: neural question generation for reading comprehension. In ACL 2017 - 55th annual meeting of the association for computational linguistics, proceedings of the conference (long papers), pp. 1342–1352 (2017). https://doi.org/10.18653/v1/P17-1123
11. Indurthi, S., Raghu, D., Khapra, M.M., Joshi, S.: Generating natural language question-answer pairs from a knowledge graph using a RNN based question generation model. In: 15th conference of the european chapter of the association for computational linguistics, EACL 2017 - proceedings of conference, pp. 376–385 (2017). https://doi.org/10.18653/v1/e17-1036
12. Chan, Y.H., Fan, Y.C.: A recurrent Bert-based model for question generation. In: MRQA@EMNLP 2019 - proceedings of the 2nd workshop on machine reading for question answering, association for computational linguistics (ACL), pp. 154–162 (2019). https://doi.org/10.18653/v1/d19-5821
13. Bousmaha, K.Z., Chergui, N.H., AliMbarek, M.S., Hadrich, L.B.: AQG: Arabic question generator. Rev. d'Intelligence Artif. **34**(6), 721–729 (2020). https://doi.org/10.18280/RIA.340606
14. Cho, K., et al.: Learning phrase representations using RNN encoder-decoder for statistical machine translation. In: EMNLP 2014 - 2014 conference on empirical methods in natural language processing, proceedings of the conference, association for computational linguistics (ACL), pp. 1724–1734 (2014). https://doi.org/10.3115/v1/d14-1179
15. Chung, J., Gulcehre, C., Cho, K., Bengio, Y.: Empirical evaluation of gated recurrent neural networks on sequence modeling (2014). https://arxiv.org/abs/1412.3555v1
16. Mikolov, T., Sutskever, I., Chen, K., Corrado, G., Dean, J.: Distributed representations ofwords and phrases and their compositionality. In: Advances in Neural Information Processing Systems, Neural information processing systems foundation (2013). https://arxiv.org/abs/1310.4546v1
17. Zhang, Y., Kamigaito, H., Aoki, T., Takamura, H., Okumura, M.: Generic mechanism for reducing repetitions in encoder-decoder models. In: International conference recent advances in natural language processing, RANLP, pp. 1606–1615 (2021). https://doi.org/10.26615/978-954-452-072-4_180

18. Cahuantzi, R., Chen, X., Güttel, S.: A comparison of LSTM and GRU networks for learning symbolic sequences. In: Arai, K. (ed.) Intelligent Computing. SAI 2023. LNCS, vol. 739, pp. 771–785. Springer, Cham (2023). https://doi.org/10.1007/978-3-031-37963-5_53
19. Bahdanau, D., Cho, K.H., Bengio, Y.: Neural machine translation by jointly learning to align and translate. In: 3rd International conference on learning representations, ICLR 2015 - conference track proceedings, international conference on learning representations, ICLR (2015). https://arxiv.org/abs/1409.0473v7
20. Vijayakumar, D.B.A.K., Cogswell, M., Selvaraju, R.R., Sun, Q., Lee, S., Crandall, D.: Diverse beam search: decoding diverse solutions from neural sequence models. In: The Thirty-Second AAAI Conference on Artificial Intelligence (AAAI-18), Association for the Advancement of Artificial Intelligence, pp. 7371–7379 (2018)
21. Nakov, P., et al.: SemEval-2016 task 3: community question answering. In: SemEval 2016 - 10th international workshop on semantic evaluation, proceedings, association for computational linguistics (ACL), pp. 525–545 (2016). https://doi.org/10.18653/v1/s16-1083
22. Nakov, P., Màrquez, L., Moschitti, A., Mubarak, H.: Arabic community question answering. Nat. Lang. Eng. **25**(1), 5–41 (2019). https://doi.org/10.1017/S1351324918000426
23. Wan, M., McAuley, J.: Modeling ambiguity, subjectivity, and diverging viewpoints in opinion question answering systems. In: 2016 IEEE 16th international conference on data mining, Institute of Electrical and Electronics Engineers (IEEE), pp. 489–498 (2017). https://doi.org/10.1109/icdm.2016.0060
24. Carneiro, T., Da Nobrega, R.V.M., Nepomuceno, T., BinBian, G., De Albuquerque, V.H.C., Filho, P.P.R.: Performance analysis of Google colaboratory as a tool for accelerating deep learning applications. IEEE Access **6**, 61677–61685 (2018). https://doi.org/10.1109/ACCESS.2018.2874767
25. Pennington, J., Socher, R., Manning, C.D.: "GloVe: global vectors for word representation. In: EMNLP 2014 - 2014 conference on empirical methods in natural language processing, proceedings of the conference, association for computational linguistics (ACL), pp. 1532–1543 (2014). https://doi.org/10.3115/v1/d14-1162
26. Papineni, K., Roukos, S., Ward, T., Zhu, W.-J.: BLEU: a method for automatic evaluation of machine translation. In: Proceedings of the 40th annual meeting of the association for computational linguistics, association for computational linguistics, pp. 311–318 (2002). https://doi.org/10.3115/1073083.1073135
27. Lavie, A., Agarwal, A.: METEOR: an automatic metric for MT evaluation with high levels of correlation with human judgments. In: Proceedings of the second workshop on statistical machine translation, Prague, Czech Republic: association for computational linguistics, pp. 228–231 (2007). https://aclanthology.org/W07-0734/
28. Lin, C.Y.: Rouge: a package for automatic evaluation of summaries. In: Proceedings of the Workshop on Text Summarization Branches Out (WAS 2004), no. 1, pp. 25–26 (2004). https://aclanthology.org/W04-1013
29. Rajpurkar, P., Zhang, J., Lopyrev, K., Liang, P.: SQuad: 100,000+ questions for machine comprehension of text. In: EMNLP 2016 - conference on empirical methods in natural language processing, proceedings, association for computational linguistics (ACL), pp. 2383–2392 (2016). https://doi.org/10.18653/v1/d16-1264
30. Lavie, A.: Evaluating the output of machine translation systems. In: AMTA 2010 - 9th conference of the association for machine translation in the Americas, China (2010). https://www.cs.cmu.edu/~alavie/Presentations/MT-Evaluation-MT-Summit-Tutorial-19Sep11.pdf

Evaluation of Machine Learning Methods for Anomaly Detection in Satellite Telemetry: Advantages and Limitations

Achraf Djerida(✉)

Space Operating Center of Telecommunication Systems (COTS),
Algerian Space Agency, Medea, Algeria
adjerida@asal.dz

Abstract. Anomaly detection is key to satellite equipment as it can lead to the detection of potential faults that can lead to great damages. So far that are various methods that have been developed ranging from simple thresholds to machine learning methods. The latter receives recently great attention due to its efficiency and capabilities in treating complex situations. However, still there are many challenges in the application of these methods due to several reasons related to the learning complexities as well as the characteristics of the processes. During this work, we evaluate some machine learning (with different architectures) methods on common NASA datasets (SMAP and MSCL) based on different parameters and criteria to have a clear vision of the methods impact. Results show that the complexities of ML methods learning still poses a challenge for the filed. Moreover, it is advised to use lightweight ML methods to get better performance in terms of time and learning complexities.

Keywords: Anomaly detection · Satellite telemetry · Machine learning

1 Introduction

Satellite telemetry is an essential source for proper control and management of its mission [8]. There are usually thousands of telemetry channels for each satellite which give information about different features, such as temperature, and different subsystems. Each subsystem has a key role with other systems and therefore advanced analysis is needed to assess its situation. The big dataset size forces engineers to look for automatic and reliable methods that can be used to assess the situation of the satellite in rapid way. Such propositions can save efforts and time especially in emergency situations where the risk is very high. Anomalies can be either point (unusual value), collective (sequence of point values) or contextual type [1]. The latter is more complex as the behavior of the telemetry is usual in terms of boundaries; however, its specific occurrence (the context) is a sign of anomaly. Therefore, this topic receives great attention with

S. Chikhi et al. (Eds.): MISC 2024, LNNS 1267, pp. 127–139, 2025.
https://doi.org/10.1007/978-3-031-82112-7_10

the application of different methods and strategies with the aim to get accurate and reliable solutions.

Traditional methods include fixed thresholds (boundaries) that are used as normal behavior description, when the telemetry exceeds the established values, an alarm is indicated [4]. This methodology is simple and effective as demonstrated in several application especially for simple alarms. The boundaries can be set by experts or statistical analysis of past data, in addition, they can be updated when necessary. However, it suffers against contextual alarms where the behavior cannot be distinguished. As an improvement of these methods, clustering-based methods [10] with different schemes have been used widely to provide better results. These methods can vary from simple to advanced techniques with the aim to extract useful patters that can be used in the prediction phase. Furthermore, dimensionality reduction techniques such as PCA [9], are used to transfer data into another feature space where the detection can be easier due to nonlinearity reduction. Methods such as nearest-neighbor [11] and similar methods have been also integrated and applied in some systems. Yet, these methods still have disadvantages concerning the complex alarms and their processing which necessitates the development of advanced methods that can tackle such alarms.

The extensive use of machine learning methods in several fields with the promising results pushes several researchers to apply different methods in detecting the different types of alarms [10]. Their usage ranges from simple machine learning methods to advanced deep learning techniques with complex structures. As data is big, adequate methods need to be taken into account to avoid heavy costs. Autoencoders has been applied to distinguish the normal and unusual behavior in several datasets due to the Encoder and Decoder parts which have a reversible work type. They are characterized with their simplicity and effectiveness in special situations. On the other hand, Long Short-Term Memory (LSTM) has been applied based on its memory and its ability to fuse past and present data for time-dependent properties [1]. This option is very critical for time-series data where time lag is an essential parameter that can affect directly the performance. Generative Adversarial Networks (GANs) includes a learning and discriminator mechanism to improve the learning mechanism and the generated data to fit the original one [4]. Although these methods show advanced performance in several applications, a detailed look can reveal several disadvantages when the method is applied in real satellite systems with different telemetry and complexity.

During this work, we study the anomaly detection scheme for some machine learning methods in public datasets where both point and contextual anomalies are present. The goal is to identify the advantages and limitations of such methods in order to investigate on reliable solutions that can be more applicable in real systems. The evaluation scheme tackles two main properties which are the window size (time lag) and the training time (number of epochs). The study examines the ability of the current methods to serve as lightweight ML methods which combines both low training resources and reasonable performance. Taking into account the need to update the model parameters, model architecture, lightweight architectures can be widely used without extra costs.

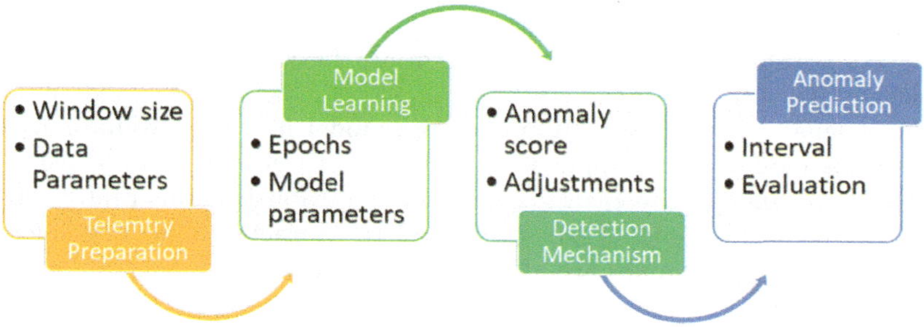

Fig. 1. The general scheme of ML methods

2 Methodology

During this section, we present the key steps during our methodology to evaluate ML methods in terms of lightweight and performance.

2.1 General Scheme

Anomaly detection in satellite telemetry using Machine Learning methods receives recently great importance due to the big data size. They can construct models which can predict in accurate ways complex alarms especially with contextual properties. Figure 1 shows the general scheme of the Machine learning methods.

The first step includes the preparation of the telemetry data to be in convenient format for the learning process. This step includes missing values processing, resampling and features selection. It is very critical step as the model uses this data to extract useful information. As time-series data is dependent on the time lag (window size), proper selection is needed as high value can lead to complex modelling and complex execution time.

The learning process depends on several parameters such as model type, structure and learning parameters. There are several machine learning types that varies in terms of the building blocks such as Long Short-Term Memory (LSTM), Generative Adversarial Networks (GAN) and simple networks such as autoencoders. These models include other parameters such as learning rates and number of epochs which cannot be identified directly and can affect both the performance and time complexity of the methods.

The detection mechanism step aims at establishing a proper anomaly score. Let us assume that a model M_t is trained on the training dataset (No alarms), a detection score S_d can be defined as:

$$S_d(x_i) = f_s(M_t(x_i)) \tag{1}$$

where f_s is a score mapping function and x_i is a telemetry at time i.

f_s is the key step in transforming the model prediction into an interesting value (score) that can be used to differentiate between normal and anomaly samples.

The anomaly prediction step applies the model into the production phase. The learned model M_t is applied on new data to get predictions which are converted into anomaly scores $S_d(x_i)$using the established function f_s. In this step, convenient thresholds TH_i are defined to characterize normal telemetry. Samples outside these values are identified as anomalies.

2.2 Exploited Methods

During this work, we select three methods with different structures to evaluate their performance.

Autoencoders are common machine learning methods and they are composed from two parts: Encoder and Decoder. The former maps the input data into different feature space (can be higher or lower in dimension) where new information about data can be exploited easily (See Fig. 2). The decoder transforms the feature space back into the original value which results in difference (reconstruction loss). The model is learned to minimize the loss so that better performance is achieved.

Let L be the loss resulted from the input and reconstructed values, the autoencoders aim at minimizing the loss value by fining the optimal architecture parameters H:

$$\arg\max_{H} L(x_i) \tag{2}$$

Their principle is almost similar with Principal Component Analysis (PCA) in the mapping of the input data into feature space and then reconstructing back the original data. However, they cannot reflect complex behavior when int is related to time lags.

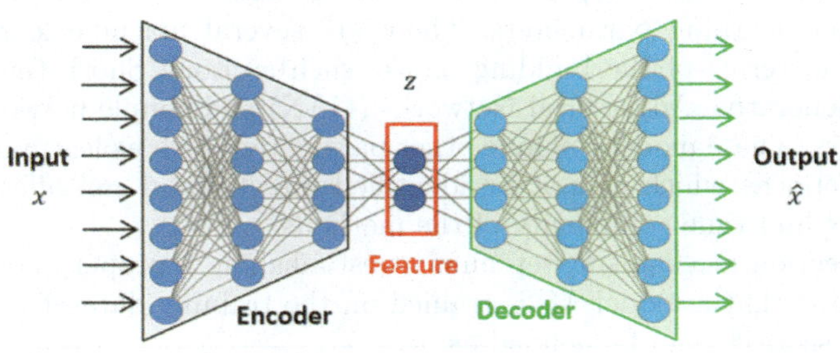

Fig. 2. Architecture of Autoencoders [5]

LSTM. Long Short-Term Memory (LSTM) receives great importance in anomaly detection in satellite telemetry. Its importance is reflected in learning the time-series properties. Its memory is used to infer the relationship between past and present data which can help in identifying useful structures. The used method is inspired by [1] which develops an LSTM model for each telemetry channel which can be more flexible. LSTM architecture (See Fig. 3) includes four parts: input gate, output gate, forget gate and Remember cell. Their importance comes from the role of these gates in controlling past and present information.

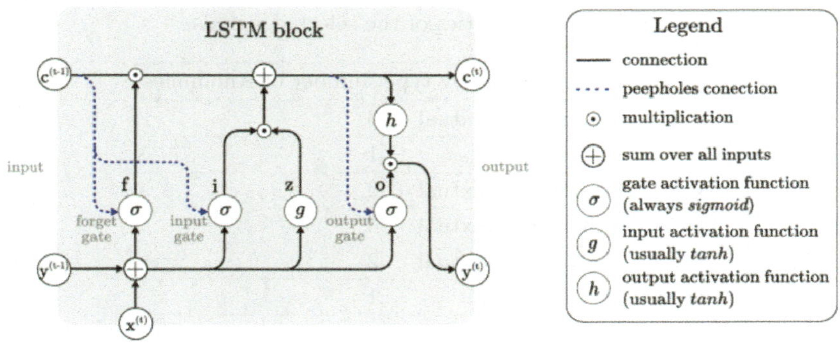

Fig. 3. The general Architecture of LSTM [6]

TadGAN. This method is inspired by Generative Adversarial Networks (GANs) [4] (Fig. 4). They are composed of three parts: Generator, Adversarial (Discriminator) and the network. The generator learns to produce data that are very close to actual one, on the other hand, the adversarial networks aim to distinguish between the generated and the actual samples.

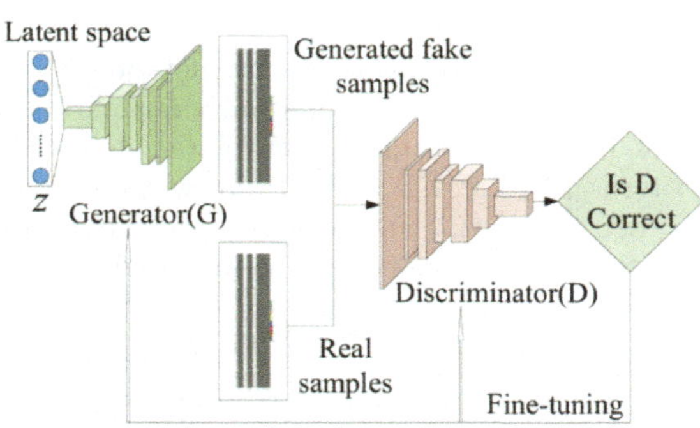

Fig. 4. The general Architecture of GAN [7]

3 Experiments

3.1 Benchmark

To evaluate the selected machine learning methods, we use six (06) signals from NASA benchmark: Soil Moisture Active Passive (SMAP) and Mars Science Laboratory (MSCL) [1]. These signals are randomly chosen to evaluate in detail the behavior of the methods in the presence of one/two and three anomalies. Table 1 shows the characterization of the selected signals.

Table 1. Properties of the selected signals

Signals	Dataset	Anomaly type	Number of Anomalies
P-1	SMAP	Contextual	3
S-1	SMAP	Point	1
E-1	SMAP	Contextual	2
F-7	MSCL	Contextual	3
T-8	MSCL	Contextual	2
D-15	MSCL	Point	1

It should be highlighted that the contextual anomalies denoted also the collective anomalies.

3.2 Implementation and Evaluation Criteria

To have equal comparisons between the selected machine learning methods, we employ the Orion library [3] which includes several commonly used methods in satellite telemetry anomaly detection. Three methods are used: dynamic LSTM (inspired by [1]), autoencoders and TADGAN [4]. Python is the main programming language and the code is implemented in Google Collaboratory with T4 GPU as the execution environment.

In order to evaluate the performance of the methods, we employ weighted contextual scoring as defined in the evaluation benchmark. This technique is preferred as it takes into account the length of the detected sequences. The commonly used criteria TN, TP, FN and FP are weighted based on the given sequence's lengths. Let S_L be the length of each detected sequence, G be the ground-truth and R the resulted segments, the weighted parameters can be formulated as:

$$TN, TP, FN, FP = C(G, R, S_L) \tag{3}$$

where C denoted the weighted confusion matrix. Once the recall and precision are calculated based on the resulted parameters, F1-score can be calculated as:

$$F1 - score = \frac{2 \times Recall \times precision}{precision + recall} \tag{4}$$

4 Results and Discussions

During this section, we assess the results of the methods in terms of the selected criteria as well as the time complexity based on the execution time.

4.1 Preliminary Results

During this section, we assess the performance of the methods using default parameters (with window size of 250 samples) except the number of epochs which is modified to 5 to have rapid assessment of their performance. Tables 2, 3 and 4 report the resulted measures for all the methods and Fig. 5 shows the time execution of each method.

The obtained results show that there are several differences in terms of the evaluated criteria. In terms of F1-score (Compromise between Recall and Precision), TadGAN produces the best score (0.33) followed by LSTM (0.27). In terms of accuracy, AER shows the best score (0.87) followed by TadGAN(0.84). These results report the complexity of the anomaly detection task especially for contextual alarms where all the methods fail to detect P-1 alarms under the selected set-up.

Table 2. Criteria scores for LSTM with 5 epochs and 250 windows size

Signal	Accuracy	Recall	precision	F1-score
P-1	0.91	0	NAN	NAN
S-1	0.95	0.19	1	0.32
E-1	0.92	0.25	0.29	0.27
F-7	0.91	0	0	NAN
T-8	0.55	1	0.13	0.22
D-15	0.70	0	NAN	NAN
Mean	0.82	0.24	0.35	0.27

Table 3. Criteria scores for TadGAN with 5 epochs and 100 window size

Signal	Accuracy	Recall	precision	F1-score
P-1	0.89	0	NAN	NAN
S-1	0.91	0.31	0.31	0.31
E-1	0.94	0	NAN	NAN
F-7	0.95	0.84	0.65	0.73
T-8	0.72	0.40	0.10	0.16
D-15	0.65	0.10	0.25	0.14
Mean	0.84	0.27	0.26	0.33

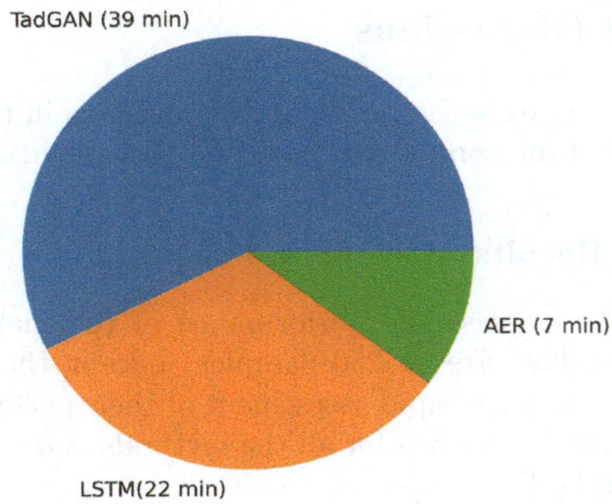

Fig. 5. Plot showing the execution time distribution for each method

Table 4. Criteria scores for AER with 5 epochs and 100 window size

Signal	Accuracy	Recall	precision	F1-score
P-1	0.91	0	NAN	NAN
S-1	0.94	0	NAN	NAN
E-1	0.93	0.059	0.25	0.10
F-7	0.92	0	NAN	NAN
T-8	0.81	1	0.26	0.41
D-15	0.70	0	NAN	NAN
Mean	0.87	0.18	0.25	0.25

LSTM fails to detect three alarms (2 contextual and 1 point), whereas TadGAN missed only two alarms (02 Contextual) which shows superiority on point alarms compared to LSTM. AER fails on 4 alarms which is revealing the complexities in the detection process. This analysis shows that advanced machine learning methods are preferable in complex alarms detection compared to simple models such as autoencoders.

On the other hand, Fig. 5 reports that TadGAN has high time complexity (39min) for 5 epochs execution and 100 window size. This remark shows that the method necessitates adequate resources and time which can be complex in situations where the training size is big and GPU resources are limited. Autoencoders report the least execution time (7 min) with 100 windows size. LSTM produces a compromise in terms of both performance and time complexity (Even though its default windows size is 250).

As the aim of this study is to provide an evaluation to get the best methods in terms of lightweight models that can produce good results without high time complexity, we need to evaluate the effect of the window size on the performance of LSTM and AER which can be more helpful to get an idea about the evolution of their criterion with different values.

4.2 Effect of the Window Size

In this section, we evaluate the effect of the time lag (window size) on the performance as well as the time complexity to have enough testing. For lightweight purposes, we select LSTM and AER. Tables 5, 6, 7, 8, 9 and 10 report the obtained results in terms of window size for 50 epochs whereas Fig. 6 shows the execution time comparisons for the different sizes.

Table 5. Criteria scores for LSTM with 50 epochs and 5 window size

Signal	Accuracy	Recall	precision	F1-score
P-1	0.90	0	0	NAN
S-1	0.92	0	0	NAN
E-1	0.93	0.08	0.23	0.12
F-7	0.92	0	NAN	NAN
T-8	0.44	1	0.12	0.19
D-15	0.63	0	0	NAN
Mean	0.79	0.18	0.07	0.16

Table 6. Criteria scores for AER with 50 epochs and 5 window size

Signal	Accuracy	Recall	precision	F1-score
P-1	0.91	0	NAN	NAN
S-1	0.94	0	NAN	NAN
E-1	0.93	0.02	0.13	0.03
F-7	0.92	0	NAN	NAN
T-8	0.81	1	0.26	0.41
D-15	0.7	0	NAN	NAN
Mean	0.87	0.17	0.19	0.22

From these results, it can be seen that both methods fail at 4 alarms with different scores. In general, autoencoders perform well at smaller window sizes however, their time complexity is higher than LSTMs. at 50 window size, AER produces 0.38 F1-score which better than the default setting produced in Table 4 at 100 lag. It is remarked that as the windows size increases, the execution time increases significantly: 5 epochs (AER 227 s /LSTM 158 s) and 50 epochs (AER 1303 s /LSTM 810 s). This result shows that it is recommended in ML methods to choose these values carefully so that to obtain optimal performance. In addition, by having comparisons with previous section, there is a compromise between execution time (5 epochs versus 50) and high scores which can result by having enough time lag and enough training time. Results also reveal the difficulties in detecting contextual alarms as well as their repeatability for point alarms.

Table 7. Criteria scores for LSTM with 50 epochs and 20 window size

Signal	Accuracy	Recall	precision	F1-score
P-1	0.91	0	0	NAN
S-1	0.93	0	0	NAN
E-1	0.89	0.11	0.1	0.1
F-7	0.92	0	NAN	NAN
T-8	0.42	1	0.1	0.19
D-15	0.7	0	NAN	NAN
Mean	0.8	0.18	0.05	0.14

Table 8. Criteria scores for AER with 50 epochs and 20 window size

Signal	Accuracy	Recall	precision	F1-score
P-1	0.91	0	NAN	NAN
S-1	0.94	0	NAN	NAN
E-1	0.92	0.08	0.18	0.11
F-7	0.92	0	NAN	NAN
T-8	0.81	1	0.26	0.41
D-15	0.7	0	NAN	NAN
Mean	0.87	0.18	0.22	0.26

Table 9. Criteria scores for LSTM with 50 epochs and 50 window size

Signal	Accuracy	Recall	precision	F1-score
P-1	0.91	0	NAN	NAN
S-1	0.94	0	NAN	NAN
E-1	0.94	0.36	0.55	0.44
F-7	0.91	0	0	NAN
T-8	0.42	1	0.1	0.19
D-15	0.7	0	NAN	NAN
Mean	0.8	0.23	0.22	0.31

Table 10. Criteria scores for AER with 50 epochs and 50 window size

Signal	Accuracy	Recall	precision	F1-score
P-1	0.91	0	NAN	NAN
S-1	0.94	0	NAN	NAN
E-1	0.94	0.34	0.48	0.4
F-7	0.92	0	NAN	NAN
T-8	0.76	1	0.22	0.36
D-15	0.7 0	NAN	NAN	NAN
Mean	0.86	0.22	0.35	0.38

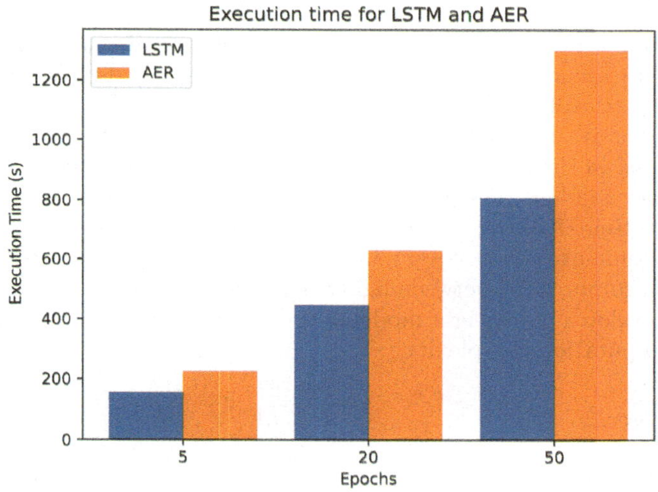

Fig. 6. Execution time comparisons between LSTM and AER

4.3 Discussion

The results show clearly the challenges as well as the advantage of machine learning methods for anomaly detection. Taking into account the big data size of satellite telemetry, the need to have intelligent and automatic solutions is very demanding. In addition, the developed method should be reliable as well as not computationally expensive to avoid extra resources and other complexities. The exploited methods show limitations against execution time as well as contextual alarms. It should be highlighted that our study investigates lightweight aspects which can be beneficial for the community and more effective in the implementation side in industrial systems.

There are several parameters that need to be taken into account for ML methods which can be both advantageous as well as limitations. From them, we could highlight the size of the training dataset, the time lag (window size) and the number of epochs. The challenges are clear on different datasets as well as different signals where a set of parameters cannot be optimal for all situations which limit the method performance. The time lag is key parameters for time series ML methods as it determines the methods complexity and reliability. On the other hand, the number of epochs is an important parameter, however, its value needs to be selected in a way to avoid overfitting/underfitting problems.

To summarize, The application of ML methods to anomaly detection in satellite telemetry shows the need for lightweight and robust methods that can be less computationally expensive as well as robust against changes in dataset size as well as their parameters.

4.4 Conclusion

The conducted works examines the performance of three machine learning methods, on satellite anomaly detection, namely: Autoencoders, LSTM and TadGAN. The methods are evaluated on six (06) signals from NASA benchmark: Soil Moisture Active Passive (SMAP) and Mars Science Laboratory (MSCL) which are commonly used in the literature. Results show that autoencoders produces good scores at lower window sizes (0.38 F1-score at 50 time lag), however, LSTM has the execution time. Although, TadGAN produces the best score (0.33) at around 5 epochs its time execution is very high. Therefore, it is remarked that still far away to get a compromise between lightweight and performance. As future work, we plan to develop a lightweight model that can be able to detect complex alarms with low computation complexity.

References

1. Hundman, K., Constantinou, V., Laporte, C., Colwell, I., Soderstrom, T.: Detecting spacecraft anomalies using LSTMs and nonparametric dynamic thresholding. In: Proceedings of the 24th ACM SIGKDD International Conference on Knowledge Discovery & Data Mining, pp. 387–395 (2018)
2. Baireddy, S., et al.: Spacecraft time-series anomaly detection using transfer learning. In: Proceedings of the IEEE/CVF Conference on Computer Vision and Pattern Recognition, pp. 1951–1960 (2021)
3. Project, T.: Orion library (2024). https://sintel.dev/Orion/index.html. Accessed 12 July 2024
4. Geiger, A., Liu, D., Alnegheimish, S., Cuesta-Infante, A., Veeramachaneni, K.L TadGAN: time series anomaly detection using generative adversarial networks. In: 2020 IEEE International Conference on Big Data (Big Data), pp. 33–43 (2020)
5. Alaghbari, K., Lim, H., Saad, M., Yong, Y.: Deep autoencoder-based integrated model for anomaly detection and efficient feature extraction in IoT networks. IoT. **4**, 345–365 (2023)
6. Van Houdt, G., Mosquera, C., Nápoles, G.: A review on the long short-term memory model. Artif. Intell. Rev. **53**(8), 5929–5955 (2020). https://doi.org/10.1007/s10462-020-09838-1
7. Dan, Y., Zhao, Y., Li, X., Li, S., Hu, M., Hu, J.: Generative adversarial networks (GAN) based efficient sampling of chemical composition space for inverse design of inorganic materials. NPJ Comput. Mater. **6**, 84 (2020)

8. Chen, J., Pi, D., Wu, Z., Zhao, X., Pan, Y., Zhang, Q.: Imbalanced satellite telemetry data anomaly detection model based on Bayesian LSTM. Acta Astronaut. **180**, 232–242 (2021)
9. Tariq, S., et al.: Detecting anomalies in space using multivariate convolutional LSTM with mixtures of probabilistic PCA. In: Proceedings of the 25th ACM SIGKDD International Conference on Knowledge Discovery & Data Mining, pp. 2123–2133 (2019)
10. Nalepa, J., Myller, M., Andrzejewski, J., Benecki, P., Piechaczek, S., Kostrzewa, D.: Evaluating algorithms for anomaly detection in satellite telemetry data. Acta Astronaut. **198**, 689–701 (2022)
11. Liu, D., Pang, J., Xu, B., Liu, Z., Zhou, J., Zhang, G.: Satellite telemetry data anomaly detection with hybrid similarity measures. In: 2017 International Conference on Sensing, Diagnostics, Prognostics, and Control (SDPC), pp. 591–596 (2017)

8. [illegible] P., [illegible] machine learning [illegible] models [illegible] (2021)
9. [illegible] learning [illegible] (2023)
10. [illegible]
11. [illegible] (2016)

Data Science and its Applications

Assessing Landslide Susceptibility Using Machine Learning and Remote Sensing Data: A Case Study of Southeastern Constantine, Algeria

Zakaria Matougui(✉), Mohamed Yacine Daksi, and Mehdi Dib

Centre de Recherche en Aménagement du Territoire (CRAT), Campus Zouaghi Slimane, Route de Ain elBey, 25000 Constantine, Algérie

zakaria.matougui@crat.dz

Abstract. This paper investigates landslide susceptibility in Constantine, Algeria, using advanced machine learning techniques integrated with Geographic Information Systems (GIS) and remote sensing data. The study employs two robust gradient boosting algorithms, XGBoost and LightGBM, to develop predictive models. These models are fine-tuned using Bayesian optimization, which enhances their accuracy. The model performance is validated through cross-validation and interpreted using SHAP values, providing a detailed analysis of the factors influencing landslide risk. The study shows high predictive accuracy, with ROC AUC scores consistently exceeding 0.83, highlighting the effectiveness of the applied methods. Key factors influencing landslide susceptibility, such as elevation, slope, and proximity to roads, are identified, offering critical insights for risk assessment and management in landslide-prone areas. This research underscores the potential of integrating machine learning and GIS for effective natural hazard prediction and urban planning strategies.

Keywords: Landslide susceptibility · Machine learning · Constantine

1 Introduction

Growing technological advances recently have pushed many fields to change their courses in order to cope with the challenges, and to benefits from this phenomenon. In this regard, Geographic Information Systems (GIS) have adopted and integrated Artificial Intelligence (AI) in order to tackle complex spatial problems [1]. In the natural hazard predictive modeling field, the use of AI for natural hazards assessment like ground movement, wildfire and floods, demonstrating the accuracy and efficiency of AI models in spatial predictions [2].

Machine learning (ML) a part of AI plays an important role in the landslide field, a variety of methodologies have been adopted for the assessment of landslide susceptibility, for example logistic regression model [3], ML adopting individual [4] or

S. Chikhi et al. (Eds.): MISC 2024, LNNS 1267, pp. 143–155, 2025.
https://doi.org/10.1007/978-3-031-82112-7_11

ensemble algorithms [5]. In order to assure the reliability of susceptibility mapping, the implementation of performant models is crucial.

In the northern part of Africa, precisely in Algeria, the complex geological and topographical context led to many challenges regarding landslide hazards. The city of Constantine, in northeastern Algeria, is known by its complex geological, morphological, and topographical context that increases the landslide susceptibility [6, 7].

In the present paper, data science and ML techniques are employed to tackle complex problem-solving in GIS through a detailed assessment of landslide susceptibility. The methodology involves the use of advanced ensemble ML algorithms, specifically XGBoost and LightGBM, to analyze remote sensing data and other environmental factors. The study area, a region heavily affected by landslides in Constantine, Algeria, serves as the focus for this assessment. The process begins with the collection of relevant data, including elevation, slope, land use, and proximity to streams, proximity to roads, aspect and curvature. These variables are then used to train the ML models, which have been optimized using Bayesian methods to improve accuracy and predictive performance. The results are interpreted using SHAP (SHapley Additive exPlanations) values, which provide insights into the relative importance of each feature in the prediction of landslide risk, ultimately aiding in the identification of high-susceptibility areas.

2 Study Area

The study area is located southwest of the city of Constantine in northeastern Algeria, approximately 430 km east of the capital Algiers (Fig. 1). It covers an area of 31.5 km^2 within the city's perimeter.

Constantine is significantly affected by land instabilities due to its geomorphological and geological characteristics, climatic and seismic conditions, and human activities. Geomorphologically, the Constantine region is part of the mountainous zone in the northeast of the Atlas Mountains, featuring a mix of mountains, hills, plains, and rivers. The altitude ranges from 400 to 800 m, decreasing from northeast to southwest.

Geologically, the region belongs to the outer domain of the Tellian Atlas chain, part of the North African Alps (Maghrebides). The study area is primarily covered by Neogene and Quaternary formations. The Neogene clays and conglomerates, which cover areas such as Boussouf, 20 August, Ciloc, Bellevue, and parts of the city center, are particularly sensitive to water and prone to instability. These areas experience medium to high plasticity due to the presence of water, leading to gravitational instability.

The climate in the Constantine region is semi-arid, with a hot and relatively dry season from June to August and a wet season from December to April. This climate contributes to the swelling and shrinking of the clays in the Neogene basin, exacerbating land instability issues.

3 Data Used

To effectively landslide susceptibility assessment, our study relies on two main types of data: a landslide inventory (Fig. 2) and related environmental factors (Fig. 3). The landslide inventory was compiled through field surveys and analysis of satellite images.

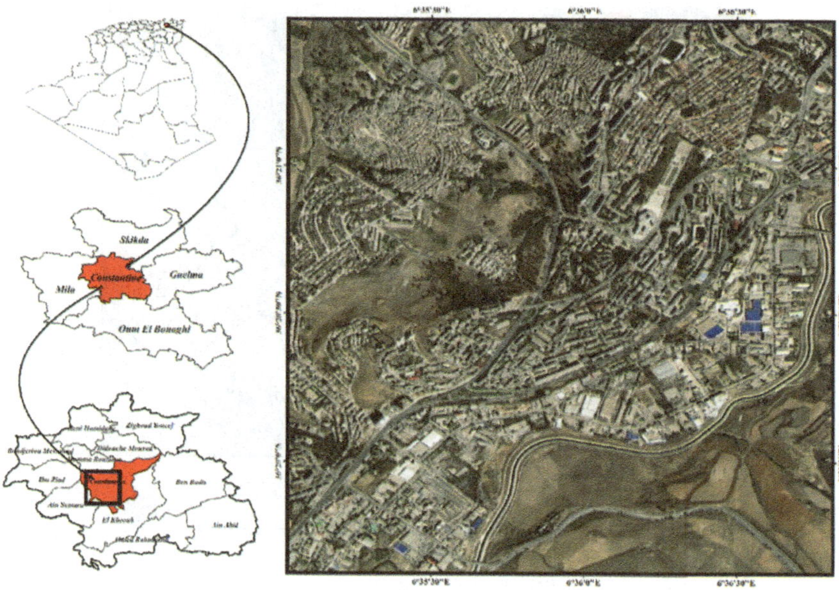

Fig. 1. Location of the study area

The environmental factors we considered include several critical datasets. For instance, we used data from ALOS PALSAR that provides elevation, ground slope, aspect, and proximity to streams with a resolution of 30 m, covering the years 2006 to 2011. This data is essential for understanding the landscape and water flow characteristics that affect landslide occurrence and can be accessed via this DOI: https://doi.org/10.5067/Z97HFCNKR6VA. We also used land use data from the ESA WorldCover project, which gives us detailed information on the types of ground cover present and is available through this DOI: https://doi.org/10.5281/zenodo.5571936. Additionally, we included road network data from OpenStreetMap, which helps us understand human alterations to the landscape; this data is current as of 2019 and is under the OpenStreetMap © license, accessible at OpenStreetMap Copyright.

The correlation matrix depicted in the heatmap (Fig. 4) is crucial in the pre-modeling phase for machine learning applications, particularly when assessing landslide susceptibility. It serves identifying relationships between different variables, which can inform feature selection. By understanding how variables relate to each other, researchers can select the most relevant predictors for the model and potentially reduce the feature space to improve model efficiency and interpretability. In addition, the matrix is instrumental in diagnosing multicollinearity, a common problem where predictors are highly correlated with one another. Multicollinearity can lead to unreliable and unstable estimates of regression coefficients, making it difficult to interpret the effect of each independent variable. In the matrix, a high correlation coefficient between predictors would suggest a potential collinearity issue.

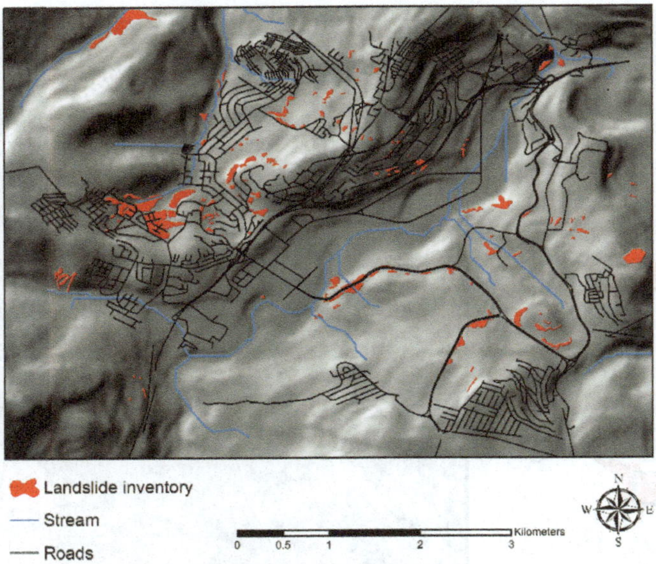

Fig. 2. Landslide inventory

As shown Fig. 4 elevation, aspect, slope, curvature and distance to stream have a weak correlation, indicating that the factors derived from the same source are not corelated. The most significant relationship is between slope and land use (−0.33), where steeper slopes are associated with different types of land use, suggesting variations in terrain usage based on slope. Overall, while some correlations are present, most factors show weak relationships, suggesting that they contribute independently to the characteristics of the landscape, important for modeling and predicting landslide susceptibility.

Considering landslides and non-landslides superposed on each individual factor map, the distribution of these events can be represented in histograms, as shown in Fig. 5. These histograms compare the distribution of various factors for locations with and without landslides.

Analyzing the histogram for aspect, we observe that landslides are more frequent in specific aspect ranges, particularly around 50° to 360° and 300° to 350°, which correspond to north-facing slopes. North-facing slopes receive less direct sunlight, leading to different moisture conditions compared to south-facing slopes. These conditions, such as retained humidity and reduced evaporation, can make north-facing slopes more susceptible to landslides.

The data indicates that landslides are more likely to occur on moderate slopes, especially those exceeding 8°. Additionally, landslides are prevalent in areas with specific types of land use, particularly near roads (within 250 m) and streams (within 400 m). This suggests that human alterations and proximity to water bodies increase landslide susceptibility.

Shrubland and built-up areas show a higher incidence of landslides, indicating that these environments might be more susceptible due to factors like less effective vegetation

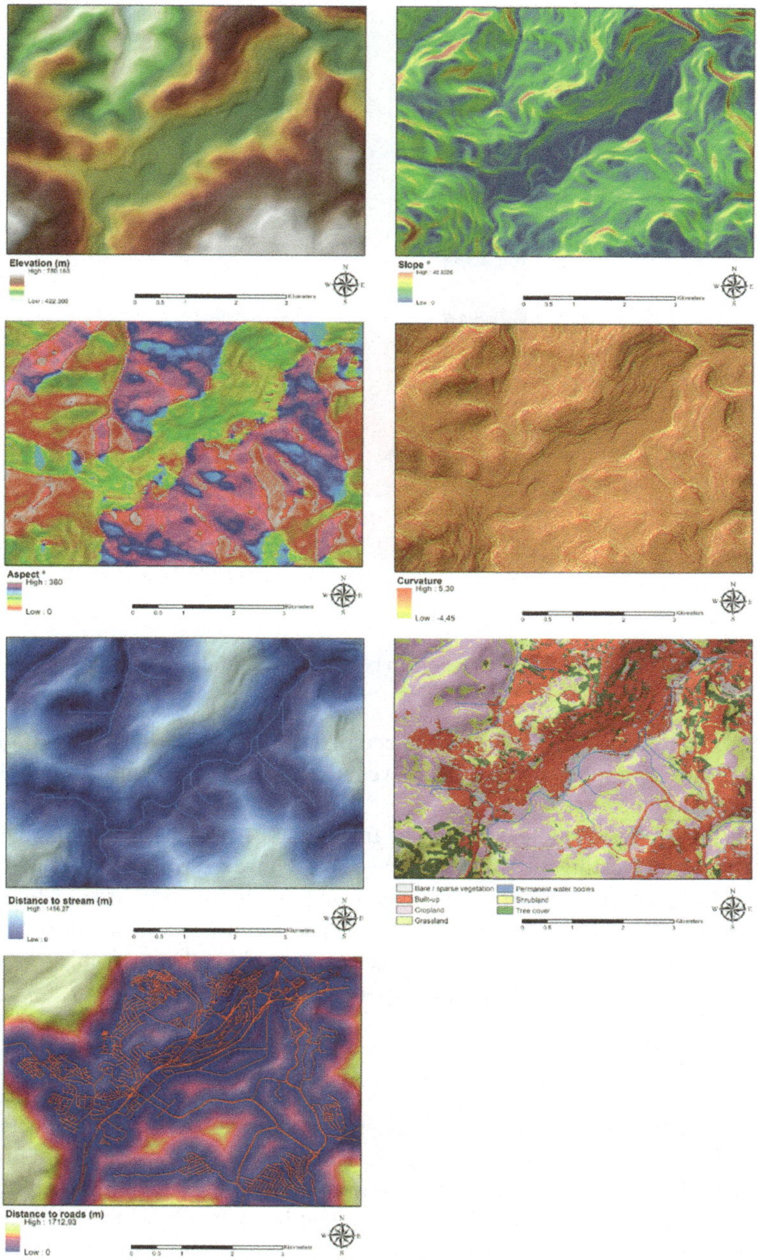

Fig. 3. Causative factors map

root systems in shrubland and landscape alterations in urban areas. In contrast, tree cover and cropland exhibit fewer landslides, suggesting that these land uses might help

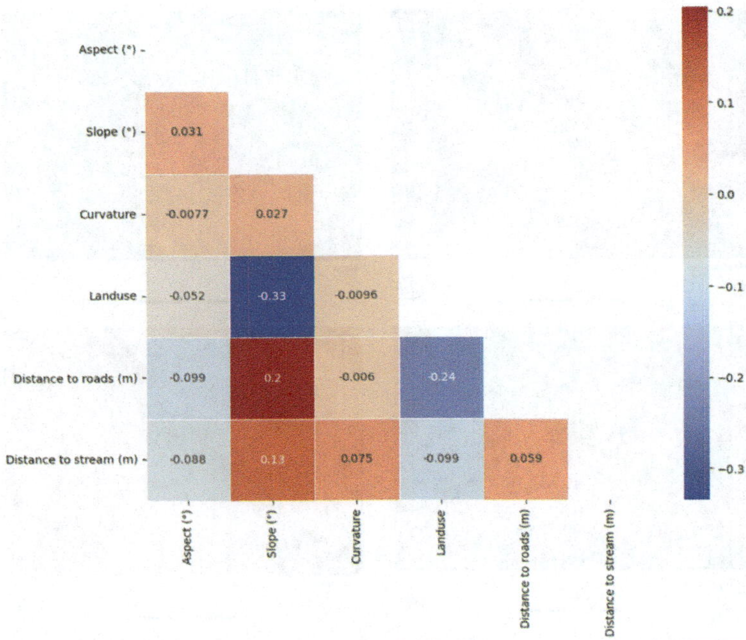

Fig. 4. Correlation heatmap between the causative factors

stabilize the soil. Grassland shows a balanced distribution, implying a neutral influence on landslide susceptibility. Bare or sparse vegetation and permanent water bodies have minimal landslide occurrences.

The histograms also indicate a higher frequency of landslides in areas with non-null curvature, suggesting that both concave and convex terrains are more prone to landslides. Considering the correlation matrix and the distribution of landslides across different factors, it is evident that all the proposed factors influence landslide occurrences. However, these factors are independent of each other, making them suitable for inclusion in machine learning models for landslide susceptibility assessment.

4 Methodology

The approach to landslide susceptibility assessment utilized previously collected raster data by converting it into a structured dataset suitable for machine learning modeling. Each raster cell was transformed into a data point with corresponding feature values. During the modeling phase, two powerful gradient boosting algorithms, XGBoost and LightGBM, were employed due to their efficiency, scalability, and ability to handle complex relationships between features and landslide occurrences [8, 9]. To enhance model performance, Bayesian Optimization was applied for hyperparameter tuning, using a probabilistic model to balance exploration and exploitation, thus allowing for a comparison of default and optimized models. After training and optimizing the models, landslide susceptibility maps of the region were presented, and the SHAP (SHapley

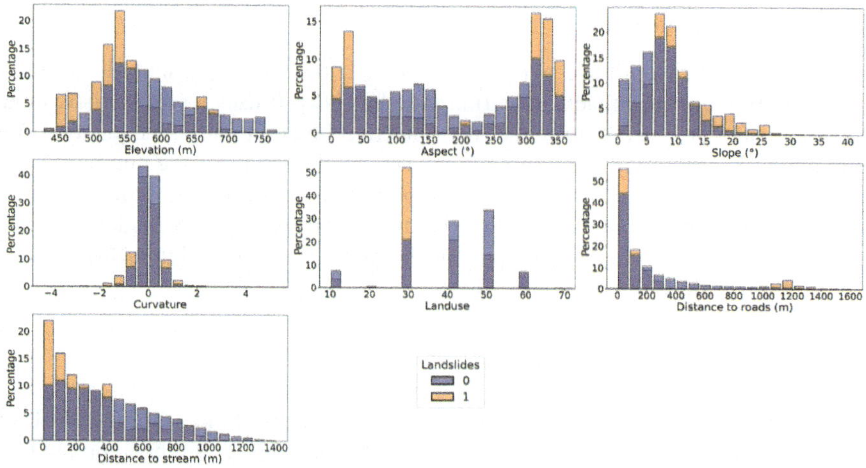

Fig. 5. Comparison of causative factors for landslide and non-landslide locations with histograms showing the distribution of aspect, slope, curvature, land use, and proximity to roads and streams. The land use categories represented include Tree cover (10), Shrubland (20), Grassland (30), Cropland (40), Built-up areas (50), Bare/sparse vegetation (60), and Permanent water bodies (70).

Additive exPlanations) method was used to interpret the results. SHAP values provided insights into the contribution of each feature to the model's predictions, elucidating the importance and impact of different factors on landslide susceptibility.

4.1 XGBoost

XGBoost, or Extreme Gradient Boosting, is a powerful and scalable machine learning algorithm used for supervised learning tasks [10]. It is designed to deliver high performance through efficient parallel processing, handling of sparse data, and regularization techniques to prevent overfitting. XGBoost's novel tree-pruning method enhances accuracy and prevents overfitting, making it a versatile tool for various applications, including classification, regression, and ranking problems. Its speed and efficiency make it a suitable choice for landslide susceptibility assessment [11].

4.2 LightGBM

Light Gradient Boosting Machine, is another efficient gradient boosting framework developed by Microsoft [12]. It employs a histogram-based learning approach and grows trees leaf-wise, resulting in faster training times and reduced memory usage. Is optimized for handling large datasets and categorical features, making it ideal for big data applications. Its ability to support parallel and distributed learning further enhances its performance, making it a strong candidate for modeling landslide susceptibility [13].

4.3 Bayesian Optimization

This method used to optimize expensive objective functions, particularly useful for hyperparameter tuning in machine learning models [14]. It uses a probabilistic model, such as a Gaussian Process, to model the objective function and an acquisition function to balance exploration and exploitation. This method reduces the number of function evaluations needed, ensuring efficient optimization and finding global optima in high-dimensional spaces. In the context of landslide susceptibility assessment, Bayesian Optimization can fine-tune the hyperparameters of XGBoost and LightGBM models to enhance their predictive performance [15].

5 Results and Discussions

To produce the landslide susceptibility models, both variants of the XGBoost and LightGBM algorithms were trained using the training dataset (70%) and then applied to the full dataset for prediction. The resulting landslide susceptibility maps, as shown in Fig. 6, illustrate the predicted susceptibility levels across the study area.

These maps reveal distinct zones of varying landslide susceptibility. High-risk areas are particularly noticeable in regions with steep slopes, close proximity to streams, and specific land use types, including shrubland and built-up areas. The maps provide a visual representation of the spatial distribution of landslide susceptibility, helping to identify and prioritize areas for mitigation efforts.

To evaluate the predictive performance of the models, Bayesian optimization was employed to fine-tune the hyperparameters, with the optimal parameters outlined in Table 1. The models were then trained on the training dataset. Subsequently, a cross-validation approach, specifically 5-fold 2 times repeated cross-validation, was used to assess the models' performance, utilizing metrics such as ROC AUC, accuracy, and F1 score. This validation was conducted on the test dataset to ensure the models' robustness and generalizability.

The results of the evaluation method are presented in Fig. 7, which showcases the performance of the XGBoost and LightGBM models, both in their default and optimized configurations. The figure provides a comparative analysis of the models based on ROC AUC scores, accuracy, and F1 scores. The ROC AUC scores demonstrate that both the default and optimized models have high predictive power, with scores consistently above 0.83, indicating a strong ability to differentiate between landslide and non-landslide areas. The accuracy scores, while slightly lower, still reflect a good level of precision in the models' predictions, particularly for the optimized versions. The F1 scores, which balance precision and recall, also show that the optimized models outperform their default counterparts, suggesting that the fine-tuning via Bayesian optimization has effectively enhanced model performance. XGBoost consistently outperforms LightGBM across all metrics, particularly after optimization. However, the performance gap is relatively small, indicating that both models are similarly effective in predicting landslide susceptibility. Among the four models, the optimized versions are more suitable for this study area, as they present a more realistic depiction of the phenomenon, being more conservative about the extent of high susceptibility areas, which is crucial for accurate risk assessment and management.

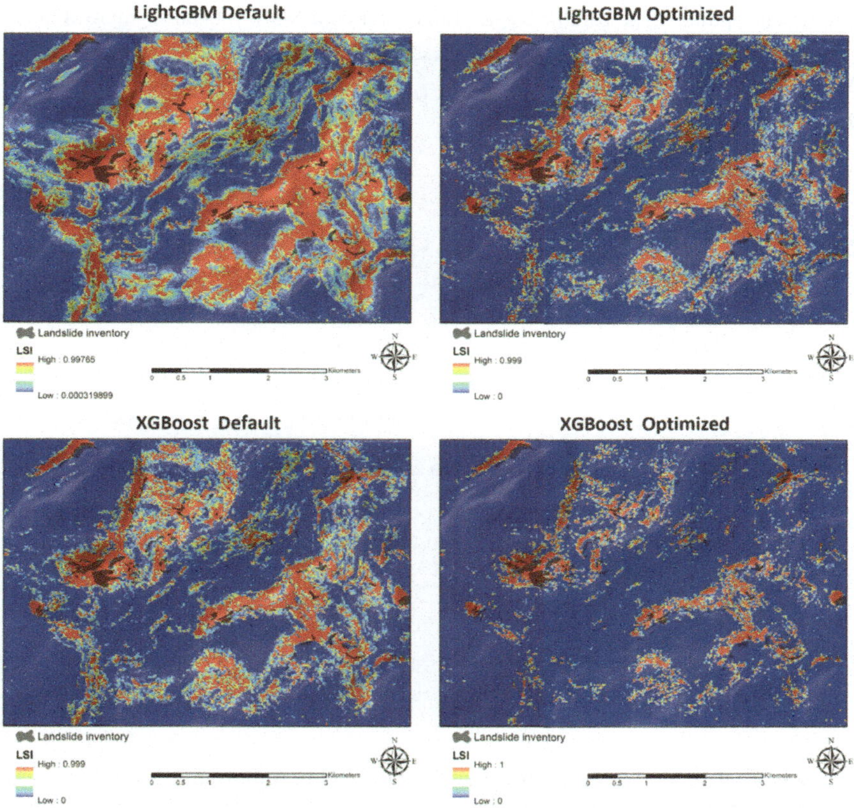

Fig. 6. Landslide susceptibility maps for default and optimized XGboost and LightGBM models

The SHAP values presented in Fig. 8 indicate the impact of each feature on the outputs of the models, with higher absolute values reflecting greater influence. The plots reveal that certain features consistently have a significant impact on landslide susceptibility predictions across both models and configurations. Notably, elevation, distance to roads and aspect are among the most influential factors. The SHAP values for these features are prominently higher, particularly in the optimized models, highlighting their critical role in the models' decision-making processes.

In both models, the optimization process appears to enhance the models' sensitivity to certain features, as evidenced by the broader range of SHAP values in the optimized plots compared to the default plots. This suggests that the optimization not only improves overall model performance but also refines the models' ability to discern the relative importance of different factors influencing landslide susceptibility.

Overall, the comparison between XGBoost and LightGBM indicates that while both models rely on similar key features, XGBoost demonstrates slightly more pronounced

Table 1. The optimal hyperparameters determined for both XGBoost and LightGBM models.

Model	Hyperparameter	Research Range	Optimal Value
XGBoost	colsample_bytree	0.5 to 1.0	1
	gamma	0 to 5	0
	learning_rate	0.01 to 1.0	1
	max_depth	3 to 10	10
	min_child_weight	1 to 10	1
	n_estimators	100 to 1000	713
	subsample	0.5 to 1.0	1
LightGBM	bagging_fraction	0.8 to 1	0.762
	feature_fraction	0.1 to 0.9	0.648
	learning_rate	0.01 to 1.0	0.425
	max_bin	20 to 90	88
	max_depth	3 to 10	9
	min_data_in_leaf	20 to 80	62
	min_sum_hessian_in_leaf	0 to 100	7.96
	num_leaves	24 to 80	55
	subsample	0.01 to 1.0	0.95

feature sensitivity. This nuanced understanding of feature importance is crucial for interpreting model predictions and for informing decision-making processes in landslide risk management.

The SHAP summary plot for the models reveals how various features influence landslide susceptibility predictions. Lower elevations, steeper slopes, and proximity to streams significantly increase the likelihood of landslides, as indicated by positive SHAP values. In contrast, higher elevations and areas far from roads generally reduce landslide risk, reflected by negative SHAP values. The aspect of slopes shows mixed impacts, depending on the direction with higher risk for north facing slopes, while land use type and terrain curvature also play roles, with certain land uses and curvatures contributing to increased susceptibility. This analysis highlights the importance of these features in predicting landslide risk, providing insights into the factors that most significantly affect model outcomes.

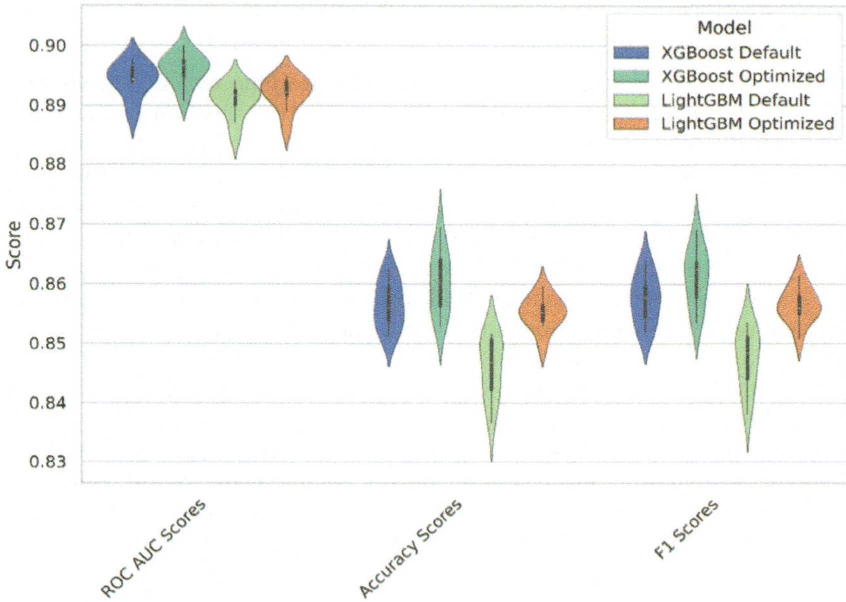

Fig. 7. Performance comparison of XGBoost and LightGBM models in default and optimized configurations using ROC AUC, accuracy, and F1 scores

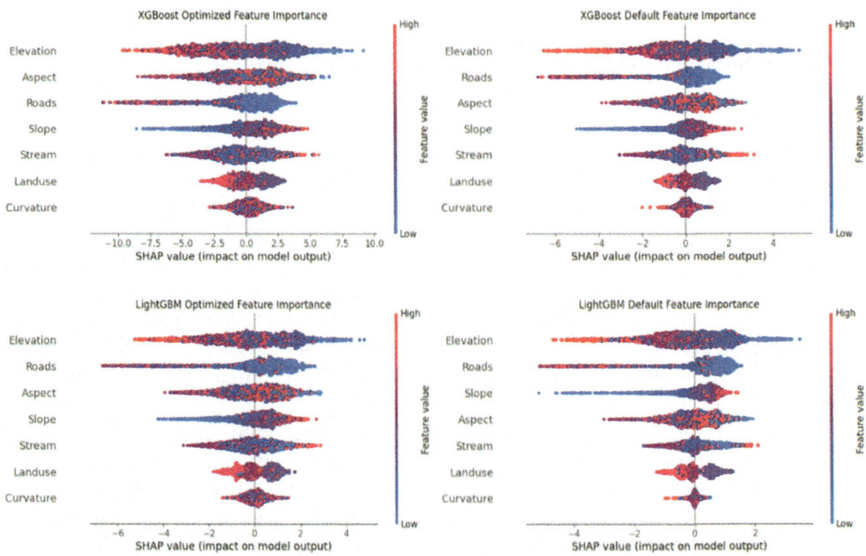

Fig. 8. SHAP (SHapley Additive exPlanations) summary plots for feature importance in XGBoost and LightGBM models

6 Conclusion

This study successfully applied advanced machine learning techniques, specifically XGBoost and LightGBM, to assess landslide susceptibility in Constantine, Algeria, a region significantly impacted by landslide hazards. Through the integration of GIS and remote sensing data, and the utilization of Bayesian optimization for model fine-tuning, All the studied model showed satisfactory performance of the proposed metrics higher than 0.8 (ROCAUC $\geq$ 0.89, F1 score $\geq$ 0.85 and accuracy $\geq$ 0.84). The factors analysis identified key factors such as elevation, slope, and proximity to roads and aspect as significant contributors to landslide risk. The SHAP values provided a detailed understanding of the relative importance of these factors in each model. The optimized models showed a more realistic representation of high susceptibility areas. This study underscores the importance of combining GIS, machine learning, and interpretative tools like SHAP for natural hazard prediction and management.

The findings are crucial for informing risk assessment and mitigation strategies, providing a robust framework that can be adapted for similar geographical regions facing landslide hazard. Future research could expand on these methods by incorporating additional environmental and anthropogenic variables to enhance predictive accuracy and reliability.

Acknowledgments. The authors did not receive support from any organization for the submitted work.

Disclosure of Interests. The authors have no competing interests.

References

1. Li, Z., Ning, H.: Autonomous GIS: the next-generation AI-powered GIS. Int. J. Digit. Earth **16**(2), 4668–4686 (2023). https://doi.org/10.1080/17538947.2023.2278895
2. Yousefi, S., Pourghasemi, H.R., Emami, S.N., Pouyan, S., Eskandari, S., Tiefenbacher, J.P.: A machine learning framework for multi-hazards modeling and mapping in a mountainous area. Sci. Rep. **10**(1), 1–14 (2020). https://doi.org/10.1038/s41598-020-69233-2
3. Budimir, M.E.A., Atkinson, P.M., Lewis, H.G.: A systematic review of landslide probability mapping using logistic regression. Landslides **12**(3), 419–436 (2015). https://doi.org/10.1007/s10346-014-0550-5
4. Marjanović, M., Bajat, B., Kovačević, M.: Landslide susceptibility assessment with machine learning algorithms. In: International conference on intelligent networking and collaborative systems, INCoS 2009, pp. 273–278 (2009). https://doi.org/10.1109/INCOS.2009.25
5. Matougui, Z., Djerbal, L., Bahar, R.: A comparative study of heterogeneous and homogeneous ensemble approaches for landslide susceptibility assessment in the Djebahia region, Algeria. Environ. Sci. Pollut. Res. **31**, 40554 (2023). https://doi.org/10.1007/s11356-023-26247-3
6. Manchar, N., Benabbas, C., Hadji, R., Bouaicha, F., Grecu, F.: Landslide susceptibility assessment in constantine region (NE Algeria) by means of statistical models. Studia Geotechnica et Mechanica **40**(3), 208–219 (2018). https://doi.org/10.2478/sgem-2018-0024

7. Achour, Y., Boumezbeur, A., Hadji, R., Chouabbi, A., Cavaleiro, V., Bendaoud, E.A.: Landslide susceptibility mapping using analytic hierarchy process and information value methods along a highway road section in Constantine, Algeria. Arab. J. Geosci. **10**(8), 1 (2017). https://doi.org/10.1007/s12517-017-2980-6
8. Sun, D., Chen, D., Zhang, J., Mi, C., Gu, Q., Wen, H.: Landslide susceptibility mapping based on interpretable machine learning from the perspective of geomorphological differentiation. Land **12**(5), 1018 (2023). https://doi.org/10.3390/land12051018
9. Yue, W., et al.: Assessment of wildfire susceptibility and wildfire threats to ecological environment and urban development based on GIS and multi-source data: a case study of Guilin, China. Remote Sens. **15**(10), 2659 (2023). https://doi.org/10.3390/rs15102659
10. Chen, T., Guestrin, C.: XGBoost : a scalable tree boosting system, vol. 13–17 (2016). https://doi.org/10.1145/2939672.2939785
11. Liu, B., Guo, H., Li, J., Ke, X., He, X.: Application and interpretability of ensemble learning for landslide susceptibility mapping along the Three Gorges Reservoir area, China. Nat. Hazards **120**(5), 4601–4632 (2024). https://doi.org/10.1007/s11069-023-06374-3
12. Ke, G., et al.: LightGBM: a highly efficient gradient boosting decision tree. In: 31st Conference on Neural Information Processing Systems (NIPS), no. Nips, pp. 1111–1116 (2017)
13. Sun, D., Wu, X., Wen, H., Gu, Q.: A LightGBM-based landslide susceptibility model considering the uncertainty of non-landslide samples. Geomat. Nat Hazards Risk **14**(1), 2213807 (2023). https://doi.org/10.1080/19475705.2023.2213807
14. Frazier, P.I.: 'A Tutorial on Bayesian Optimization BayesOpt consists of two main components : a Bayesian statistical model for modeling the objective', no. Section **5**, 1–22 (2018)
15. Sun, D., Wen, H., Wang, D., Xu, J.: A random forest model of landslide susceptibility mapping based on hyperparameter optimization using Bayes algorithm. Geomorphology **362**, 107201 (2020). https://doi.org/10.1016/j.geomorph.2020.107201

Data-Driven Bandwidth Selection in Kernel Density Estimation via Unsupervised Learning

Arsalane Chouaib Guidoum[1(✉)] and Amel Henider[2]

[1] Department of Mathematics and Computer Science, Faculty of Sciences and Technology, University of Tamanghasset, 11000 Tamanghasset, Algeria
acguidoum@univ-tam.dz

[2] Department of Finance and Actuarial, National Higher School of Statistics and Applied Economics, 42003 Kolea, Algeria

Abstract. The selection of optimal bandwidth bw in Kernel Density Estimation (KDE) is crucial for accurate density estimation. Traditional methods, such as unbiased and biased cross-validation and their variants, often suffer from performance and computational limitations. This study presents a novel, data-driven approach to bandwidth selection using unsupervised machine learning techniques. Synthetic samples are generated from the original data, a comprehensive set of statistical parameters is calculated, and bandwidth is estimated using a rule-of-thumb [2]. Feature importance is assessed through random forests (RF) to identify the most influential factors in bandwidth selection. Various unsupervised learning models are then applied to predict the optimal bandwidth for the original sample bw_{os}. The results demonstrate that this machine learning-based approach achieves superior accuracy and computational efficiency compared to traditional cross-validation methods.

Keywords: Bandwidth selection · Cross-Validation · Kernel Density Estimation · Feature Selection · Unsupervised Machine Learning

1 Introduction

Kernel density estimation (KDE) is a non-parametric method used to estimate the probability density function of a random variable. A critical aspect of KDE is the selection of the bandwidth parameter, which controls the smoothness of the resulting density estimate. The choice of bandwidth has a significant impact on the accuracy of the KDE, and thus, selecting an appropriate bandwidth is of paramount importance. Traditional methods for bandwidth selection, such as unbiased cross-validation (UCV) [6], biased cross-validation (BCV) [7], and other cross-validation techniques including maximum-likelihood cross-validation (MLCV) [15], complete cross-validation [2], modified cross-validation [16], and trimmed cross-validation [13], have been extensively studied. However, these

S. Chikhi et al. (Eds.): MISC 2024, LNNS 1267, pp. 156–170, 2025.
https://doi.org/10.1007/978-3-031-82112-7_12

methods often face limitations related to performance and computational efficiency. For example, the ucv method, as described by Bowman and Azzalini [1], minimizes an unbiased estimate of the mean integrated squared error (MISE). Similarly, the bcv method, discussed by Scott [2], adjusts the bandwidth to minimize a biased estimate of the MISE. The plug-in approach, introduced by Sheather and Jones [3], estimates the bandwidth by plugging in estimates of the unknown quantities in the asymptotic MISE expression. The mlcv method, detailed by Hall, Racine, and Li [4], selects the bandwidth that maximizes the likelihood cross-validation score. While these methods are theoretically sound, they often face challenges related to performance and computational efficiency, especially with large datasets or in the presence of complex data structures. Cross-validation methods have also been explored extensively in the literature. Stone [5] pioneered the use of cross-validation for density estimation, highlighting its potential for model selection. The Leave-One-Out Cross-Validation approach, as described by Rudemo [6], is another well-known technique where each observation is left out in turn, and the model is trained on the remaining data. Bowman [7] further expanded on these ideas, introducing smoothed cross-validation methods.

Unsupervised learning models, such as Ridge Regression [8], Lasso Regression [9], Gradient Boosting [10], Support Vector Machines (SVM) [11], and Neural Networks [12], have shown significant promise in various applications, including density estimation. Ridge Regression and Lasso Regression are popular for their ability to handle multicollinearity and perform feature selection, respectively. Gradient Boosting and SVM are powerful techniques for regression and classification tasks, while Neural Networks offer flexibility in modeling complex data structures. Leveraging the power of these machine learning techniques for bandwidth selection in KDE offers a novel and potentially more efficient approach. This study proposes a data-driven method for bandwidth selection using unsupervised machine learning models. Our approach begins with generating synthetic samples from the original sample using bootstrap resampling or perturbation techniques. For each synthetic sample, we calculate a comprehensive set of statistical parameters (see Table 1). We then use these parameters to determine the bandwidth for each synthetic sample using a theoretical formula that incorporates the kernel function. Feature importance is evaluated using random forest models to identify the most significant factors influencing bandwidth selection. The next step involves training various unsupervised learning models, such as Ridge Regression, Lasso Regression, Gradient Boosting, Support Vector Machines, and Neural Networks. The performance of these models is evaluated using metrics such as Root Mean Squared Error (RMSE) and R^2, and the best-performing model is selected for predicting the bandwidth for the original sample. Our results indicate that the machine learning-based approach not only improves the accuracy of bandwidth selection but also enhances computational efficiency compared to traditional methods.

This paper is structured as follows: Sect. 2 comprehensively details the methodology, including the generation of synthetic samples, the calculation of statistical parameters, the evaluation of feature importance, and the application of various unsupervised learning models for predicting the optimal bandwidth. Section 3 presents the simulation study, applying our approach and we compared the performance of various unsupervised learning models in predicting the optimal bandwidth bw_{os} of original sample with traditional cross-validation methods. All simulations and analyses are conducted using the R programming language [14]. Finally, Sect. 4 provides a summary of our findings and discusses potential future research directions.

2 Methodology

In this section, we describe the comprehensive methodology employed in our study for bandwidth selection in KDE using unsupervised machine learning techniques. We begin by evaluate feature importance to identify the most influential factors in bandwidth selection using a random forest model. Subsequently, we apply various unsupervised learning models to predict the optimal bandwidth bw_{os} of original sample. The performance of these models is assessed using metrics such as Root Mean Squared Error (RMSE) and R^2, and the best model is selected based on these criteria. One argument for choosing these evaluation measures lies in their complementary roles in assessing model performance:

- RMSE quantifies the magnitude of prediction errors, giving a clear measure of how far the predicted values are from actual values on average, which is particularly relevant for models where accuracy in absolute terms matters.
- R^2 measures the proportion of variance in the data explained by the model, providing insight into the model's ability to capture the underlying structure of the data.

Together, they provide a robust evaluation of both accuracy and model fit.

2.1 Determining the Most Feature Importance

To determine the optimal bandwidth bw for KDE using a data-driven approach, we begin with the first step, which involves identifying the most important factors. This process is detailed in Algorithms 1, 2, 3 and 4, which form the foundation of our overall method.

Algorithm 1 involves generating synthetic samples based on the initial dataset $(x_1, x_2, \dots, x_n)$. These synthetic samples can be created using methods such as bootstrap resampling or adding perturbations to the original data. The difference between bootstrap resampling and adding perturbations lies in how they modify the original dataset:

- Bootstrap resampling generates new datasets by randomly selecting observations from the original dataset with replacement, allowing repetitions and preserving the original data structure.

- Adding perturbations introduces small variations or noise to the original dataset, modifying the values directly without resampling. This method alters the dataset slightly but maintains the same observations, often to test model sensitivity or robustness against minor changes.

These methods serve different purposes in statistical and machine learning tasks.

Algorithm 1. Generation of m-Synthetic Samples

Require: Original sample $x_1, x_2, \ldots, x_n$, Method (1 for Bootstrap Resampling, 2 for Perturbation)
Ensure: Synthetic samples

```
procedure GENERATE_SYNTHETIC_SAMPLES
    if Method = 1 then
        for each synthetic sample do
            Bootstrap Resampling ▷ Randomly sample with replacement from the original sample
        end for
    else if Method = 2 then
        for each synthetic sample do
            Perturbation      ▷ Add random noise to the original sample to create variations
        end for
    end if
end procedure
```

Algorithm 2 is to calculate a comprehensive set of statistical parameters (factors) for each synthetic sample. These parameters serve as factors in subsequent analyses to determine the optimal bandwidth bw for KDE. The synthetic samples, generated from the original data, are used to ensure robustness and reliability of the estimations. The algorithm iterates over each synthetic sample, calculating various statistical parameters organized into several categories: central tendency, dispersion, shape, position, robustness, relative variability, and concentration. Each category includes specific measures such as mean, standard deviation, skewness, quartiles, and the Gini index, among others. These parameters are summarized in Table 1 with their corresponding mathematical formulas. The significance of this algorithm lies in its ability to provide a detailed and diversified characterization of the samples, which is crucial for the subsequent steps of the method. By systematically calculating these parameters for each synthetic sample, the algorithm ensures that all possible variations in the data are accounted for, thereby enhancing the accuracy and reliability of the final results.

Algorithm 2. Calculation of Statistical Parameters (Factors)

Require: m-Synthetic samples
Ensure: Statistical parameters for each sample

```
procedure CALCULATE_STATISTICAL_PARAMETERS
    for each synthetic sample do
        Calculate all categories (measures)   ▷ Statistical parameters as indicated in Table 1
    end for
end procedure
```

Table 1. Statistical Parameters (Factors) and Their Mathematical Formulas

Category	Parameters (Factors)	Mathematical Formula
Central Tendency	Mean	$(1/n)\sum_{i=1}^{n} x_i$
	Median	Middle value of the sorted data
	Mode	Most frequent value in the data
	Geometric Mean	$(\prod_{i=1}^{n} x_i)^{\frac{1}{n}}$
	Harmonic Mean	$n\left(\sum_{i=1}^{n}(1/x_i)\right)^{-1}$
Dispersion	Variance	$(1/n-1)\sum_{i=1}^{n}(x_i-\bar{x})^2$
	Standard Deviation σ	$\sqrt{\text{Variance}}$
	Interquartile Range	$Q_3 - Q_1$
	Range	$\max(x) - \min(x)$
	Mean Absolute Deviation	$(1/n)\sum_{i=1}^{n}\lvert x_i - \bar{x}\rvert$
Shape	Skewness	$\frac{(1/n)\sum_{i=1}^{n}(x_i-\bar{x})^3}{\left((1/n-1)\sum_{i=1}^{n}(x_i-\bar{x})^2\right)^{3/2}}$
	Kurtosis	$\frac{(1/n)\sum_{i=1}^{n}(x_i-\bar{x})^4}{\left((1/n-1)\sum_{i=1}^{n}(x_i-\bar{x})^2\right)^{2}}$
	Central moment of order $k = (3, 4, 5)$	$(1/n)\sum_{i=1}^{n}(x_i-\bar{x})^k$
Position	Minimum	$\min(x)$
	Maximum	$\max(x)$
	Quartiles	Values of Q_1 (25%) and Q_3 (75%)
	Coefficient of Variation	$\sigma/\bar{x}$
Robustness	Median Absolute Deviation	$\text{median}(\lvert x_i - \text{median}(x)\rvert)$
	Trimean	$(Q_1 + 2\text{Median} + Q_3)/4$
	Trimmed Mean (α%)	$(1/n-2k)\sum_{i=k+1}^{n-k} x_{(i)}$
	Winsorized Mean (α%)	$(1/n)\left(k\,x_{(k+1)} + \sum_{i=k+1}^{n-k} x_{(i)} + k\,x_{(n-k)}\right)$ α% of the extreme values is determined by the parameter k
Relative Variability	Dispersion Coefficient	IQR/Median
Concentration	Gini Index	$G(x) = 1 - \sum_{i=1}^{n}(2i-n-1)x_i$ for sorted x
Entropy	Shannon Entropy	$H(x) = -\sum_{i=1}^{k} p(x_i)\log p(x_i)$; where k is number of unique values in the sample and $p(x_i) = n_i/n$

Algorithm 3 aims to calculate the optimal bandwidth bw for each synthetic sample with a kernel function $K(x)$ (e.g., Gaussian, Epanechnikov, uniform, triangular, triweight, tricube, biweight, cosine,...). This calculation is based on minimizing the Mean Integrated Squared Error (MISE). The synthetic samples, generated from the original data, are used to ensure robustness and reliability of the bandwidth estimations. The algorithm begins by fixing the kernel function $K(x)$. For each synthetic sample, the algorithm calculates the sample size n, the standard deviation $\hat{\sigma}$, and the interquartile range (IQR). Using these values, the bandwidth is computed where $R(K)$ and $\mu_2(K)$ are specific to the chosen kernel function $K(x)$, where:

$$R(K) = \int_{-\infty}^{+\infty} K(u)^2\,du, \quad \text{and} \quad \mu_2(K) = \int_{-\infty}^{+\infty} u^2 K(u)\,du$$

For example, if the kernel is Gaussian $K(x) = \frac{1}{\sqrt{2\pi}}e^{-x^2/2}$, then $R(K) = \frac{1}{2\sqrt{\pi}}$ and $\mu_2(K) = 1$.

The algorithm systematically calculates the bandwidth for each synthetic sample, ensuring that variations in the data are accounted for. By doing so, it provides a reliable estimation of the optimal bandwidth for KDE, which is crucial for accurate and efficient data analysis.

Algorithm 3. Calculation of Bandwidth bw for m-Synthetic Samples

Require: Synthetic samples $\{X_1, X_2, \ldots, X_m\}$, where each X_i is a synthetic sample, the number of samples m, and the kernel function $K(x)$
Ensure: Bandwidths $\{bw_1, bw_2, \ldots, bw_m\}$ for each synthetic sample

procedure CALCULATE_BANDWIDTHS_FOR_SYNTHETIC_SAMPLES
  Kernel function $K(x)$ ▷ Any kernel function (Gaussian, Epanechnikov, triangular,...)
  for $i = 1$ **to** m **do**
    $n \leftarrow$ length of X_i
    $\hat{\sigma} \leftarrow \sqrt{\text{var}(X_i)}$ ▷ Calculate standard deviation
    IQR $\leftarrow Q3 - Q1$ ▷ Calculate interquartile range
    $bw_i \leftarrow \left(\frac{8\sqrt{\pi}R(K)}{3\mu_2(K)^2}\right)^{1/5} \times \min\left(\hat{\sigma}, \frac{\text{IQR}}{1.34}\right) \times n^{-1/5}$ ▷ Calculate bandwidth [2]
  end for
  return $\{bw_1, bw_2, \ldots, bw_m\}$
end procedure

Algorithm 4 aims to identify the most influential factors derived from Algorithm 2 affecting the bandwidth bw calculated in Algorithm 3. This is achieved by training a random forest model to evaluate feature importance. The algorithm starts by preparing the data, where the input data contains both the factors and the corresponding bw values. The data is split into features X and the target variable Y, with X containing the factors and Y containing the bw values. A random forest model is then trained using X as the input features and Y as the target variable, ensuring robustness through cross-validation. Once the model is trained, feature importance scores are extracted, indicating the significance of each factor in predicting bw. Features with importance scores above a predefined threshold are selected as the most influential. This process provides a comprehensive understanding of which factors most significantly impact the determination of bandwidth bw.

Algorithm 4. Feature Importance Evaluation for Bandwidth bw

Require: Data containing factors (from Algorithm 2) and bandwidth bw target (from Algorithm 3)
Ensure: List of important features influencing bw

procedure EVALUATE_FEATURE_IMPORTANCE
  Step 1: Prepare the data
    Split the data into features (X) and target (Y) ▷ Separate predictors and target variable
    $X \leftarrow$ Data without the bandwidth column ▷ Features (factors)
    $Y \leftarrow$ Bandwidth bw ▷ Target
  Step 2: Train a random forest model
    Set a random seed for reproducibility ▷ Ensure reproducibility of results
    Train the random forest (RF) model ▷ Fit the model using X as features and Y as the target variable
    Use cross-validation to ensure robustness of the model ▷ Validate the model
  Step 3: Evaluate feature importance
    Extract the feature importance scores from the trained model ▷ Determine the importance of each factor
    Select features with importance scores above a predefined threshold ▷ Identify significant factors
  Step 4: Return the important features
    return List of important features ▷ Output the significant factors
end procedure

2.2 Evaluation of Unsupervised ML Models for *bw* Prediction

In this section, we aim to leverage the most important factors identified in the previous section to apply various machine learning methods. The goal is to determine the most effective model for predicting the bandwidth *bw*. We will evaluate several regression methods, including linear models, tree-based methods, and neural networks, assessing their performance based on specific criteria to select the best model. The chosen model will then be used for accurate bandwidth prediction.

Algorithm 5. Evaluation of Unsupervised ML Models for Bandwidth Prediction

Require: Data constructed with bandwidths (target) obtained from Algorithm 3, and important factors obtained from Algorithm 4
Ensure: Best performing model for predicting *bw*

procedure EVALUATE_MODELS
Step 1: Prepare the data
Split the data into features (X) and target (Y) ▷ Separate predictors and target variable
Center and scale the features (X) ▷ Standardize the features (X)
Split the data into training and testing sets ▷ Use a suitable ratio, e.g., 70% training and 30% testing
Step 2: Apply regression methods
Apply the following regression methods:
1. Ridge Regression
2. Lasso Regression
3. Averaged Bayesian Ridge Regression (ABRidge)
4. Stochastic Gradient Boosting (GBM)
5. Support Vector Machine (SVM)
6. Neural Network Regression (nnet)
7. Model Averaged Neural Network (avNNet)
8. Neural Networks with Feature Extraction (pcaNNet)
9. Elastic Net Regression (enet)
10. Principal Component Analysis (PCAr)

Step 3: Evaluate model performance
Use the following criteria to evaluate the performance of each method on the testing set:
- R-squared (R^2)
- Root Mean Squared Error (RMSE)

Step 4: Select the best model
Identify the model with the lowest RMSE and highest R^2 ▷ Select the best performing model
Step 5: Return the best model
return Best performing model for predicting *bw* ▷ Output the best model for further use
end procedure

2.3 Prediction of Optimal Bandwidth bw_{os} for Original Sample

The purpose of Algorithm 6 is to leverage the best machine learning model identified in previous steps to predict the optimal bandwidth bw_{os} for the original dataset. This involves several critical steps: first, we calculate the important statistical factors for the original data. These factors are derived from the original dataset using the same statistical measures identified as important in previous steps. Once these factors are calculated, we use the best-performing machine learning model to predict the bandwidth. This predicted bandwidth is expected to be optimal for the original dataset, enhancing the accuracy of kernel density estimation applied to the data.

Algorithm 6. Prediction of Optimal Bandwidth bw_{os} for Original Sample

Require: Original sample *original_data*; Important factors from Algorithm 4; Best model from Algorithm 5
Ensure: Predicted optimal bandwidth bw_{os} for original sample
1: **procedure** PREDICT_BANDWIDTH
Step 1: Calculate Important Factors for Original Data
2: Compute all important statistical factors for the original dataset ▷ Use statistical measures identified in Algorithm 4
Step 2: Prepare Data for Prediction
3: $X_{original} \leftarrow$ Important factors calculated for original data ▷ Prepare the data with important factors
Step 3: Predict Bandwidth using Best Model
4: $bw \leftarrow$ Best model from Algorithm 5. predict(X_{original}) ▷ Predict bandwidth using the best model
Step 4: Return the Predicted Bandwidth
5: **return** bw_{os} ▷ Output the predicted optimal bandwidth
6: **end procedure**

The Fig. 1 shows the diagram of the proposed methodology for bandwidth selection in KED using unsupervised machine learning. This process improves accuracy by dynamically adapting the bandwidth bw_{os} to the original sample.

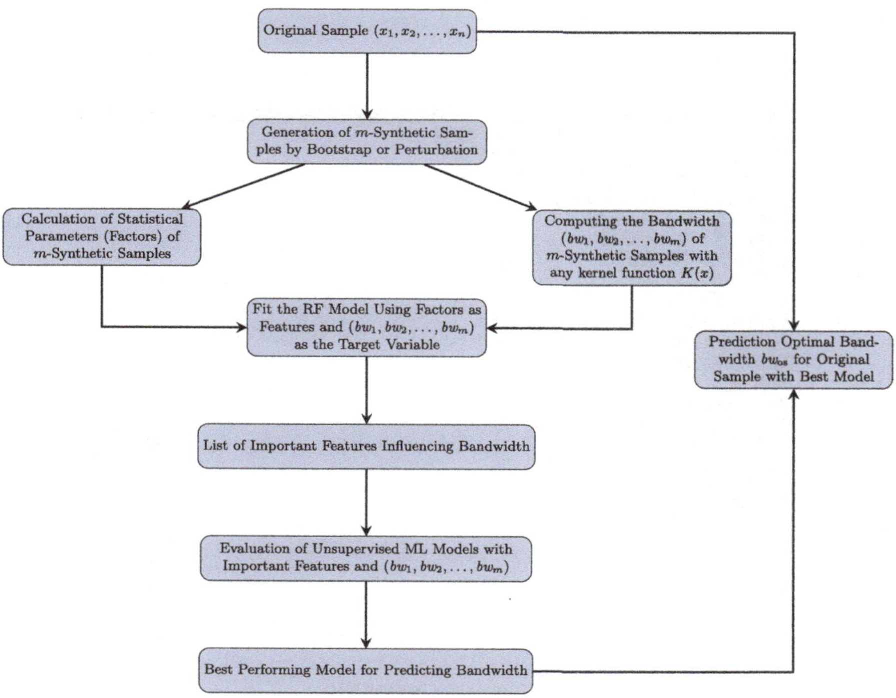

Fig. 1. Workflow for Bandwidth Selection Using Unsupervised ML.

3 Simulation Study

To evaluate the effectiveness of our proposed methodology, we conducted a simulation study using a Weibull distribution with shape parameter $\alpha = 2$ and scale parameter $\beta = 4$, where the true density function is given by:

$$f(x;\alpha,\beta) = \frac{\alpha}{\beta}\left(\frac{x}{\beta}\right)^{\alpha-1} \exp\left(-\left(\frac{x}{\beta}\right)^{\alpha}\right), \quad x \geq 0$$

An original sample of size $n = 500$ was simulated from this distribution. Following our methodology, $m = 10000$ synthetic samples were generated using bootstrap resampling method, and a comprehensive set of statistical parameters was calculated. The bandwidths $(bw_1, bw_2, \ldots, bw_m)$ for each synthetic sample were then calculated using the methods described with Gaussian kernel. Using Algorithm 4, we identified the important factors influencing the bandwidth with the threshold set at the $25th$ percentile (first quarterly Q_1) of the importance scores. The significant factors were those with importance scores greater than or equal to this threshold, as shown in Fig. 2. We also present the correlation matrix between the important factors and the bandwidth bw, represented by Fig. 3.

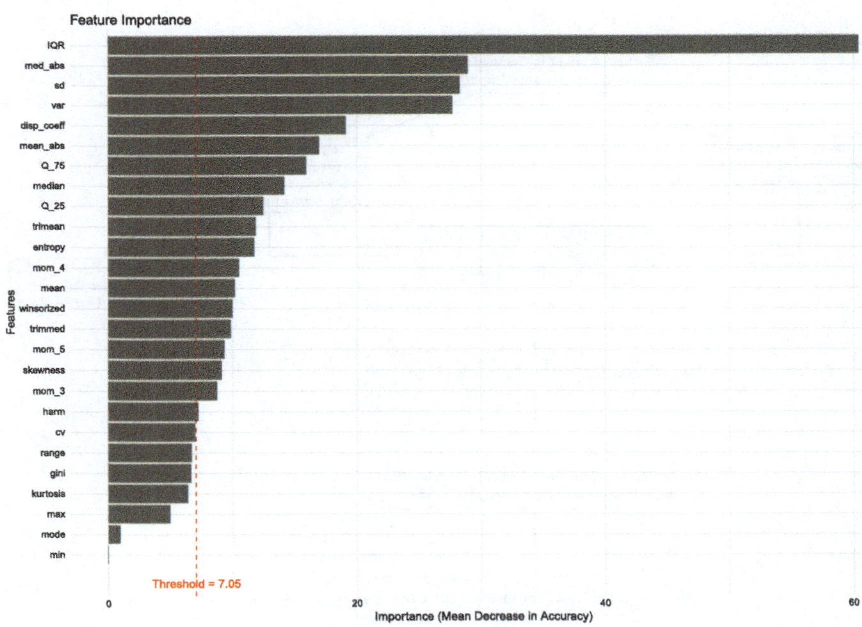

Fig. 2. Feature Importance Scores with Threshold at Q1 (25th Percentile).

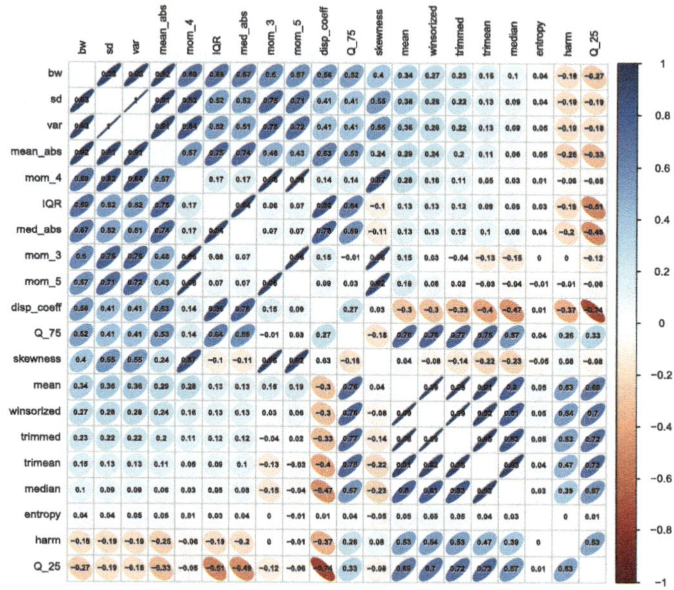

Fig. 3. Correlation Between Important Factors and Bandwidth.

After identifying the important factors, we applied various unsupervised ML models to predict the optimal bandwidth *bw*. The performance of these models was evaluated using the criteria RMSE and R^2. Our analysis revealed that the Model Averaged Neural Network (avNNet) provided the best performance (with bootstrap sampling), achieving a minimum RMSE value of 0.0007283 and an $R^2 = 0.99592$. Figure 4 illustrates the comparison of different unsupervised ML models in terms of RMSE and R^2.

After establishing that the Model Averaged Neural Network (avNNet) is the top-performing model (with bootstrap sampling), we used it to predict the bandwidth bw_{os} for the original sample. The predicted optimal bandwidth for the original sample, with the Gaussian kernel, was found to be $bw_{os} = 0.55773$. This result demonstrates the effectiveness of the avNNet model in accurately estimating the bandwidth, leveraging the important statistical factors identified earlier.

To evaluate the effectiveness of our method, we compared the predicted bandwidth for the original sample bw_{os} obtained by the best unsupervised ML (in this example we have avNNet if we use bootstrap sampling, and GBM if use perturbation sampling) with traditional methods, specifically unbiased cross-validation (ucv), biased cross-validation (bcv), Complete cross-validation (ccv), Modified cross-validation (mcv), and Trimmed cross-validation (tcv). The comparison was based on the Mean Squared Error (MSE) criterion, which is defined as follows:

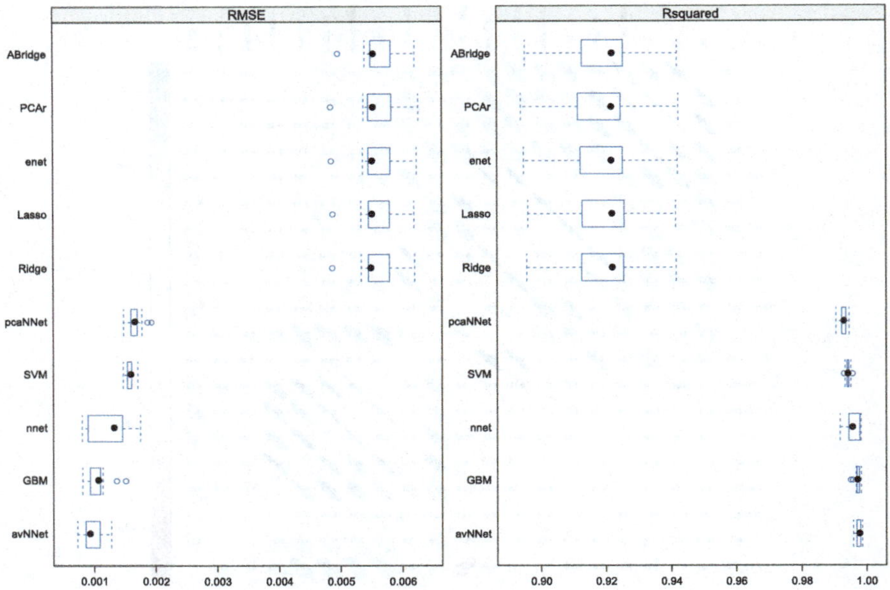

Fig. 4. Comparison of different unsupervised ML models: RMSE and R^2 for Bandwidth Prediction.

$$\text{MSE}(x; bw) = \frac{1}{n}\sum_{i=1}^{n}\left(\underbrace{\frac{1}{n\,bw}\sum_{j=1}^{l} K\left(\frac{x_i - x_j}{bw}\right)}_{\text{KDE with different } bw} - \underbrace{f(x_i; \alpha, \beta)}_{\text{true density}}\right)^2$$

where $K(\cdot)$ is the kernel function. By comparing the $\text{MSE}(x; bw)$ values, we can assess the accuracy of our method relative to these classical methods. Our findings indicate that the proposed approach (see Fig. 1) provided a lower MSE compared to traditional approaches (see Table 2), underscoring its robustness and precision in bandwidth selection.

The Fig. 5 illustrate the performance comparison of various bandwidth selection methods for the Gaussian and Triweight kernels (the bandwidths are ordered in ascending order based on their MSE values, from the smallest to the largest MSE). Figure 5a presents the comparison using Gaussian kernel (similarly, Fig. 5b with Triweight kernel), highlighting that the bandwidth selected using the avNNet model with bootstrap resampling $bw_{\text{os}}(\text{B})$ and the GBM model with perturbation $bw_{\text{os}}(\text{P})$ achieves the lowest MSE compared to classical methods. These results demonstrate the effectiveness of the proposed machine learning approach in achieving accurate bandwidth selection.

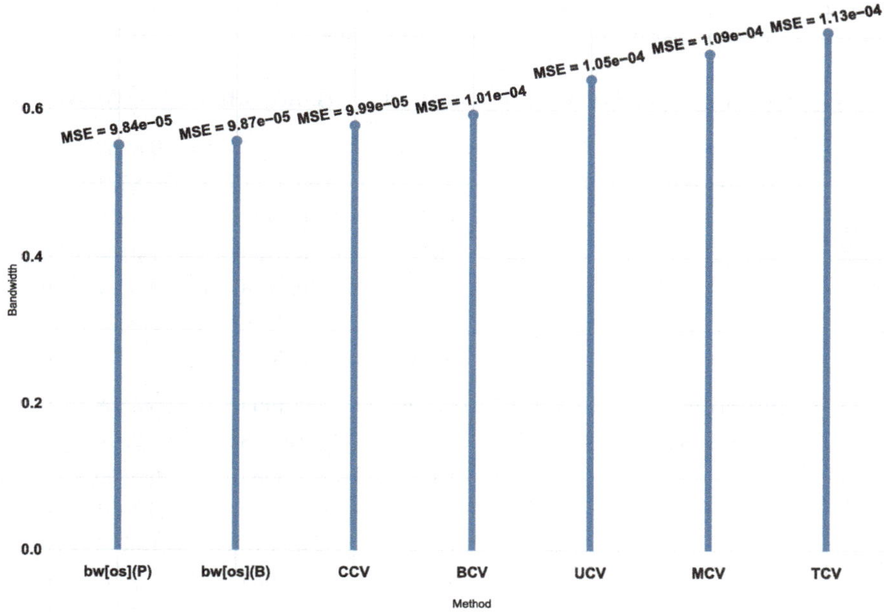

(a) Gaussian Kernel

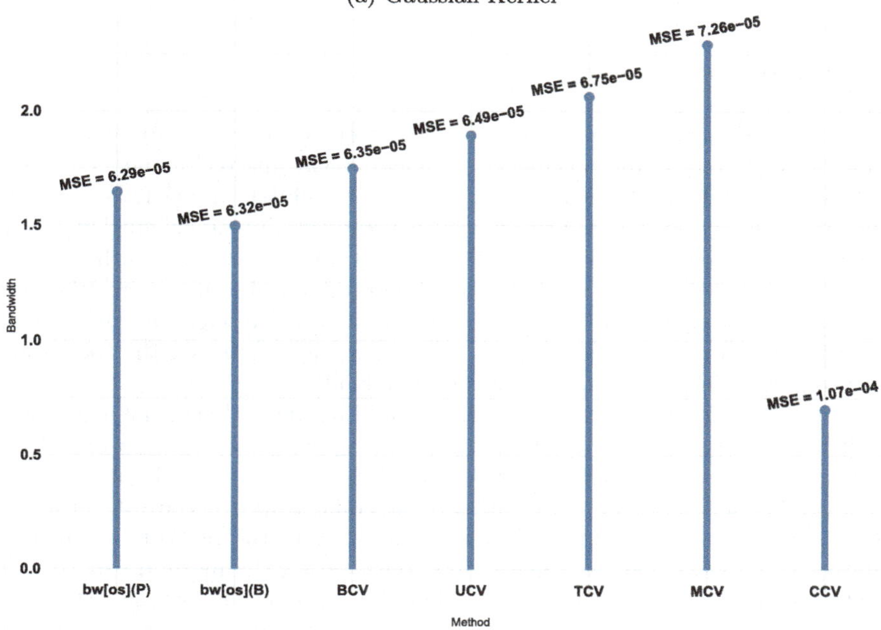

(b) Triweight Kernel

Fig. 5. Performance Comparison Using MSE of Bandwidth Selection Methods for Weibull Distribution with Shape Parameter $\alpha = 2$ and Scale Parameter $\beta = 4$.

Table 2. Performance Comparison of Bandwidth Selection Methods for Weibull distribution with shape parameter $\alpha = 2$ and scale parameter $\beta = 4$.

K(x)	**Resampling**	**Best Model**	**Method**	**Bandwidth**	**MSE**
Gaussian	Bootstrap	avNNet	$bw_{os}(B)$	0.55773	9.8666×10^{-5}
	Perturbation	GBM	$bw_{os}(P)$	0.55215	9.8423×10^{-5}
			UCV	0.64228	1.0519×10^{-4}
			BCV	0.59398	1.0103×10^{-4}
			CCV	0.57901	9.9927×10^{-5}
			MCV	0.67707	1.0898×10^{-4}
			TCV	0.70682	1.1275×10^{-4}
Triweight	Bootstrap	avNNet	$bw_{os}(B)$	1.49692	6.3158×10^{-5}
	Perturbation	GBM	$bw_{os}(P)$	1.64482	6.2940×10^{-5}
			UCV	1.89557	6.4856×10^{-5}
			BCV	1.74767	6.3451×10^{-5}
			CCV	0.70188	1.0714×10^{-4}
			MCV	2.29543	7.2597×10^{-5}
			TCV	2.06435	6.7452×10^{-5}

4 Conclusion

In this study, we have presented a novel data-driven approach to bandwidth selection in KDE by leveraging unsupervised machine learning techniques. Our approach addresses the limitations of traditional cross-validation methods for bandwidth selection, by offering improved accuracy and computational efficiency. The proposed methodology involves a series of algorithms to generate synthetic samples, compute relevant statistical features, determine optimal bandwidths, and identify the most significant factors influencing bandwidth selection. By employing a variety of unsupervised learning models, we demonstrated the robustness of our approach in predicting the optimal bandwidth.

In real-world applications, this method could be highly beneficial across various fields. For example, in finance, it can be applied to accurately model the distribution of financial returns or risks, which is essential for risk management and portfolio optimization. In healthcare, it could improve the estimation of disease prevalence or patient characteristics, aiding in the planning of medical resources or personalized treatments. In cybersecurity, the method can be used to detect anomalies in network traffic, identifying potential cyber-attacks more efficiently. In environmental science, it can help estimate the spatial distribution of pollutants, animal populations, or other environmental factors, which is critical for environmental monitoring and management.

The advantages of this approach over traditional methods, such as faster computation and better adaptability to complex data structures, make it particularly suited for big data applications, where computational resources and processing time are often limiting factors. Furthermore, it provides enhanced robustness and accuracy, especially when dealing with noisy data or datasets that have complex underlying distributions.

Future work could explore the integration of additional unsupervised learning models and the application of this methodology to other non-parametric estimation problems. Additionally, further research could investigate the performance of our approach in high-dimensional data settings and its adaptability to various types of distributions. Overall, this study contributes to the field of non-parametric statistics by providing a robust and efficient framework for bandwidth selection in KDE, leveraging the power of unsupervised machine learning to enhance density estimation accuracy.

Acknowledgments. The authors are thankful to the editor and the anonymous referees for their thorough review and highly appreciate the comments and suggestions which significantly contributed to increasing the quality of our paper.

Disclosure of Interests. The authors declare that they have no known competing financial interests or personal relationships that could have appeared to influence the work reported in this paper. All data and materials used in this study are fully disclosed in the supplementary materials to ensure transparency and reproducibility of the research.

Supplementary Material. The R code can be found on GitHub at https://github.com/acguidoum/R-code-LLNCS-2024, so that readers can replicate the results.

References

1. Bowman, A.W., Azzalini, A.: Applied Smoothing Techniques for Data Analysis: The Kernel Approach with S-Plus Illustrations. Oxford University Press, Oxford (1997)
2. Scott, D.W.: Multivariate Density Estimation: Theory, Practice, and Visualization. Wiley, Hoboken (1992)
3. Sheather, S.J., Jones, M.C.: A reliable data-based bandwidth selection method for kernel density estimation. J. Roy. Stat. Soc.: Ser. B (Methodol.) **53**(3), 683–690 (1991)
4. Hall, P., Racine, J., Li, Q.: Cross-validation and the estimation of conditional probability densities. J. Am. Stat. Assoc. **99**(468), 1015–1026 (2004)
5. Stone, M.: Cross-validatory choice and assessment of statistical predictions. J. Roy. Stat. Soc.: Ser. B (Methodol.) **36**(2), 111–147 (1974)
6. Rudemo, M.: Empirical choice of histograms and kernel density estimators. Scand. J. Stat. **9**(2), 65–78 (1982)
7. Bowman, A.W.: An alternative method of cross-validation for the smoothing of density estimates. Biometrika **71**(2), 353–360 (1984)
8. Ehsanes Saleh, A.K.Md., Arashi, M., Golam Kibria, B.M.: Theory of Ridge Regression Estimation with Applications. Wiley (2019)
9. Ranstam, J., Cook, J.A.: LASSO regression. Br. J. Surg. **105**(10) (2018)

10. Natekin, A., Knoll, A.: Gradient boosting machines, a tutorial. Front. Neurorobot. **7**(21), 1–21 (2013)
11. Sanz, H., Valim, C., Vegas, E., et al.: SVM-RFE: selection and visualization of the most relevant features through non-linear kernels. BMC Bioinform. **19**(432) (2018)
12. Abiodun, O.I., Jantan, A., Omolara, A.E., Dada, K.V., Mohamed, N.A., Arshad, H.: State-of-the-art in artificial neural network applications: a survey. Heliyon **4**(11) (2018)
13. Jones, M.C.: The performance of kernel density functions in kernel density estimation. Statist. Probab. Lett. **26**(2), 105–110 (1996)
14. R Core Team, ***R***: A Language and Environment for Statistical Computing, R Foundation for Statistical Computing, Vienna, Austria (2024). https://www.R-project.org/
15. Silverman, B.W.: Density Estimation for Statistics and Data Analysis. Chapman and Hall (1986)
16. Hall, P.: On the bias of bootstrap cross-validation. Biometrika **78**(3), 635–640 (1991)

Autonomous Data Tiering Using Reinforcement Learning for 3-Tier Hierarchical Storage

Nedjah Oussama(✉), Mokhtari Omar, Boumahdi Fatima, and Mancer Yasmine(✉)

LRDSI Laboratory, Department of Computer Science, Faculty of Sciences, University of Blida 1, Blida, Algeria
nedjahoussama.no@gmail.com, mancer.yas@gmail.com

Abstract. In today's data-driven world, organizations must efficiently store and manage massive amounts of information. As data generation keeps increasing, the demand for scalable and cost-effective storage solutions becomes critical. Data tiering appears to be a strategic solution to this problem by categorizing data based on access frequency and storing it across multiple storage tiers.

This paper presents a new solution for autonomous data tiering using reinforcement learning (RL) within a 3-tier hierarchical storage system. By utilizing the adaptability and learning capabilities of RL, our approach dynamically optimizes data placement, ensuring high-performance access for frequently used data while lowering costs for infrequently accessed data. Our solution integrates an RL agent that is trained to recognize access patterns and make data placement decisions in real-time. This dynamic tiering strategy reduces latency while maintaining low costs. Through extensive experimentation and analysis, we show the effectiveness of our approach, highlighting its potential to transform the way data is handled in large-scale storage systems.

Keywords: Data · Storage · Data Tiering · Reinforcement Learning

1 Introduction

In the modern day, when data has become the prevailing form of currency, the efficient management of data is of utmost importance. Consider the vast amount of emails, images, movies, papers, and sensor readings that are produced every second worldwide. Unseen, there exists a complex infrastructure constantly functioning to store and oversee this overwhelming amount of data. Data tiering is the solution for this situation. Data tiering is a storage management technique that categorizes and stores data in different tiers based on its value, performance requirements, and access frequency. This strategy optimizes storage resources by storing high-value and frequently accessed data on high-performance storage tiers and less critical or infrequently accessed data on lower-cost storage tiers [8].

Conventional methods of data tiering can be likened to attaching tags to the garments and depending on chance. Although they may operate well for a period

S. Chikhi et al. (Eds.): MISC 2024, LNNS 1267, pp. 171–181, 2025.
https://doi.org/10.1007/978-3-031-82112-7_13

of time, they lack adaptability. As our data consumption patterns change and our storage needs evolve, we need more intelligent and flexible solutions.

For example, an online marketplace manager gathers consumer data, product information, sales records, and other important data. Identifying the suitable storage location for this data, along with developing effective retrieval techniques, is a significant challenge. Moreover, it is not simply a question of having a sufficient amount of storage space; it is equally crucial to utilize that space intelligently.

Traditional data tiering strategies frequently prove insufficient. They fail to consider the complex nature of our data consumption or the continuous changes in our access patterns. As a result, we are left with inefficiencies, wasted resources, and slower speeds.

The contributions of this study are to accomplish the following:

- Proposed an improved approach to data tiering, one that is capable of adapting and acquiring knowledge from its environment.
- To analyze the current level of advancement, to provide a practical solution for implementation, to determine and verify the suggested solution.
- This paper proposes a solution to the data tiering problem by dynamically managing the data, achieving low latency values while maintaining low-cost storage. This is accomplished by utilizing a reinforcement learning algorithm with adequate states as input.

The paper is structured as follows: in Sect. 2, we will establish the foundation by examining the existing advancements in data tiering and reinforcement learning. Next in Sect. 3, we will begin the process of constructing by focusing on the essential details. We will provide a comprehensive explanation of the intricate aspects involved in the design and implementation of our new data tiering platform. Then in Sect. 4, we will evaluate the performance of our development. We will be putting it to a wide range of challenges, including real-world stress testing. Next, we will thoroughly examine the outcomes, distinguishing the victors from the less successful participants. Finally, we will summarize the knowledge obtained through our research. We will provide an overview of the primary elements in each section and their contribution to the overall objectives of our study. In addition, we will explore the opportunities for any further development in our research.

2 Related Works

This section serves as an introduction to our work by providing a thorough summary of the fundamentals and the corpus of previous research, including the utilization of artificial intelligence in data-tiering. By clarifying important terminology and placing our study in the larger context of similar research.

2.1 Traditional Methods in Data Tiering

Traditional methods for solving data-tiering problems rely on defining static rules that are applied to the storage depending on the frequency or recency of accessed files.

Recency-Based Policies. Recency-based cache replacement policies prioritize keeping recently accessed items in cache, assuming that items accessed recently will be accessed again in the near future [1]. Examples of these algorithms include **LRU-K** [9] (Least Recently Used) and **LIRS** (Low Inter-reference Recency Set) [10].

Frequency-Based Policies. LFU (Least Frequently Used) [4] is a cache replacement policy that removes the least frequently accessed items from the cache when more space is required for new items. Unlike recency-based policies, which prioritize recent accesses, LFU considers the frequency of accesses over time [3].

Hybrid Policies. Hybrid policies in data management are strategies that combine two or more existing policies to improve performance. These policies could be variations on existing algorithms or entirely new approaches that combine the strengths of several techniques.[2] Examples of this policy include **LFRU** (Least Frequently Used Replacement) [4] and **ARC** (Adaptive Replacement Cache) [7].

2.2 Reinforcement-Learning Based Policies

Reinforcement learning policies all share a fundamental principle. However, the specific algorithm or policy used by the agent can vary significantly. Examples of these policies are **DDPG** (Deep Deterministic Policy Gradient) [11], **A3C** (Asynchronous Advantage Actor Critic) [5], **Deep Q-Learning** [6], and **Fuzzy Rules** [12], each with their own learning methods.

3 Proposed Solution

For our proposed solution, we utilized an architecture that summarizes the research conducted in our paper. Figure 1 provides a visual representation of this architecture, while the description is as follows:

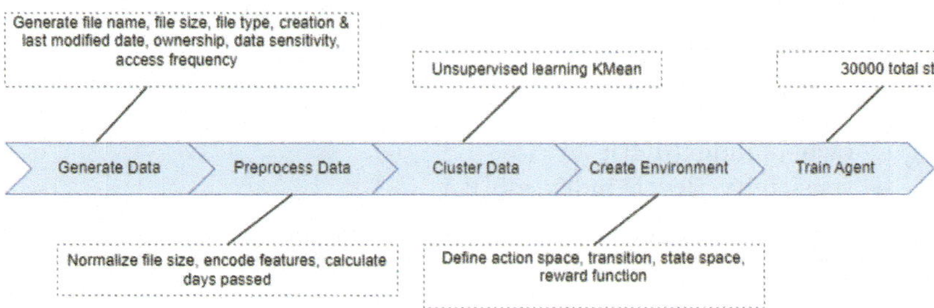

Fig. 1. Global approach for our proposed solution.

1) Files generation: This is the step in which we generate files at random with raw metadata such as file name, size, type, creation and last modified dates, ownership, sensitivity level, and access frequency. We can visualize a sample of three files in our dataset in Table 1.

Table 1. Sample of the dataset

File name	File size	File type	Creation Date	Modified Date	Owner	Sensitivity level	Access Frequency
Exam	15000 Byte	Archive File	2024-03-24	2024-05-15	admin	High	0.0032
Homework	7100 Byte	Image File	2024-01-12	2024-02-01	user1	Low	0.0002
Interrogation	9400 Byte	Spreadsheet File	2024-02-07	2024-03-17	user2	Medium	0.00057

2) Preprocess data: At this stage, we refine the metadata of the previously generated files in order to improve the learning ability of our proposed reinforcement learning agent. The file size will be normalized using the z-score method[1], and the creation and modification dates will be replaced with the number of days passed, while the sensitivity level, ownership, and file type will be one-hot encoded. We use the file name for indexing and the file access frequency for testing purposes.

The whole process can be summarized in pseudocode to make it easier to comprehend, as shown in pseudocode 1.1.

Algorithm 1.1: Preprocess data

```
Input: X
Output: np.array(X, dtype = np.float64), mean, std_dev
filetypes ← ['Text File', 'Spreadsheet File', 'Presentation File', 'Image
File', 'Audio File', 'Video File', 'Archive File', 'Executable File',
'Database File', 'Font File', 'Code File', 'Configuration File'];
owners ← ['user1', 'user2', 'user3', 'admin'];
sensitivity_levels ← ['Low', 'Medium', 'High'];
data ← np.array(X[:, 0], dtype=np.float64);
foreach a ∈ X do
    a[0], mean, std_dev ← calculate_z_score(data, float(a[0]));
    a[1] ← filetypes.index(a[1]);
    a[2] ← days_since(a[2]);
    a[3] ← days_since(a[3]);
    a[4] ← owners.index(a[4]);
    a[5] ← sensitivity_levels.index(a[5]);
return np.array(X, dtype=np.float64), mean, std_dev;
```

[1] $z = \frac{x - mean}{std}$.

Taking the example of the dataset sample Table 1, after removing the file name and file frequency, the newly generated preprocessed data is in the form of Table 2, where the current date is 2024-05-15.

Table 2. Sample of the preprocessed dataset

File size	File type	Creation Date	Modified Date	Owner	Sensitivity level
1.356	6	52	0	3	2
−1.025	3	124	104	0	0
−0.331	1	98	59	1	1

3) Clustering data: This process splits our files into clusters based on similarities using an unsupervised learning technique known as KMeans. Based on the elbow method and silhouette analysis, we chose k = 9 as our optimal value. This helps the agent learn by minimizing the agent's action.

4) Creating environment: This stage serves as the basis for our proposed solution. The task involves creating a simulation for the environment, particularly a tiered hierarchical storage system, as well as the reward function, states, and definition of dealing with agent actions.

Reward 'r':
The **reward function** would take the following form: if the storage overflows or the agent removes data from the storage, the reward is −5; otherwise, the reward is calculated as shown in Eq. 1. Storage utilization is calculated as shown in Eq. 2.

$$\text{reward} = 0.7 \cdot \text{storage latency} + 0.3 \cdot \text{storage utilization} \tag{1}$$

$$\text{Storage utilization} = 0.7 \cdot \text{Tier1 util} + 0.2 \cdot \text{Tier2 util} + 0.1 \cdot \text{Tier3 util} \tag{2}$$

State 's':

s_1: One-hot encoding for the position of each cluster (tier 1, tier 2, tier 3).

s_2: Temperature of each cluster.

s_3: Each tier capacity utilization.

The (Data management of scientific applications in a reinforcement learning-based hierarchical storage system) [12] study found a reliable way to calculate the temperature based on access frequency and file priority. To mathematically represent the temperature, we define two methods for increasing and decreasing the temperature:

Increase Temperature: When a file is accessed, we use

$$temperature = 1 - 0.5 \cdot e^{-0.2 \cdot \text{number of requests}} \tag{3}$$

the higher the number of requests, the closer the temperature to 1.

Decrease Temperature: We decrease the temperature by 0.05 if a file remains unaccessed for a period of time. The lowest value possible is 0, and we define 5 iterations to start decreasing the temperature.

Action 'a':
The agent has the ability to control every cluster in the storage by either moving up or down a tier, which makes a total of 18 actions. We added an action for the agent to do nothing, bringing the total number of actions to 19.

These elements combine to form the full picture of our reinforcement learning framework. Figure 2 provides a visual representation of this architecture.

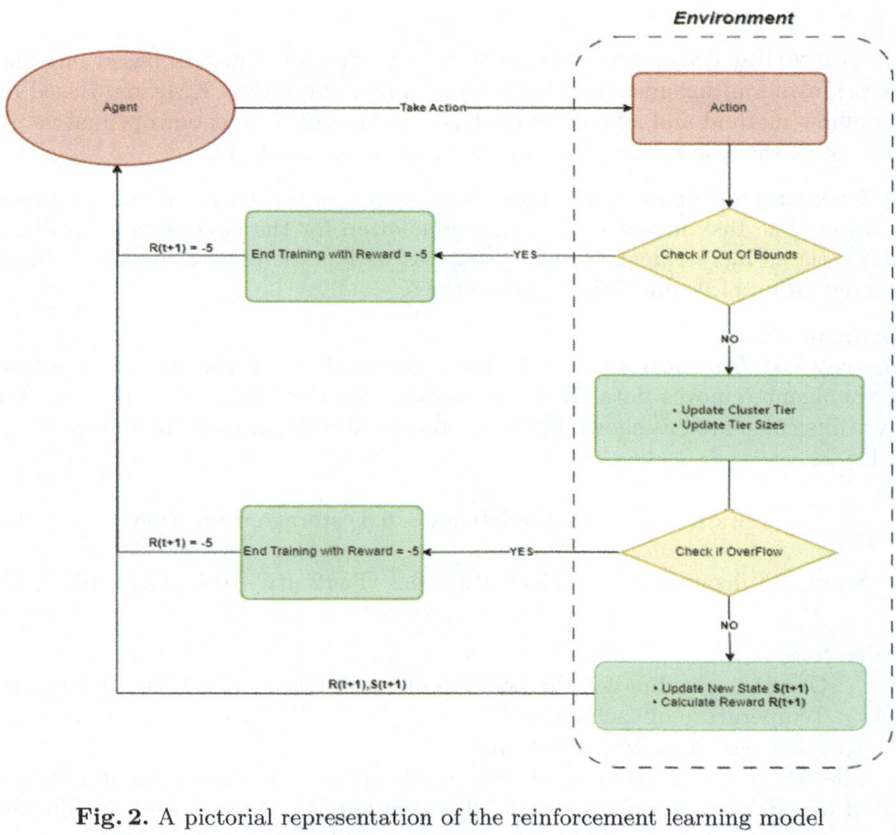

Fig. 2. A pictorial representation of the reinforcement learning model

5) Training the agent: At this stage, we chose a reinforcement-based policy algorithm, particularly the A3C algorithm, which belongs to the model-free gradient-based category with actor-critic networks. The agent then utilizes the outcomes of the previous steps to analyze patterns and gain intelligence independently. By integrating the algorithm into the environment, we will have a complete reinforcement learning model. The algorithm architecture can be visualized in Fig. 3.

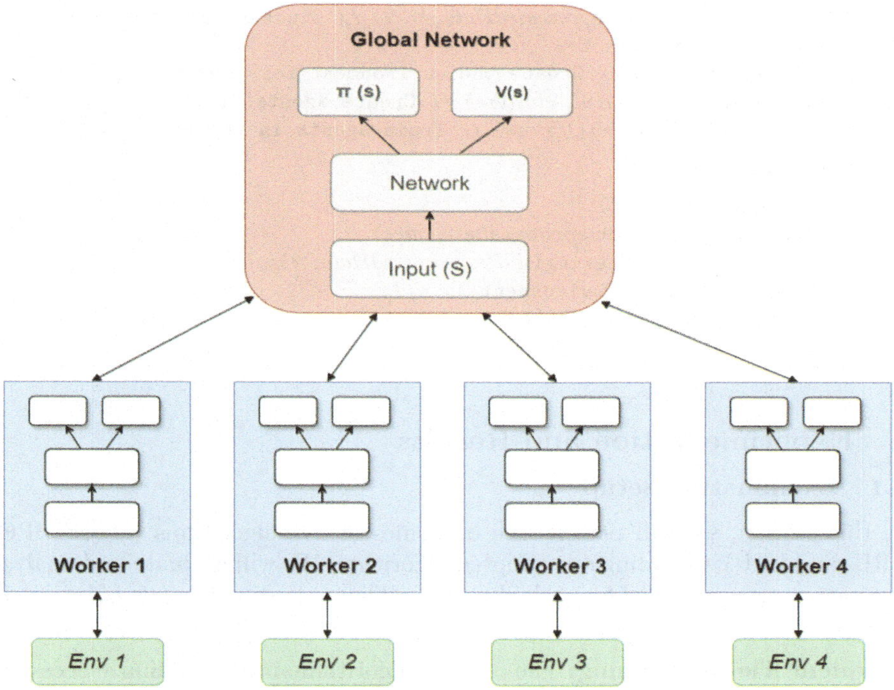

Fig. 3. The architecture of our A3C proposed solution.

Figure 3 shows that each AC worker is trained on parallel threads to obtain gradients. These gradients are then utilized for updating the parameters of the global network, which is a multi-layer perceptron. The multi-layer perceptron is composed of the following parameters:

- **Number of layers:** 2 layers.
- **Number of neurons:** 64 neurons.
- **Activation function:** Tanh[2].
- **Number of steps trained:** 30000 step.
- **Number of parallel environments:** 4 environments.

We can summarize and structure our solution implementation in the form of pseudocode as presented in Algorithm 1.2.

[2] $\tanh(x) = \frac{\sinh(x)}{\cosh(x)} = \frac{e^x - e^{-x}}{e^x + e^{-x}}$.

Algorithm 1.2: Proposed solution workflow

Result: Trained Agent
Function Definitions:

```
GenerateFiles() // Generate multiple data files
PreprocessData(files) // Preprocess the data from all files
KMeansClustering(allPreprocessedData, k=9) // Cluster the combined data
   using K-Means with k=9
CreateEnvironment() // Create the environment for agents
CreateAgents(environment, clusters) // Create agents in the environment
TrainAgents(totalNumberOfSteps) // Train agents in the environment
```

Main Algorithm:

```
files ← GenerateFiles();
allPreprocessedData ← PreprocessData(files);
clusters ← KMeansClustering(allPreprocessedData, 9);
environment ← CreateEnvironment();
CreateAgents(environment, clusters);
model ← TrainAgents(totalNumberOfSteps);
```

4 Experimentation and Results

4.1 Examination Setup

In this section, we will present the outcomes of five algorithms (DQN, PPO, LRU, and LFU), including our adopted algorithm. We will evaluate and analyze the parameters employed for each algorithm. The evaluation consists of two main analyses:

Frequent Tiers Accessing: The optimal algorithm aims to minimize access to lower tiers while maximizing access to higher tiers.

Average Latency: The algorithm should minimize the average latency of accessing files to the lowest possible value.

In order to conduct tests, we will establish predetermined latency values for each tier. As such, Tier 1 will have a latency of 1 ms, Tier 2 will have a latency of 5 ms, and Tier 3 will have a latency of 10 ms. Additionally, we will conduct a total of 5000 file requests.

4.2 Traditional Methods

The **LRU** algorithm performed the worst of the Five, with an average latency of 6.96 ms. The **LFU** algorithm was the second-worst performer, with an average latency of 6.55 ms.

4.3 Reinforcement Learning-Based Algorithms

In our reinforcement learning-based algorithms, we set the neural network architecture for all algorithms to be the same and run it for a total of 30000 steps to make the comparison unbiased. Every network contains two fully connected layers, and each layer includes 64 units with Tanh as an activation function.

With an average latency of 5.42 ms, the **DQN** algorithm came in third place out of the five. The second-best performer was the **PPO** algorithm, with an average latency of 4.51 ms, while the first and best performer was the **A3C** algorithm, with an average latency of 2.68 ms.

Table 3 can be used to summarize the evaluation and testing models, displaying the average latency and the differences between the best and selected models.

Table 3. Models evaluation result

	LRU	LFU	DQN	PPO	A3C
Average latency	6.96 ms	6.55 ms	5.42 ms	4.51 ms	2.68 ms
Difference in time	+159.7%	+144.4%	+102.2%	+68.3%	-
Tier 1 accessed frequency	15%	17%	42%	56%	75%
Tier 2 accessed frequency	23%	24%	16%	22%	17%
Tier 3 accessed frequency	62%	59%	42%	22%	8%

4.4 Comparative Analysis

The experiment conducted in this research was aimed at evaluating the performance of our algorithm (A3C) against four other algorithms, which are categorized as two traditional methods (LRU, LFU) and two reinforcement learning-based methods (PPO, DQN). We took two main problems into account in this comparison: average latency and frequent tier access.

Average Latency: The outcomes show the dominance of the A3C algorithm with an average latency of 2.68 ms, outperforming traditional methods such as the LRU with 6.96 ms and the LFU with 6.55 ms, as well as reinforcement learning-based algorithms such as DQN with 5.42 ms and PPO with 4.51 ms. These results show that traditional methods lack the dynamic adaptability to big data access patterns due to their static nature, whereas reinforcement learning-based algorithms have successfully adapted to the changes in the access pattern, resulting in lower latency values in comparison to the traditional methods, especially the A3C algorithm, which indicates that the A3C algorithm managed to learn an effective way to maintain low latency.

Frequent Tier Access: The results were also in favor of the A3C algorithm, with access frequency to Tier 1 being 75%, Tier 2 at 17%, and Tier 3 at 8%, which shows that A3C has been able to predict the access pattern successfully with low access to Tier 3. Reinforcement learning-based algorithms show a high access frequency to Tier 1 with moderate access to Tier 2 and 3, outperforming the traditional methods that show a low access frequency to Tier 1 with moderate access to Tier 2 and high access to Tier 3.

5 Conclusion

In conclusion, this paper shows that reinforcement learning-based algorithms are a better approach for data tiering, specifically the Asynchronous Advantage Actor-Critic (A3C), due to their ability to learn and adapt to new access patterns. The main goal of this research was to develop a strategy to improve the efficiency of data management by reducing latency and utilizing the maximum capacity of the higher tiers. The study findings have significant implications, such as:

A3C efficiency in data management helps reduce latency and optimize tier utilization, which shows the power of reinforcement learning in improving data management solutions. This approach can reduce the financial cost while maintaining an efficient storage system with full utilization of the storage.

The reinforcement learning approach, specifically the A3C algorithm, showed flexibility in adapting to the changing access pattern; this ability is crucial for today's unstable access data patterns, which have the characteristics of rapidity and huge amounts of data.

The positive outcome of this research can be further improved. A possible approach would be to integrate the reinforcement learning algorithms with another algorithm to form a hybrid model in order to join strengths. Another approach would be to make the reinforcement learning model more complex or train it on a diverse dataset.

References

1. Agarwal, A., Simha, R., LaMarca, A.: A unified approach to cache replacement and dynamic power management. ACM SIGMETRICS Perform. Eval. Rev. **23**(1), 242–253 (1995)
2. Chandra, S., et al.: A survey of hybrid cache replacement policies. ACM Comput. Surv. **51**(1), 1–34 (2018)
3. Denning, P.J.: The working set model for program behavior. Commun. ACM **11**(5), 323–333 (1968)
4. Lee, D., et al.: LRFU: a spectrum of policies that subsumes the least recently used and least frequently used policies. IEEE Trans. Comput. **50**(12), 1352–1361 (2001). https://doi.org/10.1109/TC.2001.970573. ISSN 1557-9956
5. Liu, M., Pan, L., Liu, S.: RLTiering: a cost-driven autotiering system for two-tier cloud storage using deep reinforcement learning. IEEE Trans. Parallel Distrib. Syst. **34**(2), 501–518 (2023). https://doi.org/10.1109/TPDS.2022.3224865
6. Liu, N., et al.: A hierarchical framework of cloud resource allocation and power management using deep reinforcement learning. In: 2017 IEEE 37th International Conference on Distributed Computing Systems (ICDCS), pp. 372–382 (2017). https://doi.org/10.1109/ICDCS.2017.123
7. Megiddo, N., Modha, D.S.: ARC: a self-tuning, low overhead replacement cache. In: 2nd USENIX Conference on File and Storage Technologies (FAST 2003), USENIX Association, San Francisco (2003). https://www.usenix.org/conference/fast-03/arcself-tuning-low-overhead-replacement-cache

8. Mehta, H.: Storage Management in Data Centers: Understanding, Exploiting, Tuning, and Troubleshooting Veritas Storage Foundation. Prentice Hall (2004). ISBN 0131014016
9. O'Neil, E.J., O'Neill, P.E., Weikum, G.: The LRUK Page Replacement Algorithm For Database Disk Buffering. Association for Computing Machinery, New York (1993)
10. Jiang, S., Zhang, X.: LIRS: an efficient low inter-reference recency set replacement policy to improve buffer cache performance. ACM SIGMETRICS Perform. Eval. Rev. **30**(1), 31–42 (2002). https://doi.org/10.1145/511399.511340
11. Wu, N., Li, P.: Phoebe: reuse-aware online caching with reinforcement learning for emerging storage models. CoRR abs/2011.07160 (2020)
12. Zhang, T., et al.: Data management of scientific applications in a reinforcement learning-based hierarchical storage system. Expert Syst. Appl. **237**, 121443 (2024). https://doi.org/10.1016/j.eswa.2023.121443. ISSN 0957-4174

Urban Change Detection from Satellite Images: Leveraging Edge AI for Advanced Insights

Meriem Baziz[1,2(✉)], Anfal Yousra Boucetta[1,2(✉)], Leila Hamdad[1(✉)], and Imed Allal[2]

[1] Ecole nationale Supérieure d'Informatique, ESI ex INI, Algiers, Algeria
{jm_baziz,ja_boucetta,l_hamdad}@esi.dz
[2] Namla Networks, Algiers, Algeria
imed.allal@namla.cloud

Abstract. Urbanization advances at unprecedented rates, resulting in negative effects on the environment and human well-being. Remote sensing has the potential to mitigate these effects by supporting sustainable development strategies with accurate information on urban growth. Deep learning-based methods have achieved promising urban change detection results from satellite images. However, these algorithms often necessitate powerful cloud-based hardware, which introduces challenges such as latency, bandwidth usage, and data privacy concerns. In the present work, we introduce an end-to-end system for detecting urban change from satellite images, utilizing Edge AI techniques to process data close to their sources and mitigate the challenges associated with cloud usage. The results of the experiments conducted during this study have been promising. The use of active learning to overcome the challenge of annotating satellite images has halved the number of samples needed and the model training time, while achieving performance similar to training on the entire annotated dataset. Furthermore, the compression technique adopted has reduced the model size by more than 50%, thereby minimizing the storage space required at the edge without significant loss of performance. The change detection achieved an IoU of 87% on the challenging Spacenet7 dataset, highlighting the effectiveness of the implemented approach.

Keywords: Urban change detection · Satellite images · Deep Learning · Edge AI

1 Introduction

Urbanization is advancing at unprecedented speeds [16]. Consequently, the global extent of urban land is expected to grow by a factor of 2 to 6 throughout the 21st century [7]. This swift expansion of urban areas brings about various adverse impacts on the environment and human well-being [11,25]. To address these

S. Chikhi et al. (Eds.): MISC 2024, LNNS 1267, pp. 182–201, 2025.
https://doi.org/10.1007/978-3-031-82112-7_14

urban transformations effectively, it is essential to develop informed and sustainable urban change detection systems.

Remote sensing is an effective tool for monitoring the Earth's surface [3]. Urban change detection aims to distinguish between changed and unchanged pixels in bi-temporal or multi-temporal satellite images taken over the same geographical area at different times. Traditional change detection methods use arithmetic operations to derive change features from bi-temporal image pairs. Various arithmetic methods, such as image differencing, image regression, and change vector analysis, have been developed to extract change features from optical images [18]. These features are then classified into changed or unchanged pixels or objects using various classification algorithms, including machine learning algorithms. In recent years, deep learning has increasingly replaced traditional change detection methods. Deep learning models often outperform traditional methods when datasets are large, making this approach particularly useful for satellite image analysis applications. Traditionally, deep learning solutions for satellite image analysis are deployed in the cloud. While this offers virtually unlimited storage and computing power, it raises concerns about data security, latency, and network bandwidth. To address these issues, a new paradigm known as 'Edge AI' has emerged. This approach deploys artificial intelligence algorithms directly on devices at the edge of the network, rather than relying on centralized cloud servers. In the context of satellite imagery, Edge AI plays a crucial role by enabling localized processing of the vast amounts of data generated by satellites. Integrating processing capabilities directly on satellites or in ground stations reduces processing delays, minimizes data transmission, and enhances confidentiality. In this paper, we propose an end-to-end urban change detection system utilizing Edge AI techniques. By integrating Edge AI, we can process data near the point of generation, thereby reducing latency and network dependency.

The paper is organized as follows: Sect. 2 provides a review of related work on various change detection techniques. Section 3 details the proposed solution, including the general architecture, change detection methodology, and the MLops pipeline utilized. Section 4 describes the dataset and preprocessing steps, while Sect. 5 covers the training process, including initial training, active learning applications, and model compression. Section 6 presents the experimental results, and Sect. 7 concludes with a discussion of the final system results.

2 Related Works

Over the years, multiple deep learning-based bi-temporal change detection methods have been proposed, with ongoing development of newer methods. These approaches offer significant advantages, providing precise change information that is less influenced by external factors such as atmospheric interferences [2]. Researchers have argued that DL approaches outperform traditional methods, enhancing the accuracy of change detection [6,34]. DL-based frameworks for change detection are typically classified into single-stream, dual-stream, and integrated multi-model approaches [26]. Single-stream frameworks use direct classification and transformation-based methods. Direct classification merges multi-temporal data for classification into changed or unchanged categories. [19] used this on datasets from Bern, Mexico, and Ottawa, achieving 0.90 precision, while [14] achieved 0.78 precision on Ottawa data. [8] applied it to San Francisco SAR images with 0.90 precision. Transformation-based methods make heterogeneous data comparable for change detection [33]. [20] used this on Landsat7 EMTM (SAR) data, achieving 0.96 precision, and [17] with ERS and SPOT data, achieving 0.87 precision. Dual-stream frameworks include Siamese and post-classification structures. The Siamese structure employs identical networks to extract features from different temporal images, generating change maps [26]. [4] used this on ZY-3 and GF-2 datasets, achieving 0.97 precision. [10] used it with attention on WHU and EV-CD datasets, attaining 0.92 precision. Post-classification uses separate classifiers for each time period, comparing classification maps to identify changes, [22] used this with IRS-1B and IRSP6 data, achieving 0.85 precision. [31] used it on Landsat data, achieving 0.92 precision, and [1] achieved 0.94 precision on Dubai and Las Vegas data. Integrated multi-model frameworks combine spatial and temporal features, enhancing change detection performance through diverse data sources and techniques [26]. [15] used this on QuickBird data, achieving 0.78 precision, while [4] achieved 0.87 precision on WHU data. To the best of our knowledge, there has been no prior work that proposes an end-to-end system for urban change detection from satellite images integrating deep learning and Edge AI techniques, which is precisely what we are presenting in this study.

3 Proposed Solution

3.1 General Architecture

The proposed urban change detection solution incorporates a cloud-edge architecture as illustrated in Fig. 1.

- **Edge Layer:** The edge layer is pivotal in our change detection system, strategically located near the data source, either onboard the satellite (Scenario 1) or at the ground station (Scenario 2), the edge layer handles several key functions. It stores change maps locally and periodically uploads them to the cloud for long-term storage. It performs image preprocessing and post-processing to ensure efficient data handling. Inference on satellite images is conducted at this layer, producing segmentation masks directly without needing cloud-based training. It also generates change maps from these masks and issues alerts based on predefined thresholds to notify relevant parties. Additionally, the edge layer supports the deployment of the frontend application, allowing users to interact with and visualize the change detection results.

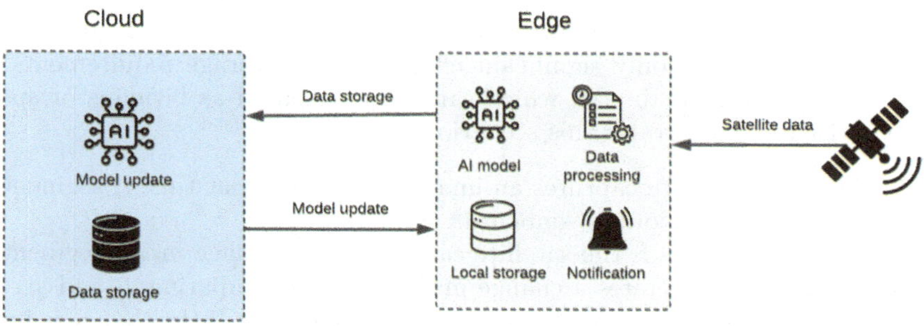

Fig. 1. General system architecture cloud-edge.

- **Cloud Layer:** The cloud layer provides extensive storage resources for long-term data preservation. It receives change maps generated by the edge layer and maintains them for future reference. Additionally, the cloud layer is responsible for retraining the deep learning model with newly acquired data to ensure its accuracy and relevance. Once retrained, the model is validated and redeployed to the edge layer to improve the precision of inferences.

3.2 Edge MLOps Pipeline

MLOps encompasses best practices for deploying and maintaining machine learning models efficiently in production, emphasizing automation, collaboration, and reproducibility [13].

We applied MLOps principles to streamline development, deployment, and continuous improvement through effective Monitoring and Continuous Training, utilizing Continuous Integration (CI) and Continuous Delivery (CD). CI automates component creation and testing to ensure pipeline stability and quality, while CD deploys the trained model to edge devices for real-time predictions. Monitoring involves collecting performance metrics, analyzing resource usage (CPU, RAM), and assessing input changes and segmentation confidence. We also implemented alert triggers to notify us if segmentation confidence drops below a set threshold, enabling timely updates and retraining to address data drift, with manual oversight to manage resource efficiency. This integrated approach ensures efficient deployment and ongoing adaptation in the production environment.

3.3 Change Detection Methodology

We propose a change detection methodology using the post-classification method, effective for urban changes with high accuracy and manageable complexity. Two scenarios are considered: Scenario 1 involves onboard edge processing with the satellite, while Scenario 2 involves edge processing at the ground station.

Scenario 1. Utilizes onboard processing to reduce ground communication bandwidth by transmitting only significant change maps. Storage requirements are based on the change map size, which can be represented as bitmaps or sparse matrices. Figure 2 illustrates this scenario.

- **(1)–(3):** The satellite captures an image m_t of area A at time t, segments it to produce s_t, and stores s_t onboard.
- **(4)–(6):** At time $t+1$, the satellite captures a new image m_{t+1}, segments it to get s_{t+1}, and generates a change map $C_{t,t+1}$ by comparing s_t and s_{t+1}.
- **(7)–(10):** Calculate the percentage of change in $C_{t,t+1}$. If the change is below a threshold, delete $C_{t,t+1}$ and s_{t+1}. If above, send $C_{t,t+1}$ to the ground station and retain s_{t+1} to replace s_t.

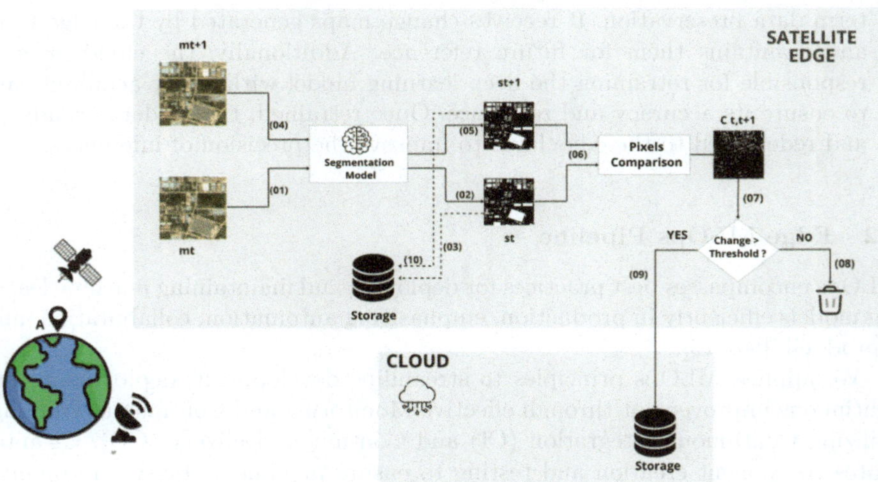

Fig. 2. Change Detection Methodology - Scenario 1

Scenario 2. Uses edge processing at the ground station, simplifying implementation and maintenance with reduced costs compared to onboard processing. It also lowers cloud dependency by handling processing locally, which limits sensitive data sent to the cloud. This scenario is illustrated in Fig. 3.

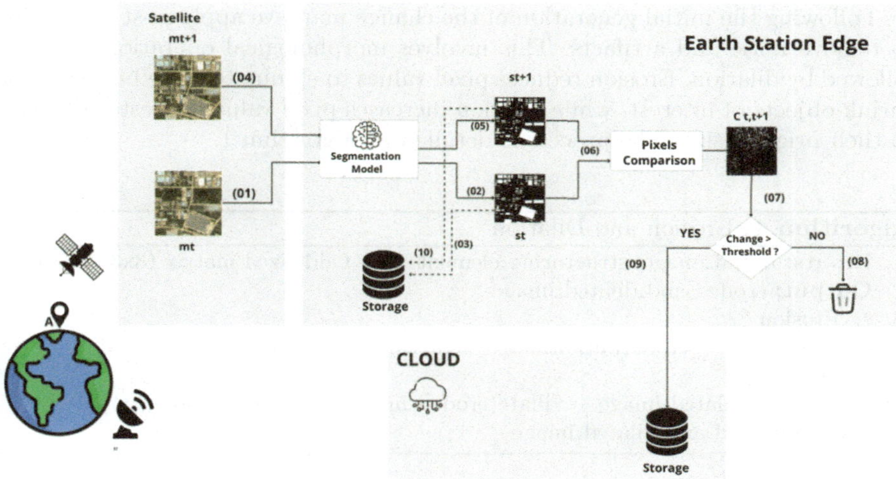

Fig. 3. Change Detection Methodology - Scenario 2

Both scenarios protect data confidentiality by minimizing sensitive information stored or processed in the cloud.

Change Maps Generation Process. Change maps are generated by comparing binary segmentation masks from two instants T_1 and T_2 pixel by pixel. Each pixel is labeled '1' for a change and '0' for no change, using the formula:

$$Y = |X_2 - X_1| \quad (1)$$

where X_2 is the image at T_2 and X_1 is the image at T_1.

After generating the change map between X_1 and X_2, we assign labels (change/no change) to the difference mask using a threshold of 11 pixels. This threshold is based on the median building size in the SpaceNet7 dataset, which is 11 pixels. Therefore, we indicate a change if the largest polygon in the change map is 11 pixels or more, meaning at least one building is identified. The threshold can be adjusted as necessary based on the specific use case. Its determination should be conducted by subject matter experts, whose expertise is crucial for accurately evaluating the unique characteristics and requirements of each application. These experts can offer valuable insights into selecting an optimal threshold that effectively delineates urban changes, taking into account factors such as building dimensions, spatial density, and the specific analytical objectives at hand.

Following the initial generation of the change map, we apply post-processing to remove noise and artifacts. This involves morphological operations: erosion followed by dilation. Erosion reduces pixel values to eliminate noise but may also shrink objects of interest, while dilation increases pixel values to restore objects to their original size. The process is detailed in Algorithm 1.

Algorithm 1. Erosion and Dilation

```
1: Inputs: initial_image, structuring_element    ▷ Odd-sized matrix (3x3, 5x5, etc.)
2: Output: eroded_and_dilated_image
3: /* Erosion */
4: eroded_image ← erode(initial_image, structuring_element)
5: /* Dilation */
6: eroded_and_dilated_image ← dilate(eroded_image, structuring_element)
7: return eroded_and_dilated_image
```

4 Dataset and Preprocessing

Comparative Analysis of Datasets. The analysis is structured around essential criteria that directly influence the relevance and utility of datasets for urban change detection. These criteria include:

- **Types of Annotations:** The analysis of the different types of available annotations (pixel-level annotations, one-hot encoding, segmentation masks, change masks) is crucial for defining the nature of the computer vision tasks to be undertaken. Since our solution focuses on building segmentation as a first step before calculating changes, we pay particular attention to datasets providing segmentation masks.
- **Classes Provided by Labels:** Categories such as buildings and roads are particularly relevant to our urban change detection project.
- **Multitemporal Data:** The availability of images taken at different times for the same area is crucial for tracking evolution and calculating changes from segmentation maps.
- **Geographic Coverage:** The geographic coverage of datasets is an important criterion to ensure the generalization of models.

Table 1 summarizes the characteristics of each dataset evaluated for our case study, based on the defined criteria:

Table 1. Comparative table of available datasets

Dataset	Types of Annotations	Multi-temporal	Size	Coverage	Classes
Esri2020	Pixel-level semantic labeling	No	Single global map	Global	Water, Trees, Grass, Vegetation, Crops, Built-up Area, Bare Land, Snow, Clouds
Change-Net	Pixel-level semantic labeling	Yes	31,000 image pairs	100 different cities in China	Building, Agricultural Land, Bare Land, Water, Road
CNAM-CD	Pixel-level semantic labeling	No	2503 GeoTiff image pairs	12 different areas in China	Background, Bare Land, Vegetation, Water Bodies, Other
Dynamic EarthNet	Pixel-level semantic labeling	No	54,750 satellite images	75 locations worldwide	Water, Soil, Agriculture, Snow and Ice, Forests and Other Vegetation
QFabric	Change polygons	Yes	/	504 locations in 100 cities	Change, No Change
DSIFN	Change masks	Yes	3940 image pairs	Beijing, Chengdu, Shenzhen, Chongqing, Wuhan, and Xian	Change, No Change
OSCD	Change masks	Yes	24 image pairs	Brazil, USA, Europe, Middle East, and Asia	Change, No Change
SpaceNet7 (Optical)	Segmentation masks	Yes	2389 images	Over 100 regions from 6 continents	Building, No Building
SpaceNet1	Segmentation masks	No	6940 images	Rio de Janeiro, Brazil, Las Vegas, Paris, Shanghai, Khartoum	Building, No Building

This comparison has allowed us to formulate several key observations:

- **Geographic Diversity:** It appears that some datasets, such as CNAM-CD and DSIFN, although they offer multitemporal data, suffer from a lack of geographic diversity. This limitation makes them less suitable for diverse urban contexts, thereby reducing their generalization potential.
- **Presence of Multitemporal Data:** Datasets such as Esri2020 and Dynamic EarthNet stand out due to their extensive geographic coverage. However, the absence of multitemporal data limits their utility for analyses requiring temporal comparisons in the same areas.
- **Suitability of Annotations:** Several datasets, including OSCD, DSIFN, QFabric, and ChangeNet, provide multitemporal data that includes change maps. However, they do not fully align with our approach, which requires segmentation masks.

In conclusion, SpaceNet7 stands out as the dataset most suited to the specific needs of our solution. Its richness in terms of geographic diversity and multi-temporal data, combined with precise annotations for semantic segmentation, makes it a strategic choice for our solution.

4.1 Dataset

The Spacenet7 dataset [28], also known as the Urban Development Multi-Temporal SpaceNet (MUDS), is particularly compelling for our study. It consists of 101 sequences of satellite images captured by Planet Labs' Dove satellite constellation. These sequences were selected for their geographic diversity and significant urbanization changes observed over a two-year period. The dataset includes a total of 2389 images, with approximately 11 million annotated buildings. Each image has a spatial resolution of 4 m per pixel and spectral resolution of 4 bands including RGB+NIR. The image size is 1024 × 1024 pixels, covering approximately 18 km^2.

Each dataset region consists of 24 consecutive monthly mosaic images[1]. Each image is accompanied by a GeoJSON file containing polygons outlining each building present in the image. Examples of dataset images and their corresponding masks are illustrated in Fig. 4.

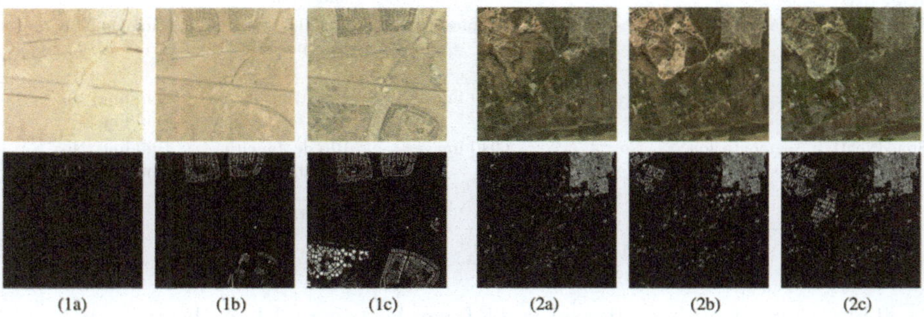

Fig. 4. Images from the SpaceNet7 dataset [28].

4.2 Data Preprocessing

In this section, we outline the preprocessing steps for training the segmentation model, as shown in Fig. 5.

Initially, we converted the segmentation masks from GeoJSON format into binary masks to facilitate their use in the model. To reduce redundancy, we retained approximately one image out of every five due to the high similarity among images within the same scene. To address the challenge of detecting small buildings; We then addressed the challenge of detecting small buildings by

[1] A mosaic is a composite of assembled images, often created to minimize cloud cover.

applying oversampling, which enlarged image features by a factor of 2 in each direction, increasing the image resolution from 1024×1024 to 2048×2048 pixels. Large images were cropped into smaller sub-images of 256×256 pixels to accommodate larger batch sizes and enable deeper model architectures during training ,this process resulted in a dataset comprising 4670 images of 256×256 pixels .For dataset partitioning, we designated 25% of the cropped images (1170 images) for testing and utilized the remaining 75% (3500 images) for training.To enhance model generalization and reduce overfitting, we incorporated various data augmentations during training, including horizontal and vertical flipping, random rotation, and modifications to contrast and brightness.

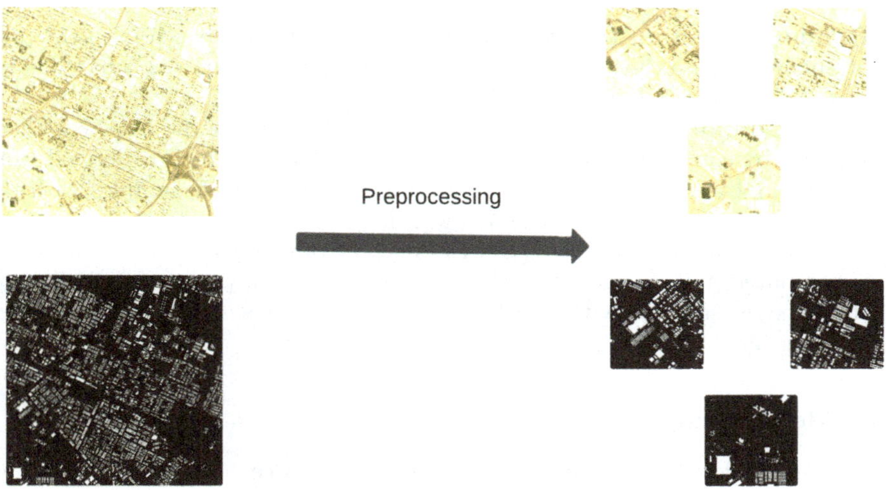

Fig. 5. Images Before and After Applying Various Preprocessing Steps

5 Training Process

The training process involves several critical steps. We begin by selecting various segmentation models, which are subsequently fine-tuned on the SpaceNet7 dataset. In this evaluation phase, we explored two primary convolutional neural network (CNN) architectures: U-Net [23] and DeepLabV3+ [5]. U-Net was tested with multiple backbones, including ResNet34 [9], MobileNetV2 [24], and HRNet [29], while DeepLabV3+ was evaluated with the MobileNetV2 backbone. To improve annotation efficiency, we incorporated active learning techniques, which aim to reduce the effort required for annotation. Additionally, we applied model compression techniques, such as quantization and pruning, to facilitate deployment in environments with resource constraints, thereby achieving a more compact model size.

5.1 Active Learning

Active learning is a technique used to reduce reliance on annotated data for deep learning models. This is particularly relevant for annotating satellite images for segmentation, which is challenging due to their complexity and high dimensionality [21].

The core of an active learning system involves selecting informative samples for the model. For binary segmentation, several methods focus on predicting model uncertainty for each pixel to select images with the highest uncertainties. The difference between these methods lies in how uncertainty measures are defined. In this context, we test different mechanisms to capture and quantify this uncertainty:

1. **Random Selection (RAND)**: Samples are randomly chosen from the pool of unlabeled data during each iteration. This method serves as a baseline for comparison with other strategies.
2. **Least Confidence (CONF)**: This technique identifies the least confident samples by computing the minimum confidence score, which is derived from the sigmoid probability of the class of interest, then the K samples with the lowest scores are selected for labeling [21].
3. **Entropy-Based Selection (ENT)**: The entropy of each pixel in an image is computed and summed to obtain the overall entropy score for the image, then the K samples with the highest entropy scores are then selected for labeling [21].

5.2 Model Compression

Deploying models on resource-constrained edge devices necessitates effective compression techniques. We explored three methods: Post-Training Quantization (PTQ), Quantization-Aware Training (QAT), and a hybrid approach combining pruning with quantization.

- **Quantization:** This technique reduces the precision of weights, biases, and activations to lower memory usage and model size. Typically, weights are stored as 32-bit floating-point values, but quantization converts them to more compact formats like FP16 or lower-bit integers (INT8, INT4), balancing memory efficiency with accuracy [27]. We tested two techniques: Quantization-Aware Training (QAT), which retrains the model with quantized parameters [30], and Post-Training Quantization (PTQ), which applies quantization without retraining, using a calibration dataset to find optimal parameters [30].
- **Pruning:** Pruning removes redundant or less significant connections in a model, which can be done in a structured or unstructured manner [32].

The third hybrid approach first prunes the model to reduce complexity and then applies PTQ to maintain accuracy.

6 Tests and Results

This section details the evaluation of our models, including the testing environment, segmentation model training, and performance metrics. We also assess the impact of active learning strategies, improvements from post-processing, and various model compression techniques to optimize performance and deployment readiness.

6.1 Testing Environment

The models were evaluated using a testing environment with the following hardware specifications: an Intel Xeon CPU operating at 2.30 GHz, an NVIDIA T4 GPU, 16 GB of RAM, and 100 GB of disk space.

It is important to note that this hardware was used exclusively for proof-of-concept purposes. In actual space deployment scenarios, hardware specifically designed for space applications will be utilized to ensure optimal performance and reliability.

6.2 Segmentation Models Training

We trained the models on the Spacenet7 dataset using transfer learning for 300 epochs. All segmentation models were trained using the Adam optimizer [12] with a fixed learning rate of 1×10^{-4}. We employed a combined loss function, integrating Binary Cross-Entropy (BCE) and Dice loss, defined as follows:

$$l_{combined} = l_{BCE} + 0.5 \cdot l_{Dice} \tag{2}$$

Given the high class imbalance in SpaceNet 7, we used F1 score, Precision, Recall, and Intersection over Union (IoU) for evaluation. IoU measures the overlap between predicted and ground truth masks, Recall assesses the model's ability to detect all building pixels, Precision evaluates prediction accuracy, and the F1 score provides a balance between Precision and Recall.

Training results, including times and performance metrics, are shown in Table 2:

Table 2. Model Performances and Training Times.

Model	F1	Precision	Recall	IoU	Training Time
Unet-Resnet34	0.934	0.941	0.927	0.93	6 h14 m
Unet-MobileNetV2	0.888	0.889	0.887	0.84	5 h56 m
DeeplabV3-MobilenetV2	0.634	0.638	0.63	0.47	5 h27 m
Unet-HRnet	0.918	0.94	0.925	0.91	7 h23 m

For segmentation, U-Net with ResNet34 achieved the best performance with a high F1-score, indicating a strong balance between precision and recall. U-Net with HRNet also performed well but had a higher complexity and training time. U-Net with MobileNetV2 provided a good balance of performance and efficiency. In contrast, DeeplabV3+ with MobileNetV2 underperformed, highlighting U-Net's superior ability to capture fine details and maintain boundary accuracy, which is crucial for building segmentation at low spatial resolution.

6.3 Active Learning Training

We chose the ResNet34 model due to its strong performance and investigated the effects of different active learning strategies. The Unet-ResNet34 model was retrained with each strategy, selecting 150 images per cycle over 10 cycles, totaling 1500 samples, or nearly half of the dataset. The results of this iterative training process are summarized in the Table 3 and Fig. 6:

Table 3. Active Learning Results based on Random, Confidence-based, and Entropy-based Selection.

AL Algorithm	IoU	Training Time
RAND	0.61	45 mins
CONF	0.84	1 h43 mins
ENT	0.90	1 h45 mins

The analysis of the results reveals significant differences in the performance of the selection strategies. As shown in Fig. 6, both uncertainty-based methods (CONF and ENT) significantly outperform random selection from the first cycle. Furthermore, the entropy-based method surpasses the confidence-based method by 6% in the final cycle, achieving an average IoU of 0.90. This indicates that annotating half of the objects in the images yields almost the same average IoU of 0.92, suggesting that annotation effort can be reduced by half with minimal loss in model precision. The results in Table 3 also show a nearly 50% reduction in training time to achieve similar results.

This experimentation demonstrates that using a confidence-based active learning strategy can halve the annotation effort and training time of models, with negligible loss in precision.

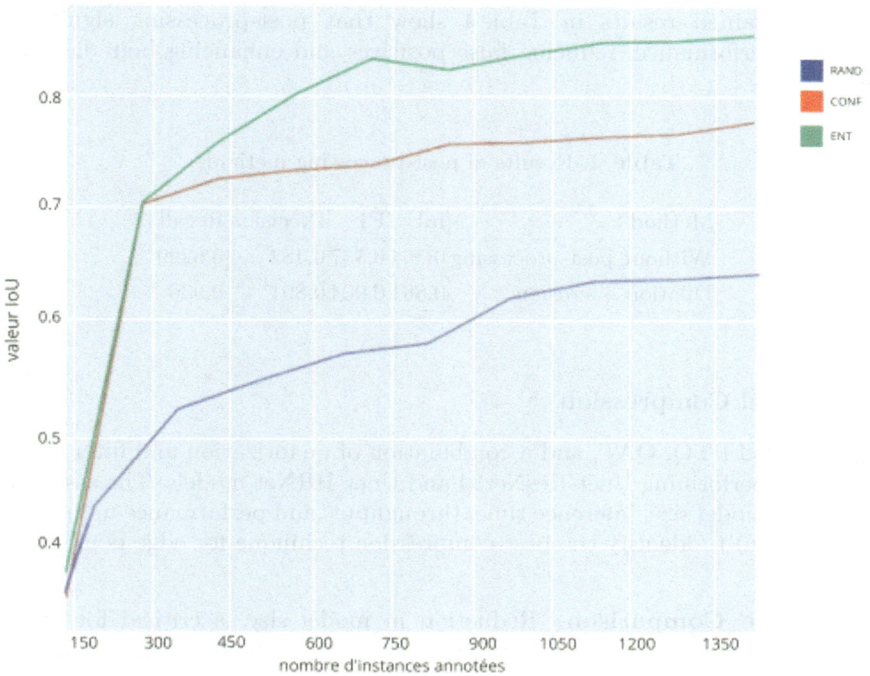

Fig. 6. Results of Active Learning Evaluation.

6.4 Change Detection

To evaluate change detection, we utilized pairs of multi-temporal images from the SpaceNet7 dataset to create a validation dataset. Change maps were generated using ground truth segmentation masks, providing a reliable reference for assessing our approach.

Initially, the change maps derived from segmentation masks exhibited lower accuracy, with reduced precision and F1 scores primarily due to false positives, especially around object boundaries. To enhance accuracy, we employed post-processing techniques, specifically dilation and erosion. The results, detailed in Table 4 and illustrated in Fig. 7, demonstrate the effectiveness of these techniques.

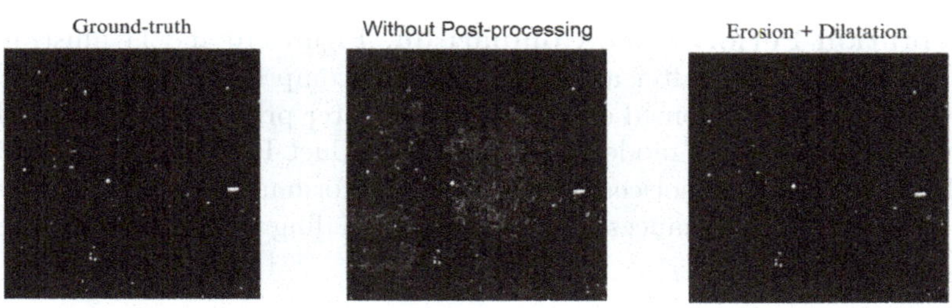

Fig. 7. Post-processing results.

The obtained results in Table 4 show that post-processing significantly improved performance, reducing false positives and enhancing boundary accuracy.

Table 4. Results of post-processing methods.

Method	IoU	F1	Precision	Recall
Without post-processing	0.464	0.557	0.482	0.6299
Dilation + erosion	0.861	0.904	0.891	0.936

6.5 Model Compression

We evaluated PTQ, QAT, and a combination of quantization and filter pruning on the top-performing Unet-ResNet34 and Unet-HRNet models. The assessment focused on model size, inference time, throughput, and performance metrics (IoU and F1 score) to identify the best compression technique for edge deployment.

Model Size Comparison. Reduction in model size is critical for efficient deployment. Figures comparing model sizes after different compression techniques are shown in Figs. 9 and 8.

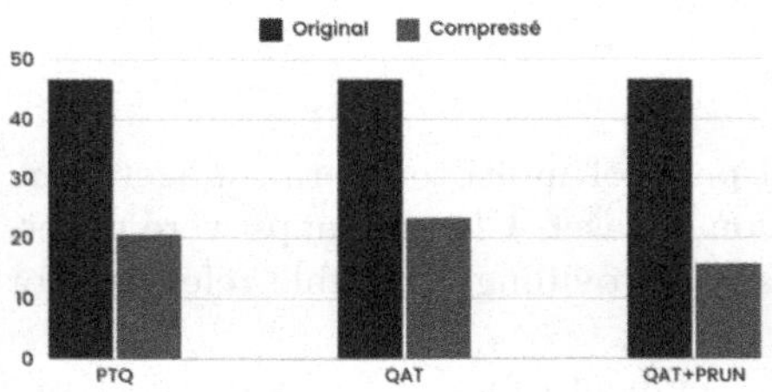

Fig. 8. Comparison of Unet-Resnet model sizes after different compression techniques.

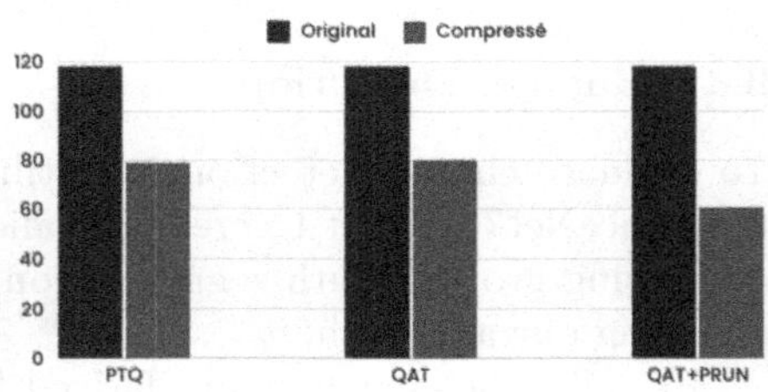

Fig. 9. Comparison of Unet-HRNet model sizes after different compression techniques.

Segmentation Performance Comparison. Figures 10 and 11 illustrate segmentation performance after applying the three compression techniques. Based on the analysis, the combined quantization and filter pruning technique achieves the greatest reduction in model size: 66.13% for Unet-ResNet34 and 48.53% for Unet-HRNet. Despite a noticeable impact on performance metrics (IOU and F1 score), this technique enhances inference speed, leading to improved throughput.

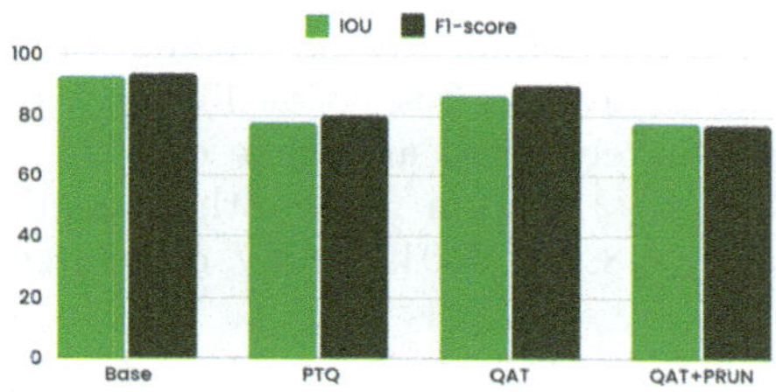

Fig. 10. Performance of the Unet-Resnet model after different compression techniques.

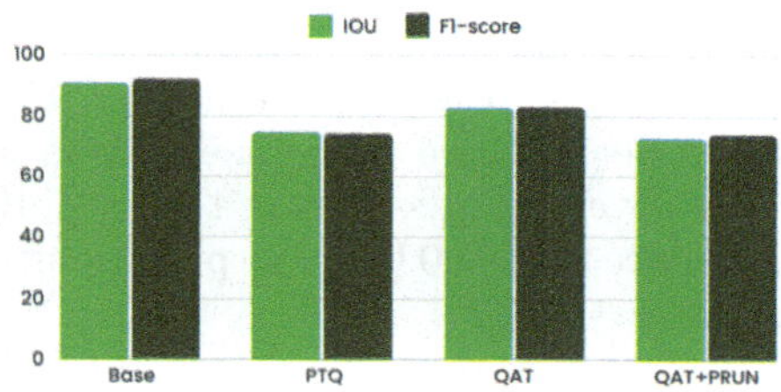

Fig. 11. Performance of the Unet-HRNet model after different compression techniques.

In selecting a compressed model for deployment, balancing size reduction with acceptable performance degradation is crucial. For practical purposes, we use quantization-aware training on the Unet-ResNet34 model. This method achieves a similar model size (approximately 23 MB) as the third compression technique, while maintaining satisfactory segmentation performance.

7 Final System Results

In this section, we present the final system's performance in terms of segmentation accuracy, processing capacity, storage requirements, and energy consumption.

7.1 Segmentation Performance and Processing Capacity

The final segmentation model used is Unet-ResNet34, which has been compressed using the training-aware quantization technique with 8-bit floating point precision. Table 5 illustrates the model's performance for urban change detection, detailing the change IoU, model size, inference time per image, and throughput.

Table 5. Performance of the final model.

Model	IoU	Size (MB)	Inference Time (s)	Throughput (image/s)
Unet-ResNet34	0.87	23.36	0.04	25.00

The results in Table 5 show that the Unet-ResNet34 model achieves a high IoU of 0.87, demonstrating effective urban change detection. Its compact size of 23.36 MB and quick inference time of 0.04 s per image allow for efficient processing with a throughput of 25 images per second, making it suitable for resource-constrained environments.

Using the model's inference time given in the table, we can calculate the processing capacity of the model. Assuming an average resolution like that of SpaceNet7, where each pixel covers approximately $4\,\mathrm{m}^2$, an image of 1024×1024 covers $1024 \times 1024 \times 4\,\mathrm{m}^2 = 4,194,304\,\mathrm{m}^2 \approx 4.19\,\mathrm{km}^2$. Given the model's inference time of $0.04\,\mathrm{s}$, the processing capacity is $8,640,000\,\mathrm{km}^2/\mathrm{day}$, calculated as follows:

$$\left(\frac{86400}{t}\right) \times a \tag{3}$$

where t is the inference time and a is the area per image, which is $4.19\,\mathrm{km}^2$ for SpaceNet7 images.

7.2 Storage Requirements

We store only change maps, which are binary and thus require just one bit per pixel. A 1024 × 1024 binary mask occupies approximately 131 KB uncompressed, compared to 1 GB for the original image. Further compression in PNG format reduces this to about 24 KB. This approach significantly reduces storage requirements by focusing solely on significant changes.

7.3 Energy Consumption

Energy consumption is calculated using the formula 4:

$$E_{tot} = (F \times E_{FLOP}) + (M \times E_{mem}) \tag{4}$$

In this equation, F represents the number of floating-point operations (FLOPs), M denotes the model size in megabytes (MB), E_{FLOP} is the energy cost per floating-point operation, and E_{mem} is the energy cost associated with retrieving one megabyte of data from DRAM.

Therefore, based on the U-net Resnet34 model results as shown earlier, we will have the results shown in the Table 6:

Table 6. Energy consumption for U-net Resnet34 models

Model	Size (M)	FLOPs (F)	Energy Consumption
U-net Resnet34 (uncompressed)	47 MB	5,999,393,792	13×10^9 pJ
U-net Resnet34 compressed	23 MB	3,043,471,860	7×10^9 pJ

The results shown in Table 6 illustrate that by compressing the model, our solution achieves a significant reduction in energy consumption, with a gain of approximately 46.15%.

8 Conclusion

This study presented an end-to-end urban change detection system leveraging Edge AI to process satellite images close to their source, reducing latency, bandwidth usage, and enhancing data privacy. The system employs active learning to significantly cut down on annotation effort while maintaining high accuracy and uses model compression techniques to facilitate deployment on resource-constrained edge devices. Key contributions include Edge AI Integration, processing images at the edge minimizes data transmission and ensures data confidentiality; Active Learning, halves the number of samples needed for training without significant performance loss; and MLOps Integration, implements continuous integration, delivery, and monitoring, ensuring efficient deployment and continuous adaptation. Future work will include testing other high-potential models, such as vision transformers, to achieve even better accuracy. Additionally, integrating active learning into the MLOps pipeline and incorporating more data sources will enhance the system's robustness. This system provides a scalable and efficient solution for sustainable urban monitoring and development.

References

1. Aldossary, A.: Analysis of urban change detection techniques in desert cities using remote sensing. Ph.D. thesis, California State University, Northridge (2012)
2. Asokan, A., et al.: Change detection techniques for remote sensing applications: a survey. Earth Sci. Inform. **12**, 143–160 (2019)
3. Ban, Y., Yousif, O.: Change detection techniques: a review. Multitemporal Remote Sens. 19–43 (2016)
4. Chen, H., et al.: Change detection in multisource VHR images via deep Siamese convolutional multiple-layers recurrent neural network. IEEE Trans. Geosci. Remote Sens. **58**(4), 2848–2864 (2019)
5. Chen, L.C., Zhu, Y.A., Papandreou, G., Schroff, F., Adam, H.: Encoder-decoder with atrous separable convolution for semantic image segmentation. In: Proceedings of the European Conference on Computer Vision (ECCV) (2018)
6. Fan, J., Lin, K., Han, M.: A novel joint change detection approach based on weight-clustering sparse autoencoders. IEEE J. Sel. Topics Appl. Earth Obs. Remote Sens. **12**(2), 685–699 (2019)
7. Gao, J., O'Neill, B.C.: Mapping global urban land for the 21st century with data-driven simulations and shared socioeconomic pathways. Nat. Commun. **11**(1), 2302 (2020)
8. Geng, J., et al.: Saliency-guided deep neural networks for SAR image change detection. IEEE Trans. Geosci. Remote Sens. **57**(10), 7365–7377 (2019)
9. He, K., Zhang, X., Ren, S., Sun, J.: Deep residual learning for image recognition. In: Proceedings of the IEEE Conference on Computer Vision and Pattern Recognition, pp. 770–778 (2016)
10. Jiang, H., Ma, A., Wu, C., Liu, L., Yu, Y.: PGA-siamnet: pyramid feature-based attention-guided Siamese network for remote sensing orthoimagery building change detection. Remote Sens. **12**(3), 484 (2020)
11. Johnson, M.P.: Environmental impacts of urban sprawl: a survey of the literature and proposed research agenda. Environ. Plan. A **33**(4), 717–735 (2001)

12. Kingma, D.P., Ba, J.: Adam: a method for stochastic optimization. arXiv preprint arXiv:1412.6980 (2014)
13. Kreuzberger, D., Kühl, N., Hirschl, S.: Machine learning operations (MLOPS): overview, definition, and architecture. IEEE Access (2023)
14. Li, Y., et al.: A deep learning method for change detection in synthetic aperture radar images. IEEE Trans. Geosci. Remote Sens. **57**(8), 5751–5763 (2019)
15. Liu, R., et al.: Remote sensing image change detection based on information transmission and attention mechanism. IEEE Access **7**, 156349–156359 (2019)
16. Liu, X., et al.: High-spatiotemporal-resolution mapping of global urban change from 1985 to 2015. Nat. Sustain. 1–7 (2020)
17. Liu, Z., et al.: Change detection in heterogeneous remote sensing images via homogeneous pixel transformation. IEEE Trans. Image Process. **27**(4), 1822–1834 (2017)
18. Lu, D., Mausel, P., Brondizio, E., Moran, E.: Change detection techniques. Int. J. Remote Sens. **25**(12), 2365–2401 (2004)
19. Lv, N., et al.: Deep learning and superpixel feature extraction based on contractive autoencoder for change detection in SAR images. IEEE Trans. Ind. Inf. **14**(12), 5530–5538 (2018)
20. Ma, W., et al.: Change detection in remote sensing images based on image mapping and a deep capsule network. Remote Sens. **11**(6), 626 (2019)
21. Monarch, R., Manning, C.D.: Human-in-the-Loop Machine Learning: Active Learning and Annotation for Human-Centered AI. Sherlter Island, NY (n.d.)
22. Raja, R.A., et al.: Wavelet based post classification change detection technique for urban growth monitoring. J. Indian Soc. Remote Sens. **41**, 35–43 (2013)
23. Ronneberger, O., Fischer, P., Brox, T.: U-net: convolutional networks for biomedical image segmentation. In: Medical Image Computing and Computer-Assisted Intervention–MICCAI 2015: 18th International Conference, Munich, Germany, 5–9 October 2015, Proceedings, Part III 18, pp. 234–241. Springer (2015)
24. Sandler, M., Howard, A., Zhu, M., Zhmoginov, A., Chen, L.C.: Mobilenetv2: inverted residuals and linear bottlenecks. In: Proceedings of the IEEE Conference on Computer Vision and Pattern Recognition (2018)
25. Sarkodie, S.A., Owusu, P.A., Leirvik, T.: Global effect of urban sprawl, industrialization, trade and economic development on carbon dioxide emissions. Environ. Res. Lett. **15**(3), 034049 (2020)
26. Shi, W., et al.: Change detection based on artificial intelligence: state-of-the-art and challenges. Remote Sens. **12**(10), 1688 (2020)
27. Surianarayanan, C., Lawrence, J.J., Chelliah, P.R., Prakash, E., Hewage, C.: A survey on optimization techniques for edge artificial intelligence (AI). Sensors **23**(3), 1279 (2023)
28. Van Etten, A., Hogan, D.: The spacenet multi-temporal urban development challenge. arXiv preprint arXiv:2102.11958 (2021)
29. Wang, J., et al.: Deep high-resolution representation learning for visual recognition. IEEE Trans. Pattern Anal. Mach. Intell. **43**(10), 3349–3364 (2020)
30. Weng, O.: Neural network quantization for efficient inference: a survey. arXiv preprint arXiv:2112.06126 (2021)
31. Wu, C., et al.: A post-classification change detection method based on iterative slow feature analysis and Bayesian soft fusion. Remote Sens. Environ. **199**, 241–255 (2017)
32. Xu, S., et al.: Convolutional neural network pruning: a survey. In: 2020 39th Chinese Control Conference (CCC). IEEE (2020)

33. Zhang, P., et al.: Change detection based on deep feature representation and mapping transformation for multi-spatial-resolution remote sensing images. ISPRS J. Photogramm. Remote. Sens. **116**, 24–41 (2016)
34. Zhang, W., Lu, X.: The spectral-spatial joint learning for change detection in multispectral imagery. Remote Sens. **11**(3), 240 (2019)

Artificial Intelligence in Healthcare

Enhanced Glioma Classification Through Multi-modal Deep Learning: Integrating Histopathological and MRI Data

Linda Ait Mohammed(✉), Fatiha Alim-Ferhat, and Farid Talbi

Division Architecture et Système Multimédia, Centre de Développement des Technologies Avancées, Alger, Algeria
{laitmohamed,falim,ftalbi}@cdta.dz

Abstract. The workload of pathologists is increasing due to the continuous rise in cancer cases. To classify tumors and assess their level of aggressiveness, pathologists must analyze a large number of pathological images, sometimes hundreds or thousands, which is both costly and time-consuming and does not necessarily guarantee perfectly accurate results. To address these challenges, automating the analysis of tissue slices mounted on glass slides using microscopes is essential. Computer-assisted techniques, particularly artificial intelligence, offer significant potential to improve tumor classification. We propose to develop a new multi-input convolutional neural network architecture, leveraging both MRI and histological data to refine glioma classification. This approach will be validated using data from the Radiology-Pathology Challenge (CPM: RAD-PATH 2020). Our results show a precision of 0.75 for MRI image classification, 0.80 for pathological image classification, and 0.83 when combining pathological and MRI images.

Keywords: Glioma Classification · Deep Learning · Multi-Modal Imaging · Histopathological · MRI · CNN · Fusion

1 Introduction

Intracranial tumors, whether benign or malignant, develop within the cranial cavity and can affect the brain as well as adjacent anatomical structures such as the meninges, cranial nerves, and pituitary gland. Among these tumors, gliomas, which account for nearly half of all brain tumor cases, originate from the brain's glial cells. Gliomas are classified into several types, including astrocytomas, oligodendrogliomas, gangliogliomas, and ependymomas [1], based on their cellular origin and location. According to the World Health Organization, these tumors are graded based on their malignancy, a classification crucial for guiding diagnosis and therapeutic management [2].

The diagnostic process for gliomas often involves a biopsy followed by microscopic analysis of tissue samples by pathologists. The samples, stained with hematoxylin and eosin, are examined to determine the tumor grade, but this process is time-consuming, complex, and sometimes prone to errors. In some cases, the mixed characteristics of gliomas make classification particularly challenging.

S. Chikhi et al. (Eds.): MISC 2024, LNNS 1267, pp. 205–211, 2025.
https://doi.org/10.1007/978-3-031-82112-7_15

The advent of computer vision and artificial intelligence techniques presents a promising opportunity to automate and enhance this process. By leveraging radiological and histological images, it is possible to develop machine learning models capable of classifying gliomas with increased accuracy [3]. Recent studies have demonstrated the potential of convolutional neural networks (CNNs) and other deep learning techniques in this area [4]. For example, approaches combining MRI and histological images have shown significant improvements in glioma classification accuracy.

Among recent research, Pei et al. [5] have developed a two-step method for classifying brain tumors. The first step involves segmenting regions of interest (ROIs) from MRI images using a U-Net convolutional neural network. These ROIs are then classified into three types of gliomas astrocytoma, oligodendroglioma, and glioblastoma using a 3D CNN. This method was validated with data from the radiology-pathology challenge (CPM: RAD-PATH 2020) [6]. Meanwhile, Rathore et al. [7] utilize a complementary approach combining radiological and pathological images. They extract a comprehensive set of technical features from tumor regions in radiological images, including intensity, histogram, and texture descriptors, primarily quantified using GLCM [8] and GLRLM matrices [9]. Features extracted from images of 107 patients are then analyzed using SVM for classification and survival outcome prediction.

Ma and Jia [10] also developed an approach for classifying glioma grades using convolutional neural networks, securing first place in the CPM: RAD-PATH 2019 challenge. Their method integrates both MRI and histological images to build two distinct prediction models. The first model, based on ResNet, focuses on classifying histological images by extracting 512×512 pixel patches from each slide. To optimize the selection of the most representative cancer cells, specific criteria are applied to the samples, and preprocessing is performed using OpenSlide software [11]. The second model uses 3D DenseNet for multimodal classification of MRI images, and the final decision is made by combining the predictions from both models.

Chan et al. [12] propose a CNN-based approach by resizing images to 244×244 pixels to fit their GPUs. They also eliminate patches where the sum of the RGB channel standard deviations is below a predefined threshold (100 or 50). Using VGG 16, they extract 4096 parameters from each image in the CPM: RAD-PATH 2019 database, while ResNet 50 extracts 2048 parameters from the Gvg-Pool layer. These numerous parameters help classify the different grades of gliomas using a Random Forest classifier.

Xue et al. [13] established a pipeline for the automatic classification of gliomas into three subtypes: astrocytoma, oligodendroglioma, and glioblastoma, using both multimodal radiological images and histological data from RAD-PATH 2019. They developed two distinct models using DenseNet architecture (Dense Connectivity Network DCN) and conducted a series of tests for each modality in the database. They conclude that combining different modalities significantly improves classification results. This research aims to explore and refine these automated approaches for glioma classification, utilizing multimodal data and innovative neural network architectures. The goal is to overcome current diagnostic challenges, reduce human errors, and enhance patient care through more reliable and effective diagnostic tools.

2 Méthodologie

2.1 Preprocessing

Digitizing pathological tissue samples stained with hematoxylin and eosin (H&E) produces WSIs of large size, nearly 1 GB each. Processing these large images is complex. To simplify the analysis, we select representative patches of cells due to numerous non-stained and damaged areas in the images. Preprocessing is carried out on each patch (512×512 pixels) based on its mean grayscale value, relying on the hypothesis that cancer cells appear lighter in color compared to other cells. This preprocessing step is particularly time-consuming, taking about 84 min per sample. The dataset used for this task is from the Radiology-Pathology Challenge (CPM: RAD-PATH 2020), which provides the images needed for analysis.

2.2 Image Classification Using a Multi-input Convolutional Neural Network

Convolutional Neural Networks (CNNs) currently exhibit promising performance not only in image classification but also in various other applications. Their effectiveness is attributed to their locally connected layers, which excel at capturing local features compared to other types of neural networks. Figure 1 illustrates the implementation of a CNN for image classification using a multi-input approach. This model is designed to classify three subtypes of gliomas: Astrocytomas (A), Oligodendrogliomas (O), and Glioblastomas (G). Each pathology is incorporated into the CNN architecture using both histological images and corresponding MRIs, as depicted in the diagram below.

In the work of Ruimin et al. [14], a deep learning model called TM-CNN is used to predict traffic speed across multiple lanes. We drew inspiration from this architecture to develop a multimodal approach for glioma classification. The proposed model features two convolutional layers: one for histological images and the other for the corresponding MRIs. This dual-flow approach is particularly advantageous as it enables the simultaneous integration of pathological information at the cellular level and image characteristics such as texture, shape, and color. This multimodal approach enhances the model's performance compared to a single-modality method. To combine the two flows, a concatenation operation is applied, where the outputs of both flows are flattened and concatenated before entering the fully connected (FC) layers. Concatenation is preferred over addition or multiplication because it preserves distinct structures for each flow while providing the flexibility to adjust each flow's structure independently.

Figure 2 illustrates the overall architecture of the model, detailing the dimensions of the inputs and outputs at each layer. Each input data sample is structured in a dimension of $14 \times 20 \times 3$, representing the specific characteristics of the processed image.

For the model, three hidden convolutional layers were chosen after a rigorous process of trial and error. This process revealed that using three convolutional layers provides better performance compared to using only one or two layers. Convolutional layers are crucial for extracting complex features from the input data, and adding additional layers allows for capturing finer details and more complex structures. To reduce the risk of overfitting during the training process, a dropout layer is integrated between the fully connected (FC) layers and the output layer. This dropout layer, with a dropout

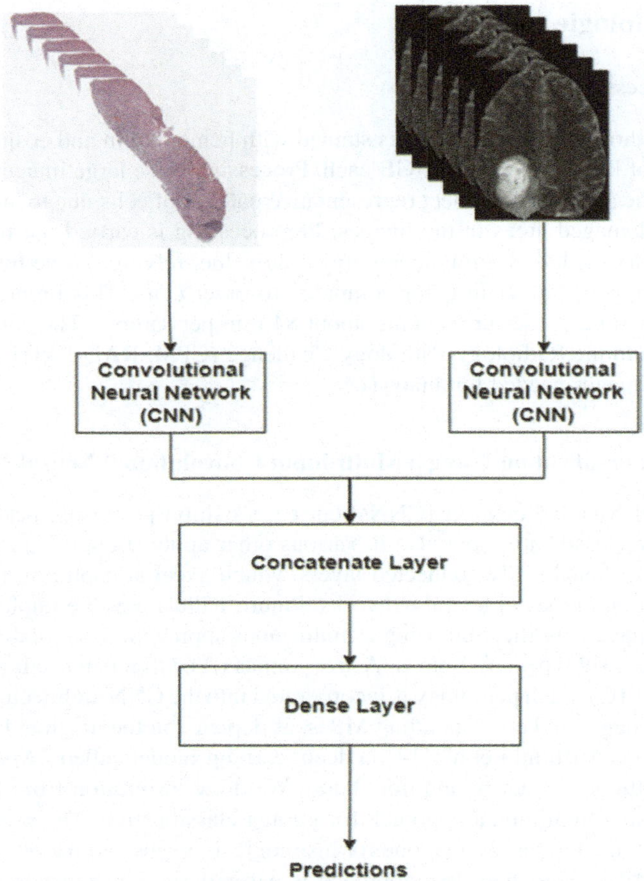

Fig. 1. Diagram of a multi-input convolutional neural network.

rate of 0.25, helps prevent the model from fitting excessively to the training data by randomly disabling a fraction of neurons during each training iteration. This improves the model's generalization by making it less sensitive to fluctuations in the training data and promoting more robust representations.

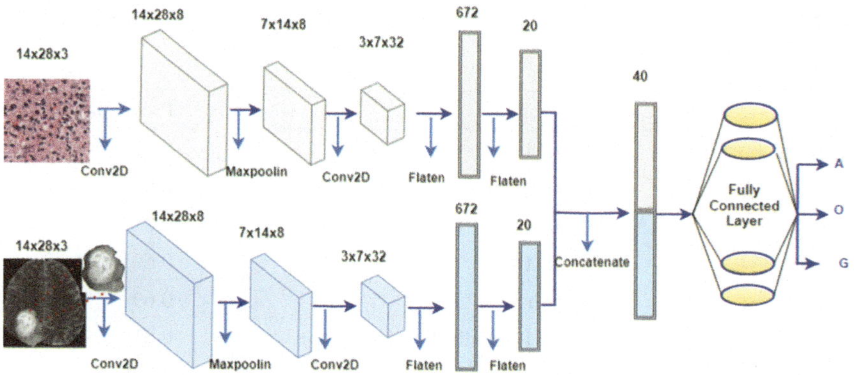

Fig. 2. Proposed pipeline for integrating histological and MRI data to classify gliomas.

3 Results and Evaluation

In evaluating segmentation and classification performance, a range of methodologies has been established. In the current study, precision was utilized as the principal metric for assessment. All metrics were evaluated using the Rad-Path 2020 dataset to ensure consistency and comparability.

Precision, as a classification metric, is determined by the ratio of the number of true positive predictions to the total number of predictions made. This metric effectively measures the model's capability to correctly identify positive cases, thereby reflecting the accuracy of the classification process.

$$Precision = \frac{TP}{TP + FP} \quad (1)$$

With TP (True Positives) and FP (False Positives).

We achieved improved validation precision scores: 0.75 for MRI image classification, 0.80 for pathological image classification, and 0.87 for the combined classification of pathological and MRI images. The fusion of both modalities significantly enhanced the model's performance. All performance parameters are detailed in Table 1.

Table 1. Results on the CPM-RadPath-2020 Validation Set

Approach	Precision
IRM	75%
Path	80%
IRM & Path	83%

We also compared our results with several recent state-of-the-art methods. Table 2 summarizes the outcomes of our approach.

Table 2. Results of Comparison with Existing Methods for Glioma Subtype Classification Using Rad and Path Data Fusion

Work	Approach	Precision
Ma and Jia. [10]	IRM	0.711
	Path	0.833
	IRM & Path	0.889
Xue et al. [13]	IRM	0.69
	Path	0.69
	IRM & Path (Random crop)	0.81
W. Chan et al. [12]	Path	0.72
	IRM & Path	0.78
Our approach	IRM	**0.75**
	Path (Random crop)	**0.80**
	IRM & Path(Random crop)	**0.83**

The results shown in Table 2 validate the effectiveness of our proposed method for glioma subtype classification. Ma and Jia [10] utilized two convolutional neural networks (CNNs), ResNet and 3D DenseNet, with a regression of the two predictions, achieving a precision of 0.889. While their results surpass ours, their performance in MRI classification is considerably lower, with a precision of 0.711. Our results exceed those of W. Chan et al. [12], despite their use of a GPU. Xue et al. [13] obtained similar results to our approach using DenseNet (Densely Connected Network DCN), but we outperform them in the classification of both MRI and pathological images.

4 Conclusion

We developed an innovative approach to brain tumor classification by combining advanced deep learning techniques. Using a dual-flux convolutional neural network, we analyzed WSI and MRI images from patients with glioblastoma. We employed three distinct methods: one based on WSI images, another on MRI, and a combined approach integrating both modalities. The results demonstrate that fusing WSI and MRI data significantly enhances classification accuracy compared to each method used alone.

Future directions include exploring additional deep learning techniques to further optimize precision and robustness, as well as extending this approach to other types of brain tumors, thereby providing more accurate diagnoses and more effective treatments.

Conflicts of Interest.. We know of no conflicts of interest associated with this publication, and there has been no significant financial support for this work that could have influenced its outcome. As Corresponding Author, I confirm that the manuscript has been read and approved for submission by all the named authors.

References

1. Louis, D.N., et al.: The 2016 world health organization classification of tumors of the central nervous system: a summary. Acta Neuropathol. (Berl.) **131**(6), 803–820 (2016). https://doi.org/10.1007/s00401-016-1545-1
2. Barrington, A.: Pathologically-Validated Tumor Prediction Maps in MRI. Masters Theses 2009 (2019) [Online]. Available: https://epublications.marquette.edu/theses_open/563
3. Ait Mohamed, L., Cherfa, A., Cherfa, Y., Belkhamsa, N., Alim-Ferhat, F.: Hybrid method combining superpixel, supervised learning, and random walk for glioma segmentation. Int. J. Imaging Syst. Technol. **31**(1), 288–301 (2021). https://doi.org/10.1002/ima.22499
4. Alim-Ferhat, F., Mohammed, L.A., Abdelaziz, M.: Integration of radiographic and histological images for the diagnosis of glioblastoma. Int. J. Biomed. Eng. Technol. **43**(4), 329–337 (2023). https://doi.org/10.1504/IJBET.2023.135395
5. Pei, L., Vidyaratne, L., Hsu, W.-W., Rahman, M.M., Iftekharuddin, K.M.: Brain tumor classification using 3D convolutional neural network. In: Crimi, A., Bakas, S. (eds.) Brainlesion: glioma, multiple sclerosis, stroke and traumatic brain injuries. Lecture Notes in Computer Science, pp. 335–342. Springer International Publishing, Cham (2020). https://doi.org/10.1007/978-3-030-46643-5_33
6. Farahani, K., et al.: Computational Precision Medicine Radiology-Pathology challenge on Brain Tumor Classification 2020 (2020). https://doi.org/10.5281/zenodo.3718894
7. Rathore, S., Iftikhar, M.A., Gurcan, M.N., Mourelatos, Z.: Radiopathomics: Integration of radiographic and histologic characteristics for prognostication in glioblastoma (2019). arXiv: arXiv:1909.07581. https://doi.org/10.48550/arXiv.1909.07581
8. Haralick, R.M., Shanmugam, K., Dinstein, I.: Textural features for image classification. IEEE Trans. Syst. Man Cybern. **SMC-3**(6), 610–621 (1973). https://doi.org/10.1109/TSMC.1973.4309314
9. Tang, X.: Texture information in run-length matrices. IEEE Trans. Image Process. **7**(11), 1602–1609 (1998). https://doi.org/10.1109/83.725367
10. Ma, X., Jia, F.: Brain tumor classification with multimodal MR and pathology images. In: Crimi, A., Bakas, S. (eds.) Brainlesion: glioma, multiple sclerosis, stroke and traumatic brain injuries. Lecture Notes in Computer Science, pp. 343–352. Springer International Publishing, Cham (2020). https://doi.org/10.1007/978-3-030-46643-5_34
11. Goode, A., Gilbert, B., Harkes, J., Jukic, D., Satyanarayanan, M.: OpenSlide: a vendor-neutral software foundation for digital pathology. J. Pathol. Inform. **4**, 27 (2013). https://doi.org/10.4103/2153-3539.119005
12. Chan, H.-W., Weng, Y.-T., Huang, T.-Y.: Automatic classification of brain tumor types with the MRI scans and histopathology images. In: Crimi, A., Bakas, S. (eds.) Brainlesion: glioma, multiple sclerosis, stroke and traumatic brain injuries. Lecture Notes in Computer Science, pp. 353–359. Springer International Publishing, Cham (2020). https://doi.org/10.1007/978-3-030-46643-5_35
13. Xue, Y., et al.: Brain tumor classification with tumor segmentations and a dual path residual convolutional neural network from MRI and pathology images. In: Crimi, A., Bakas, S. (eds.) Brainlesion: glioma, multiple sclerosis, stroke and traumatic brain injuries. Lecture Notes in Computer Science, pp. 360–367. Springer International Publishing, Cham (2020). https://doi.org/10.1007/978-3-030-46643-5_36
14. Ke, R., Li, W., Cui, Z., Wang, Y.: Two-stream multi-channel convolutional neural network for multi-lane traffic speed prediction considering traffic volume impact. Transp. Res. Rec. **2674**(4), 459–470 (2020). https://doi.org/10.1177/0361198120911052

Automated Medical Labels Detection and Text Extraction Using Tesseract

Imène Ait Abderrahim(✉) and Youcef Benrached(✉)

Khemis Miliana University, Ain Defla, Algeria
{i.aitabderrahim,mi19.y.benrached}@univ-dbkm.dz

Abstract. Medical labels are widely employed in the field of pharmacology to convey critical information about medications, including their names, prices, and other relevant details. In Algeria, the absence of an approved standardized model by the Minister of Health has resulted in the lack of automated identification systems for medical labels. Tesseract has proven to be highly effective in OCR and text extraction tasks, making it a suitable choice for this application. This study introduces an artificial intelligence-based method designed to automatically recognize and extract information from medical labels using Tesseract. Experimental results demonstrate the overall effectiveness of the system, with Tesseract proving particularly effective in text extraction tasks from medical labels images.

Keywords: Image Processing · Text Extraction · Tesseract · Optical Character Recognition (OCR) · LSTM · Medical labels · Healthcare Data Management

1 Introduction

In the contemporary digital age, data is predominantly processed and managed automatically, yet Algerian pharmacies have not fully leveraged this potential regarding medication labels. Currently, these labels are used primarily to identify medication names, while barcodes serve as the primary management tool for selling medicines. The submission of medication information to the National Social Insurance Fund (CNAS) involves attaching medication stickers to the back of prescription sheets, followed by manual verification of the labels against the delivery note (le bordereau). This manual process is time-consuming and prone to errors, prompting the question of how to automate the entire workflow from label detection to data extraction and verification.

Despite significant advancements in optical character recognition (OCR) technology, the recognition of medication labels remains unresolved. This project aims to develop an AI-based OCR system specifically designed for this purpose. By automating the detection and segmentation of information on medication labels, the content can be extracted and stored in a text file, which can then be automatically compared with other documents, such as the delivery note. This

S. Chikhi et al. (Eds.): MISC 2024, LNNS 1267, pp. 212–225, 2025.
https://doi.org/10.1007/978-3-031-82112-7_16

automation would streamline the interaction between pharmacies and CNAS, enhancing efficiency and accuracy in medication management. In conclusion, our proposed system leverages AI to fully utilize medication stickers, transforming the current manual process into an automated, efficient, and reliable system. This advancement not only addresses existing challenges but also sets the stage for future improvements in healthcare data management in Algeria.

2 Related Work

Medical labels in several countries provide general information about medications, such as the medicament's name, the number of capsules, the price, and the manufacturing company. However, there is no standardized format for this information. Some labels feature only a barcode, others combine a barcode with the medicament's name, and some display only the medication details. Due to this inconsistency, France opted to eliminate the use of medical labels in 2012, replacing them with barcodes that contain all necessary information[1]. Conversely, Algeria has its own format for medical labels. To enhance the utilization of this information, we propose leveraging advancements in computer vision and image processing to detect and recognize the content of medical labels automatically.

Computer vision operations, such as image identification and segmentation, are fundamental tasks that have recently experienced significant advancements. These tasks involve locating objects within an image, recognizing them, and segmenting them into meaningful sections. Various methods have been proposed for image detection and segmentation, ranging from traditional techniques to advanced deep learning approaches [4]. Traditional methods often involve the application of clustering or segmentation algorithms followed by the extraction of manually crafted features, such as edges, textures, and color information. While these methods have demonstrated reasonable performance, they are limited by their reliance on manually generated features, which may not adequately capture complex patterns [8,11,14,16].

Conversely, deep learning-based methods have emerged as the dominant paradigm for image segmentation and detection. Convolutional Neural Networks (CNNs) have revolutionized artificial intelligence by automatically learning hierarchical features from raw image inputs [9]. Among the prominent deep learning architectures for image identification and segmentation, the YOLO (You Only Look Once) framework stands out, attracting significant attention [12]. YOLO's single-shot detection method is highly efficient for real-time applications, directly predicting bounding boxes and class probabilities by dividing the image into a grid, demonstrating impressive accuracy and speed.

Instance segmentation approaches have also gained popularity for identifying distinct object instances within an image. Mask R-CNN [1], a well-known system, extends the Faster R-CNN object detection architecture with a branch

[1] https://sante.gouv.fr/soins-et-maladies/medicaments/professionnels-de-sante/suppression-de-la-vignette-pharmaceutique-questions-reponses-a-l-attention-des/.

for generating pixel-level segmentation masks, enabling precise instance segmentation crucial for tasks requiring object separation and counting.

Another critical area in computer vision and pattern recognition is text recognition from images, commonly known as optical character recognition (OCR). Text recognition aims to automatically extract and interpret textual information from images, facilitating the conversion of printed or handwritten text into machine-readable formats. This technology has numerous applications, including document analysis, information retrieval, digital archiving, and text-to-speech synthesis.

Text recognition techniques have significantly advanced due to developments in machine learning, deep learning, and image processing. Traditional approaches often combine preprocessing procedures, feature extraction, and classification algorithms. Preprocessing steps like image enhancement, noise reduction, and binarization improve input image quality. Feature extraction techniques, such as connected component analysis or stroke extraction, identify individual text components and their spatial relationships. Classification methods, including template matching or neural networks, are then used to recognize and interpret the extracted text. In recent years, deep learning approaches, particularly CNNs and Recurrent Neural Networks (RNNs) [2,5], have demonstrated remarkable performance in text recognition tasks [15].

Another prominent area of study in computer vision and pattern recognition is text recognition from photographs, sometimes referred to as optical character recognition (OCR). To enable the conversion of printed or handwritten text into machine-readable formats, text recognition aims to automatically extract and interpret textual information inherent within photographs. Document analysis, information retrieval, digital archiving, and text-to-speech synthesis are just a few of the many uses of this technology.

Techniques for text recognition have also advanced significantly on their turn as a result of developments in machine learning, deep learning, and image processing. Traditional approaches to text recognition frequently combined feature extraction, preprocessing procedures, and classification algorithms. To enhance the quality of the input image, preprocessing processes including image enhancement, noise reduction, and binarization are used. Methods for identifying individual text parts and their spatial relationships, such as linked component analysis or stroke extraction, are called feature extraction techniques. To detect and analyze the retrieved text, classification methods like template matching or neural networks are used. on the other hand, in recent years, deep learning approaches [2], particularly Convolutional Neural Networks (CNNs) and Recurrent Neural Networks (RNNs) [5], have shown remarkable performance in text recognition tasks [15].

In addition to these advanced algorithms, Tesseract-based methods remain important in the field of OCR. Tesseract, an open-source OCR engine first created by Hewlett-Packard and later upgraded by Google, takes a classical approach to text recognition. It uses adaptive thresholding, connected component analysis, and pattern matching to detect and recognize characters[13]. To detect and recognize characters, Tesseract uses a multistage approach that includes adaptive thresholding, linked component analysis, and pattern matching. Recent developments have integrated LSTM (Long Short-Term Memory) networks to handle line recognition, making it stronger for diverse text recognition tasks [13]. Tesseract has been shown to be effective in a variety of applications, including identifying text in historical texts and extracting information [6,7]. Tesseract's inclusion into broader image processing processes remains valuable, particularly when combined with standard techniques for preprocessing and segmentation [3]. This hybrid method guarantees that Tesseract remains a robust and adaptable tool for OCR jobs, supplementing the more complex deep learning-based systems.

3 Tesseract-Based Approach

In this section, we describe our OCR-based approach using the Tesseract from Meghatria et al. [10].

3.1 Dataset Collection and Organization

Given the absence of an Algerian Medical Labels dataset, we compiled an image dataset by photographing labels from a local pharmacy called "Dis Mebarek" in Khemis Miliana, Ain Defla, Algeria. This dataset consists of over 1,000 images of 300 different medications. The labels were photographed from various angles and positions, as illustrated in Fig. 1.

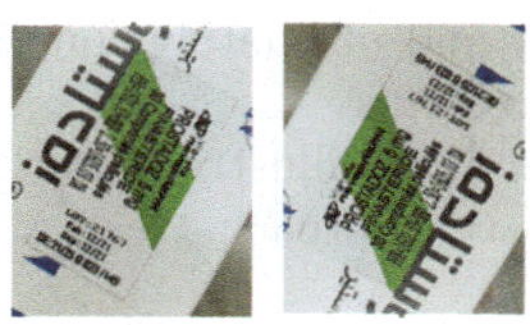

Fig. 1. Medical label Image

Following data collection, the labels were organized into a tabular format to facilitate later exploration as shown in Fig. 2). This table includes details such as the image ID, the number of textual sections on the medical label, its color (red, green, or white), and the text content of each section.

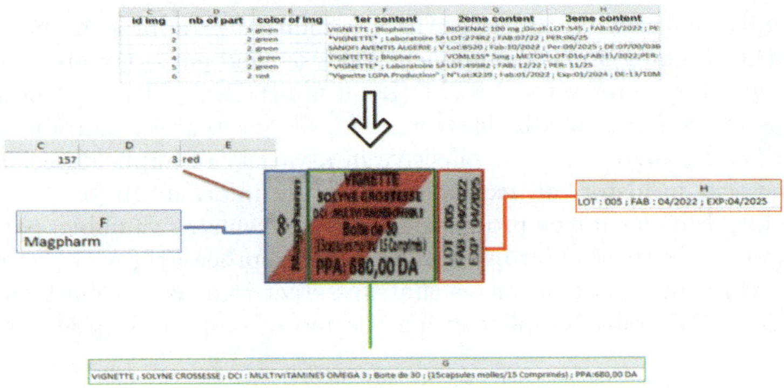

Fig. 2. Medical labels tabular documentation

3.2 Tesseract-Based Approach

The Tesseract-based method begins with the manual identification of the four corners of the medical label and proceeds using a traditional semi-intelligent approach. Histogram analysis is subsequently employed to detect the lines of text within these sections, as well as their vertical and horizontal components. Character recognition algorithms are then utilized to extract meaningful information.

This technique leverages conventional detection and segmentation methods, such as histogram thresholding for image segmentation. The aim is to achieve precise and reliable identification and segmentation of medical labels. The identified segments are subsequently passed to the Tesseract recognition module, which processes each line separately and ultimately compiles the lines to generate the recognized text from the image.

The following flowchart in Fig. 3 illustrates the steps involved in text recognition using the Tesseract-based approach.

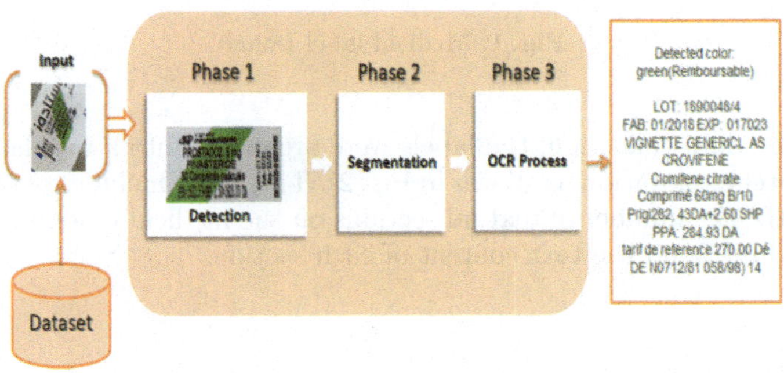

Fig. 3. Architecture of the Tesseract-based approach

Phase 1: Medical Label Detection

This phase constitutes the initial step of the Tesseract-based method. It involves manually selecting the four corners of the medical label by clicking with a mouse. After completing the selection, the system applies a corrective rotation to properly align the label.

Fig. 4. Phase1: Detection in Tesseract-based approach

Phase 2: Segmentation Phase

Medical labels contain a horizontal section with details like the manufacturer's name, medication name, dosage, and price, and a vertical section with the lot number, expiration date, and manufacturing date. To analyze and differentiate these sections, the image is converted to gray-scale. The gray-scale histogram is used to identify gaps, allowing for the segmentation of the image. Vertical sections are rotated counterclockwise for alignment, and text lines within these sections are separated using a horizontal histogram. The positions of the remaining segments are recorded and applied to the original color image for accurate segmentation. This process is repeated for each part of the label.

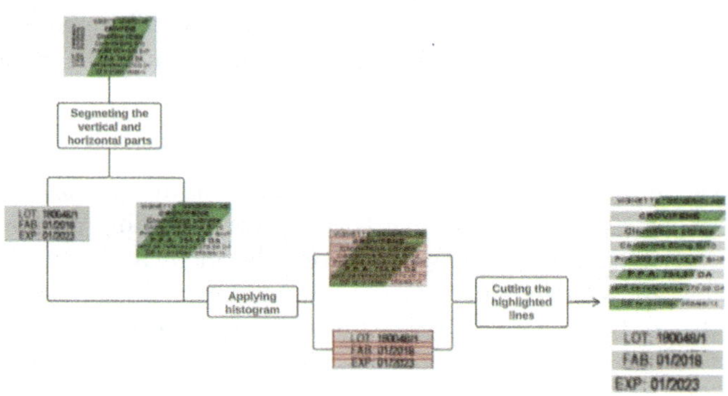

Fig. 5. Phase2: Segmentation in Tesseract-based approach

Phase 3: Character Recognition

In this stage, each line of text undergoes immediate optical character recognition (OCR) using the PyTesseract OCR model individually.

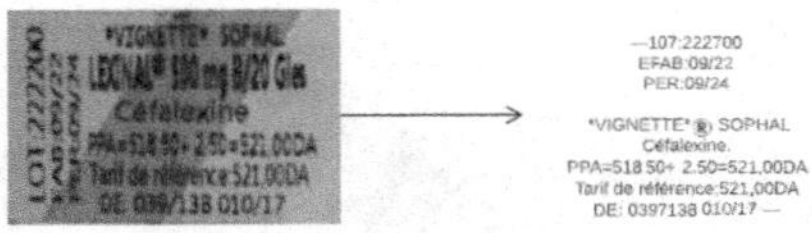

Fig. 6. Tesseract approach output

4 Proposed Fine-Tuned Tesseract Approach

One of the issues that the Tesseract approach in [10] encounters is that it does not correctly detect the text within the medical labels due to their bad printing quality and the lack of a standard model. In this section, we propose fine-tuning the model from [10] using a separate dataset that we manually created for this purpose of tuning. To do so, the fine-tuning process is mainly composed of two phases, the first phase consists of the data creation and the second phase refers to the fine-tuning process. Furthermore, to evaluate the quality of the model's performance where two metrics have been considered: *Accuracy* and *loss* (BCER). All the steps are described in more detail as follows:

4.1 Dataset Creation

Fine-tuning the data is a crucial step to enhance the performance quality of the Tesseract approach. In the context of Tesseract, combining text images, ground truth, and box files into a Long-Short Term Model (LSTM) file for fine-tuning refers to a specific process in preparing and optimizing the OCR model for better performance.

Data Preparation

In this process and as shown in Fig. 7, we first generate a dataset comprising images of text in *.tif* format along with their corresponding ground truth text files. This involves extracting text lines from the medical labels, creating text files containing the actual text for each line, and ensuring that each image and its associated text file share the same name. Then, in the next step, we would be creating the box files for each of these images using the ground truth text files which serve as the correct reference for the OCR system to learn from, see Fig. 8.

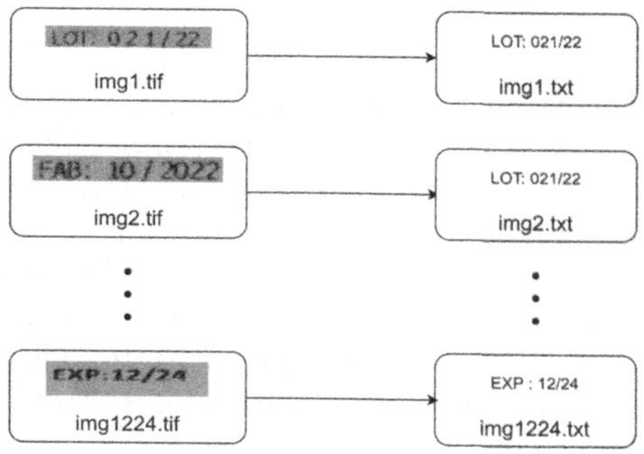

Fig. 7. Dataset preparation illustration example

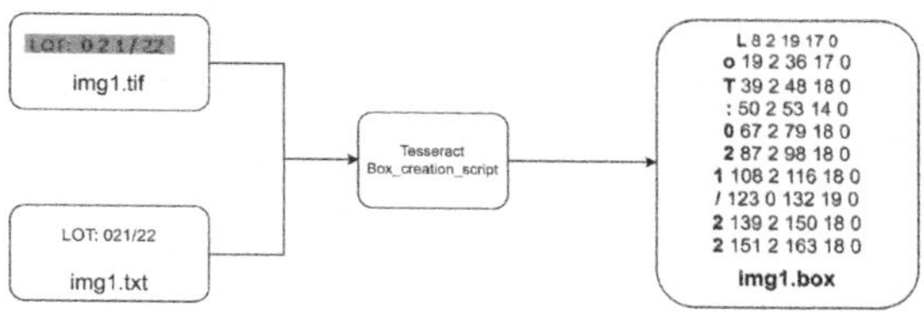

Fig. 8. Box File Creation

Training Data Creation

In this step, the text images, ground truth, and box files are combined into a training dataset that Tesseract's training process can use. This dataset is used to train the LSTM model on specific types of text images.

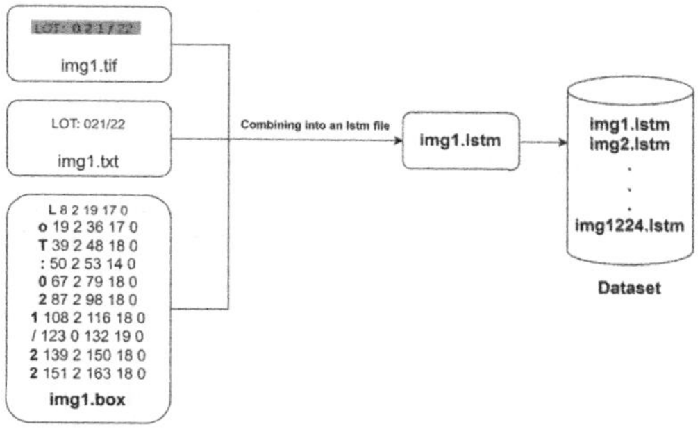

Fig. 9. LSTM File Creation

4.2 Fine Tuning Process

During fine-tuning, the existing OCR model is further trained on this prepared dataset. The process begins by extracting the pre-trained base LSTM neural network from the standard Tesseract OCR engine. This extracted LSTM serves as the foundation for our training process. Following the extraction, the model is provided with the *.traineddata* file, which contains the essential trained data required for the OCR engine to recognize text. This includes the neural network parameters, character set, and language-specific data. The *.traineddata* file can be accessed via the official Tesseract repository on GitHub[2]. In addition to the *.traineddata* file, we specified our custom dataset with the LSTM files, as seen in Fig. 10. This custom dataset is used to train and fine-tune the basic LSTM model.

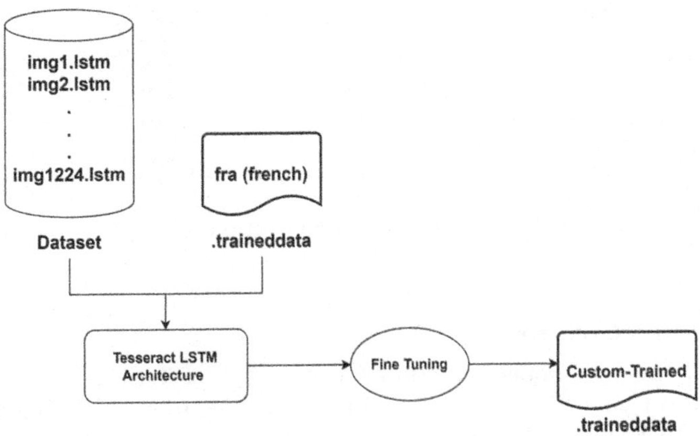

Fig. 10. Tesseract Fine Tune: Fine Tuning Process

4.3 Fine-Tuning Evaluation

The presented graphs illustrate the progress of the Tesseract OCR fine-tuning.

Training Accuracy Evaluation

The graphic illustrated in Fig. 11 tracks the model's accuracy improvement over 350 iterations. The x-axis represents the number of iterations, while the y-axis displays accuracy as a percentage. The graphic shows a steady increase in accuracy from approximately 80% to 85%. This upward trend indicates that the model is progressively learning and improving its ability to accurately recognize and interpret medical labels.

[2] https://github.com/tesseract-ocr/tessdata_best/blob/main/fra.traineddata.

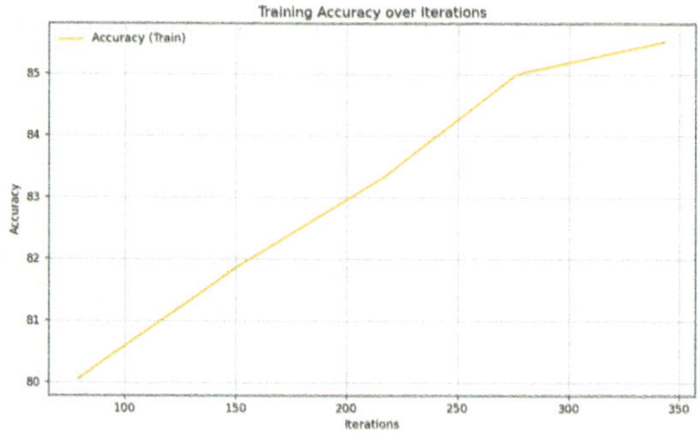

Fig. 11. Accuracy Graph during Fine-Tuning

Training Loss (BCER) Evaluation

The graphic in Fig. 12 illustrates the reduction in the model's training loss over a series of iterations. The x-axis represents the number of iterations, while the y-axis displays the Binary Cross-Entropy Loss (BCER). BCER is a common loss function for binary classification problems, measuring the performance of a classification model whose output is a probability between 0 and 1. It shows a consistent decrease in BCER from approximately 20 to 15. This downward trend in training loss indicates that the model's predictions are becoming more accurate and better aligned with the actual labels, reflecting effective learning from the training data.

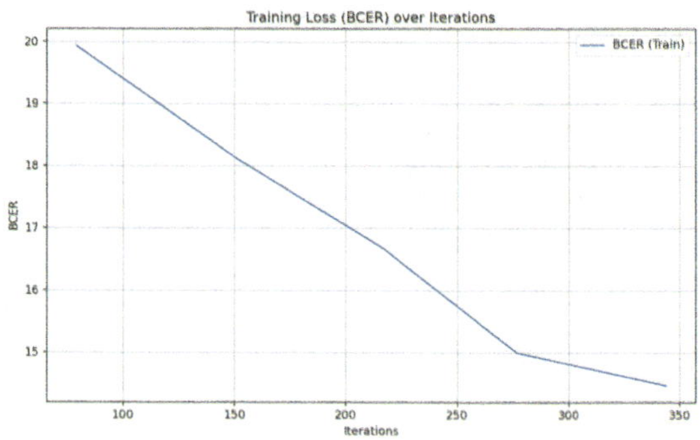

Fig. 12. Accuracy Graph during Fine-Tuning

Upon completion of the fine-tuning process, the updated Tesseract *.traineddata* file, now better adapted to the specific characteristics of the text in the medical labels, is saved. This refined model is then compiled into a new *.traineddata* file, which integrates the original base model data with the newly acquired features. The resulting *.traineddata* file is now ready for use by Tesseract for

OCR tasks, offering enhanced accuracy and reliability tailored to the specific use case for which it was fine-tuned.

5 Experimental Results

Preliminary tests on 40 images using the standard Tesseract OCR model, without fine-tuning, showed promising performance overall. Despite this, some letters were misrecognized, particularly those with distinctive features like the initial letters in medication names, which are often decorated with large capital letters. Additionally, characters with poor printing quality posed recognition challenges. These minor discrepancies provide valuable insights for future refinement and optimization of the model. Therefore, Fine-tuning the standard Tesseract OCR model was necessary to improve its character recognition accuracy in contexts where it initially struggled. By training the model on a dataset specifically targeting issues such as decorated large capital letters and poor printing quality, recognition accuracy can be enhanced. The fine-tuned model aims to better accommodate these unique text characteristics and handle variability in text presentation more effectively. This process is expected to reduce misrecognition rates and enhance overall performance, particularly in scenarios where the standard model was insufficient.

For comparison, figure Fig. 13 illustrates the recognized texts of the same medical label using each of the Tesseract models, standard and fine-tuned, respectively. This figure shows some visual contrast and assessment of the performance of each OCR engine.

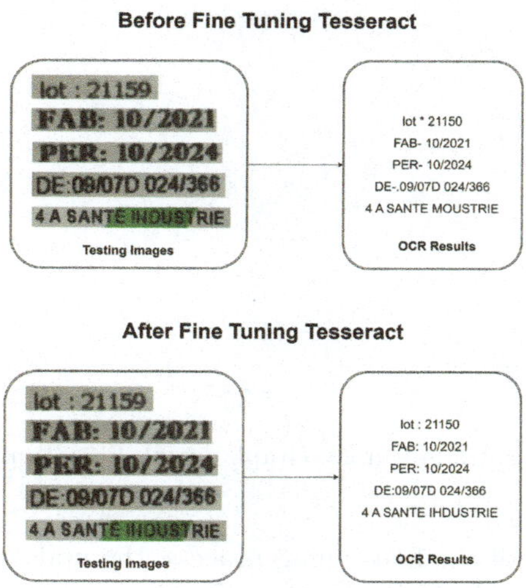

Fig. 13. Tesseract results comparison before and after the fine tuning

Additional experiments were conducted on forty new medical label images, distinct from those in the training and validation sets, to assess the effectiveness of the proposed approach using the standard model. The accuracy and reliability of the text identification were evaluated by comparing the recognized text to the original text on the medical labels, utilizing the following formula (1):

$$accuracy = \frac{NCCR \times 100}{TNC} \tag{1}$$

where NCCR stands for "Number of Characters Correctly Recognized" and TNC stands for "Total Number of Characters in the medical label".

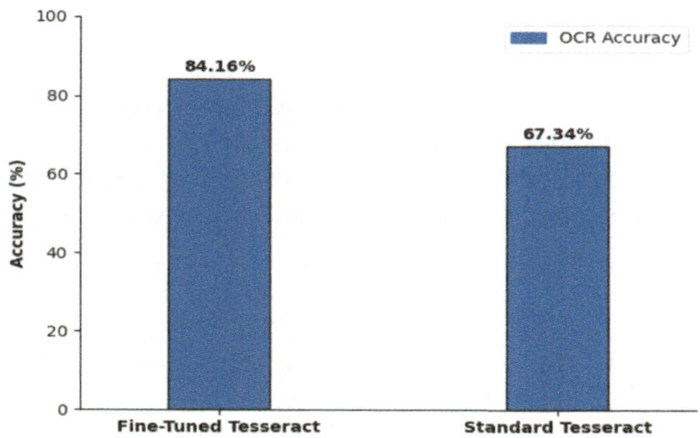

Fig. 14. Comparison study results

As shown in Fig. 14, the standard Tesseract OCR achieved a text recognition accuracy of 67.34% on 40 medical labels, which, while acceptable, suggests a need for further improvement. In contrast, the fine-tuned Tesseract model attained a significantly higher accuracy of 84.16% on the test set. This improvement highlights that fine-tuning enhances the OCR model's ability to accurately recognize text from medical labels and adapt to specific text characteristics. The results underscore the value of customized training for achieving greater accuracy and reliability in real-world applications.

These performance outcomes from the standard Tesseract model can be attributed to several factors. Primarily, the use of the Standard Tesseract OCR model in its default configuration, without any fine-tuning or additional training, impacted the results. The Standard Tesseract version 4.0 employs a neural network for text detection but lacks specialized customizations beyond its general training.

Conversely, the superior performance of the Fine-Tuned Tesseract is due to its targeted training on a specialized dataset designed to address the specific challenges of recognizing medical labels. This fine-tuning process customizes the OCR model to better handle unique text characteristics and contexts, leading to significantly improved accuracy. The enhanced results highlight the advantages of customizing OCR models for specific text recognition tasks, providing more reliable and precise outcomes compared to the standard model.

6 Conclusion

This study presents an AI-based system for text extraction from medical labels in Algeria, addressing the gap in standardized recognition systems. By evaluating a Tesseract-based approach and a fine-tuned variant of the Tesseract approach solutions, our experiments reveal that the fine-tuned Tesseract model achieved a notable accuracy of 84.16% in label detection and content extraction. This performance improvement underscores the benefits of customizing OCR models for specific text characteristics. The fine-tuned Tesseract model offers significant advantages for pharmacies, insurance providers, and CNAS by enhancing medication handling and streamlining procedures. The research demonstrates the potential of AI technologies to improve healthcare outcomes nationwide. Future work will focus on further fine-tuning Tesseract and exploring additional AI approaches to advance medical label detection and text extraction.

References

1. He, K., Gkioxari, G., Dollár, P., Girshick, R.: Mask r-cnn. In: Proceedings of the IEEE International Conference on Computer Vision, pp. 2961–2969 (2017)
2. Hossain, M.Z., Sohel, F., Shiratuddin, M.F., Laga, H.: A comprehensive survey of deep learning for image captioning. ACM Comput. Surv. (CsUR) **51**(6), 1–36 (2019)
3. Kaundilya, C., Chawla, D., Chopra, Y.: Automated text extraction from images using ocr system. In: 2019 6th International Conference on Computing for Sustainable Global Development (INDIACom), pp. 145–150. IEEE (2019)
4. Khan, W.: Image segmentation techniques: a survey. J. Image Graph. **1**(4), 166–170 (2013)
5. Koutnik, J., Greff, K., Gomez, F., Schmidhuber, J.: A clockwork rnn. In: International Conference on Machine Learning, pp. 1863–1871. PMLR (2014)
6. Kumar, S., Sharma, N.K., Sharma, M., Agrawal, N.: Text extraction from images using tesseract. Deep Learn. Tech. Autom. Ind. Appl. 1–18 (2024)
7. Kumar Garai, S., Paul, O., Dey, U., Ghoshal, S., Biswas, N., Mondal, S.: A novel method for image to text extraction using tesseract-ocr. Am. J. Electron. Commun. **3**(2), 8–11 (2022)
8. Lakshmi, S., Sankaranarayanan, D.V., et al.: A study of edge detection techniques for segmentation computing approaches. IJCA Special Issue on "Computer Aided Soft Computing Techniques for Imaging and Biomedical Applications" CASCT, pp. 35–40 (2010)
9. LeCun, Y., et al.: Handwritten digit recognition with a back-propagation network. Adv. Neural Inf. Process. Syst. **2** (1989)
10. Meghatria, R., Ait Abderrahim, I.: Detection and recognition approaches for medical labels in Algeria. In: 2024 8th International Conference on Image and Signal Processing and their Applications (ISPA), pp. 1–6 (2024). https://doi.org/10.1109/ISPA59904.2024.10536801

11. Nadernejad, E., Sharifzadeh, S., Hassanpour, H.: Edge detection techniques: evaluations and comparisons. Appl. Math. Sci. **2**(31), 1507–1520 (2008)
12. Redmon, J., Divvala, S., Girshick, R., Farhadi, A.: You only look once: unified, real-time object detection. In: Proceedings of the IEEE Conference on Computer Vision and Pattern Recognition, pp. 779–788 (2016)
13. Smith, R.: An overview of the tesseract ocr engine. In: Ninth International Conference on Document Analysis and Recognition (ICDAR 2007), vol. 2, pp. 629–633. IEEE (2007)
14. Succi, A., Torre, V.: A new approach to image segmentation. In: Image Analysis and Processing: 8th International Conference, ICIAP'95 San Remo, Italy, September 13–15, 1995 Proceedings 8, pp. 17–22. Springer (1995)
15. Ye, Q., Doermann, D.: Text detection and recognition in imagery: a survey. IEEE Trans. Pattern Anal. Mach. Intell. **37**(7), 1480–1500 (2014)
16. Zaitoun, N.M., Aqel, M.J.: Survey on image segmentation techniques. Procedia Comput. Sci. **65**, 797–806 (2015)

Mobile AR for Learning Skeleton Anatomy in Algerian Undergraduates

Mohamed Amine Guerroudji[1(✉)], Samir Benbelkacem[1], Leila Boudine[2], Kahina Amara[1], Samya Taright[2,3], Nesrine Toubaline[4], Imene Issolah[4], Rayane Gacem[4], Nadia Zenati[1], Fatima-Zohra Touia[2], Mostefa Masmoudi[1], and Ali Benkacimi[3]

[1] Center for Development of Advanced Technologies, Baba Hassen, Algiers, Algeria
mguerroudji@cdta.dz
[2] Faculty of Medicine, Benyoucef Benkhedda University of Algiers 1, Algiers, Algeria
[3] CHU Mustapha Bacha, Algiers, Algeria
[4] LRDSI Laboratory, Computer Science Department, Saad Dahlab University - Blida 1, Blida, Algeria

Abstract. The field of augmented reality (AR) is currently booming, encompassing various areas including medicine and diagnostic assistance. AR allows the development of three-dimensional virtual environments for human interaction. This work aims to design and implement an AR-based system to well support medical undergraduate students in learning and training skeleton human anatomy, conforming to Algeria's medical school requirements. Students, through their mobile devices, can experience various parts of the human anatomy, like the clavicle and scapula, in a realistic and engaging way by superimposing 3D anatomical models on the skeleton. With the help of our suggested system, undergraduate students at Algerian medical faculties will be able to manipulate virtual anatomical models, zoom, rotate, and examine them from various angles without having to directly access cadavers or traditional learning resources which are frequently unavailable due to the large number of undergraduate students which in some faculties exceeds a thousand students. Tests and evaluations were conducted to measure the effectiveness and usefulness of our AR learning system on mobile devices.

Keywords: 3D Model · Augmented Reality · Scapula · Clavicle · Human Skeleton Anatomy

1 Introduction

Augmented reality (AR) is experiencing rapid growth and finding applications in various domains, including medical training [3] and diagnostic aid [1]. Over the previous 20 years, there have been significant changes to anatomy education at the undergraduate level. This is due to a number of issues, such as cost and the difficulty of accessing human cadaver dissection. According to a recent study

S. Chikhi et al. (Eds.): MISC 2024, LNNS 1267, pp. 226–239, 2025.
https://doi.org/10.1007/978-3-031-82112-7_17

examining medical school anatomy laboratory pedagogies [4], computer-based and alternative teaching strategies are equally effective at producing similar results as dissection in anatomy classes. Thus, learning spatial knowledge can be facilitated by using three-dimensional anatomical visualizations than by two-dimensional images. Numerous studies on the application of AR technology in education revealed that the technology has a favorable impact on learning [13, 21].

Prior to beginning our proposal, we have comprehensively surveyed works relevant to our research, which are summarized in a comparative table (see Table 1). The comparison criteria used for this synthesis include : (1) platform, (2) anatomy, (3) software, (3) Marke/Markerless interaction, and (4) interactivity, based on the study conducted in [6,7], and [11]. We also included additional criteria, which reflected the expectations that clinicians had expressed: (5) AI, (6) captioning, (7) audio SDK, and (8) evaluation.

Table 1. Summary of work on AR for education in the literature.

	Date	Platform	Anatomy	Software	AI	Marker/ Markerless	Interaction	SDK
Kurniawan & al. [10]	2018	Mobile	Internal & External Organs	Unity	No	Marker	Yes	Vuforia
Jamali & al. [5]	2015	Mobile	Heart, Skeleton	Unity	No	Marker	Yes	Vuforia, ARToolkit
Bakar & al. [9]	2021	Mobile	Gross anatomy	Unity	No	Markerles	Yes	VuForia, ARToolKit ,Wikitude
Andayani & al. [12]	2019	Web, Mobile, Desktop	Lung Anatomy	/	No	Marker	No	Vuforia
Rita & al. [7]	2018	Site Web/TV Applications	Lung, Larynx, Bronchus, Bronchioles, Alveoli	/	No	Marker	Yes	/
Hossain & al. [16]	2021	Mobile	The Skeletal System	Unity	Markerles	Yes	Yes	EasyAR
Amara & al. [15]	2023	Smartphone	Brain tumour	Unity	phantom head	/	Yes	Vuforia
Loucif & al. [17]	2019	Mobile	Heart, Lungs, Liver, Stomach And Kidney	Unity	Natural Marker	Yes	Yes	/
Basurco & al. [18]	2022	Tablet/ Hololens 2	Skull	Unity	Natural Marker	/	Yes	Vuforia
Guerroudji & al. [22]	2024	Smartphone	Brain	Unity	phantom head	/	Yes	Vuforia

Table 1 leads us to the following conclusion:

- The reviewed works have successfully enhanced the understanding of complex anatomical subjects. However, they share a common limitation in terms of the narrow anatomical range studied. According to the platform criteria, three distinct types of platforms stand out: mobile applications, web applications, and HoloLens 2 tablets/headsets.

- Mobile applications are widely appreciated for their accessibility, convenience, flexibility, and interactive experience. Some applications have added additional features, such as supplementary resources (text and audio) and multilingual support (e.g., Arabic and English) as seen in [17], or self-assessment quizzes to test user memory based on the application's content, as noted in [5]. However, these applications face limitations in terms of technical requirements and software capabilities, as highlighted by [5], and [8]. Web applications are praised for their cost-effectiveness and visual enhancements but suffer from internet dependency and complex interactions compared to mobile applications. HoloLens 2 tablets/headsets excel in spatial understanding and real-time visualization but are expensive and complex, requiring specialized equipment and expertise for implementation.
- Each platform type has its strengths and weaknesses, and the choice depends on specific needs regarding accessibility, interactivity, cost-effectiveness, and visual quality.

The purpose of this research is to improve knowledge about the human skeleton for Algerian undergraduate students through the utilization of augmented reality, this, by experiencing clavicle and scapula bones in a first stage. We designed a dedicated AR learning system which principally overlay 3D anatomical bone data on the human skeleton, enabling students to experience a realistic and immersive skeleton models, without the need for direct access to cadavers or traditional, costly teaching tools. We aim to improve student understanding and learning while reducing cognitive resource demands for computerized medical education in Algeria. Thus, our proposed AR learning system will be integrated into a mobile application accessible to all students and teachers. Our choice was made following discussions with researchers and university hospital academics as well as other members of the national research project team, who emphasized the need for an accessible and cost-effective end-product. The remaining part contained in the present paper is organized as follows. After briefly reviewing the related works on the anatomy training with AR in Sect. 1, Sect. 2 presents our proposed method. Section 3 describes our developed training environment, while Sect. 4 provides preliminary results. Finally, we highlight the results obtained in Sect. 5.

2 Materials and Method

The outline of the proposed AR skeleton visualization and interaction method is described in Fig. 1. Students and teachers can access our application which via a camera to scan the environment. Then, the 3D model of the whole skeleton was rendered and displayed in mobile devices. Users have the possibility to visually interact, explore and manipulate each bone and obtain detailed information in 3D forms.

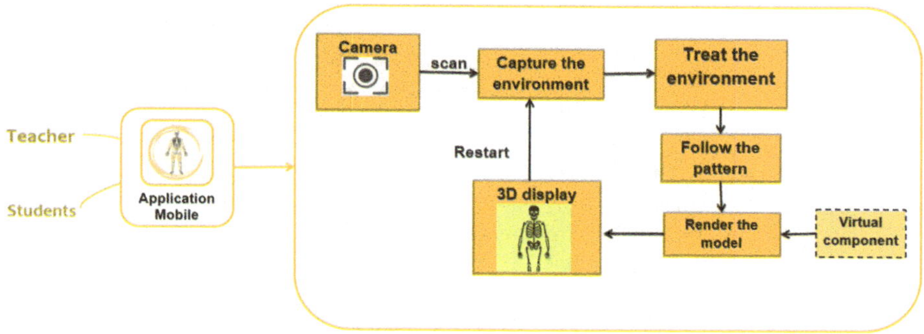

Fig. 1. Architect of the proposed solution.

In our case study, we experienced our method on learning scapula and the clavicle, utilizing teaching materials supplied by Algerian medical faculty members.

2.1 Analyse of Requirements

In the first stage, we defined the requirements of our system and identified the actors in the system.

Functional Requirements. We have conducted a variety of studies and interviews that could help us in more accurately determining the features of our system. We gathered information from the clinicians and specialists at Mustapha Pasha Hospital as well as the professors of the Algiers Faculty of Medicine. Relevant data about body skeleton, like clavicle and scapula, was provided by the laboratory of Clinical Anatomy from University Alger 1. Our goal is to accurately determine how anatomical knowledge should be incorporated into our system. Our goal is to precisely pinpoint the most effective approach for integrating anatomical data into our system and to provide direction to elaborate our own evaluation procedure. Undergraduate students and medical teachers can use their mobile smartphone to be able to visualize and interact with 3D bone models.

Identification of Actors. An external actor is a system user, another system, or any other entity that interacts with the system and has the ability to view or change its state. Three actors (teachers, students, and administrators) have been identified as potential users of our system. The role of each actor is described in Table 2.

Table 2. Role of system actors.

Actor	Role
Administrator	- Add an account for a user. - Edit an account for a user. - Delete a user account
Students	- Authenticate. - Scan the environment. - Select the plant. - Insert 3D model. -Manipulation of the 3D model (Rotation, Scale, Move). - Selecting a component of the 3D object (Highlight). - Inspection of the selected component (Rotation, Scale, Move, read information, Listen to information)
Teacher	- Authenticate. - Scan the environment. - Select the plant. - Insert 3D model. - Manipulation of the 3D model (Rotation, Scale, Move). - Selecting a component of the 3D object (Highlight). - Inspection of the selected component (Rotation, Scale, Move, read information, Listen to information). - Edit content

Technical Requirements From a technical point of view, we separate the requirements into two categories: software and hardware needs. The software requirements include:

- An Unreal Engine 3D game graphics engine.
- ARCore plugin for Unreal enginer.
- A C++ and Bleuprint development environment.
- Playfab plugin.

Unreal Engine: Unreal Engine of Epic Games is a powerful and efficient system for developing high-quality interactive experiences, video compilations, simulations, architectural visualizations, and excellent real-life applications [2]. These advanced graphics capabilities, blueprints, and C++ programming allow you to develop advanced environments and create new mechanics. Unreal Engine proposes new interfaces, documentation, didactic files, and a local community.

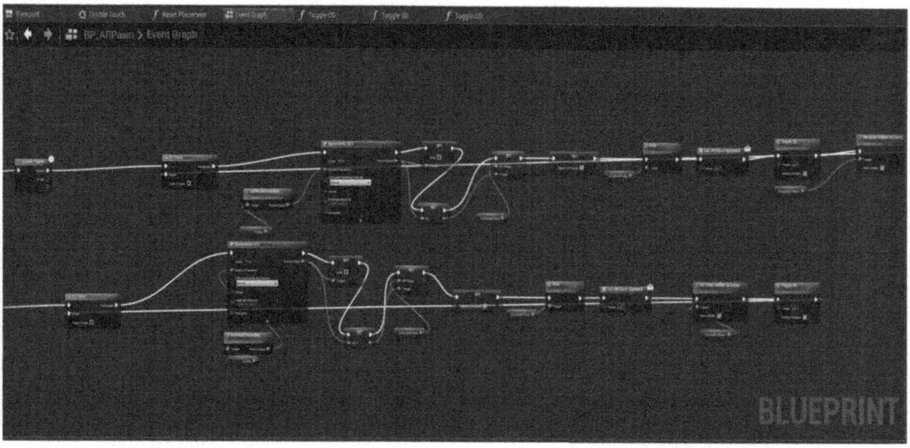

Fig. 2. Blueprint script example.

Blueprint: is a programming language based on visual scripting [19, 20] to define classes or objects in the engine. Figure 2 show an example of Blueprint.

On the other hand, we identified material requirements in order to take into account the software guidelines mentioned above:

- Smartphone and/or tablet supporting AR Library.
- Gamer type workstation.
- Wireless Internet network.

Therefore, we implemented our system and conducted experiments using a gamer workstation and smartphones. Their specifications are listed in Tables 3.

Table 3. Characteristics of the device used.

Device	Processor	Ram	GPU
Station Acer Nitro	Intel (R) Core^{TM} i9-10900 GHz CPU @ 2.80 GHz 2.81 GHz	64.00 GB 3600 MHz	-NVIDIA Ge Force RTX3080 Ti32GB -Dedicated GPU 12 GB
Redmi note 9 (M2003J15SG)	Octa-core (2 × 2.0 GHz Cortex-A75 & 6 × 1.8 GHz Cortex-A55)	4 GB	Mali-G52 MC2

3 Experimental Results

Authentication and Registration: Upon launching the application with a smartphone, users (students and teachers) are prompted to enter their email and password. When a new user hope experience the application, he/she navigate to the registration interface to obtain his/her own account. Figure 3 show the first window of the application' user interface.

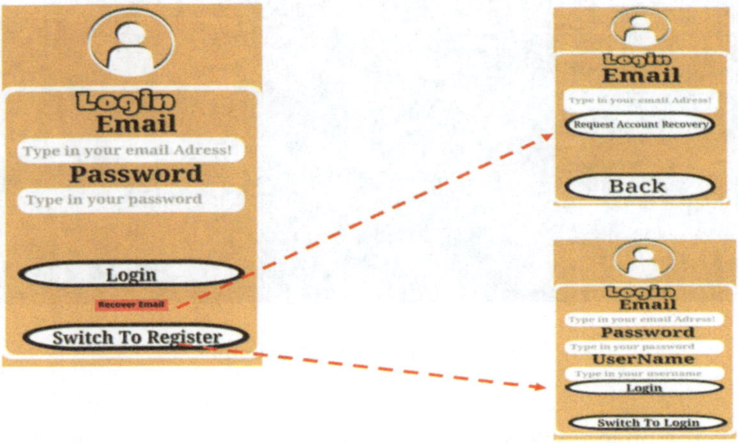

Fig. 3. Connection interface.

After authentication, the the user is directed to the main menu as shown in Fig. 4.

Fig. 4. Main Menu Interface.

The smartphone's camera is activated to capture the surrounding environment. The environment analysis process is then started. Upon analysing the environment, the application highlights suitable spaces for placing the 3D model (Fig. 5).

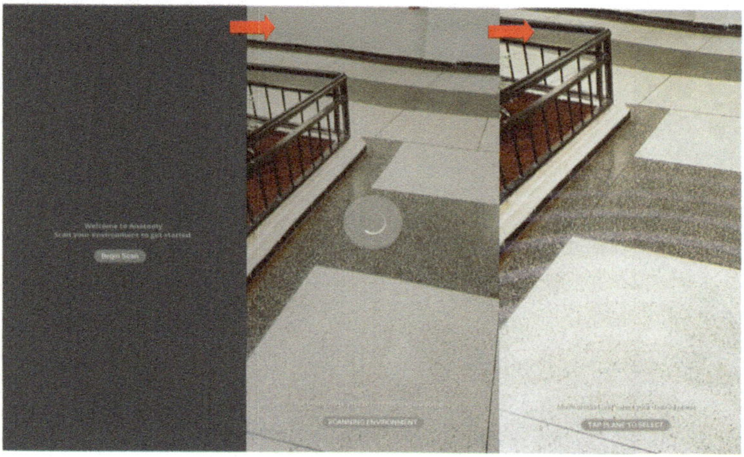

Fig. 5. Scan the environment.

The user click on the smartphone to display the skeletal structure (Fig. 6).

Fig. 6. Inserting the 3D model into the environment.

To enhance user interaction, the application includes an interactive menu that features a clear visual tutorial on basic manipulation gestures for virtual objects (Fig. 7).

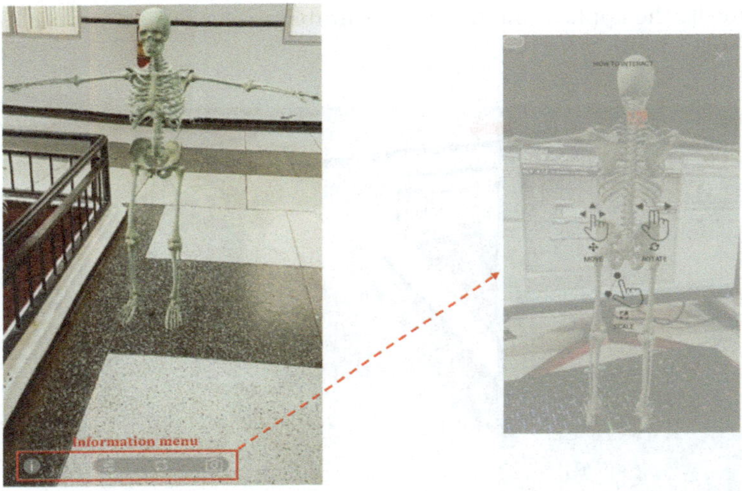

Fig. 7. Information menu.

Once the model is displayed, users can select a component for better visualization and interaction. Double-clicking on a visualized 3D organ provides text and audio support for enhanced understanding (Figs. 8).

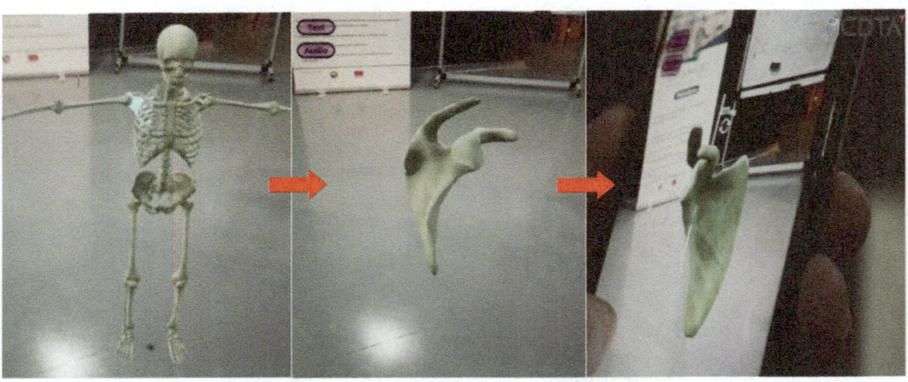

Fig. 8. Selection of a 3D Component.

If the user clicks the "Option" button on the main menu, several alternatives are provided (Fig. 9), offering choices between axial, appendicular, or zonal skeleton 3D views.

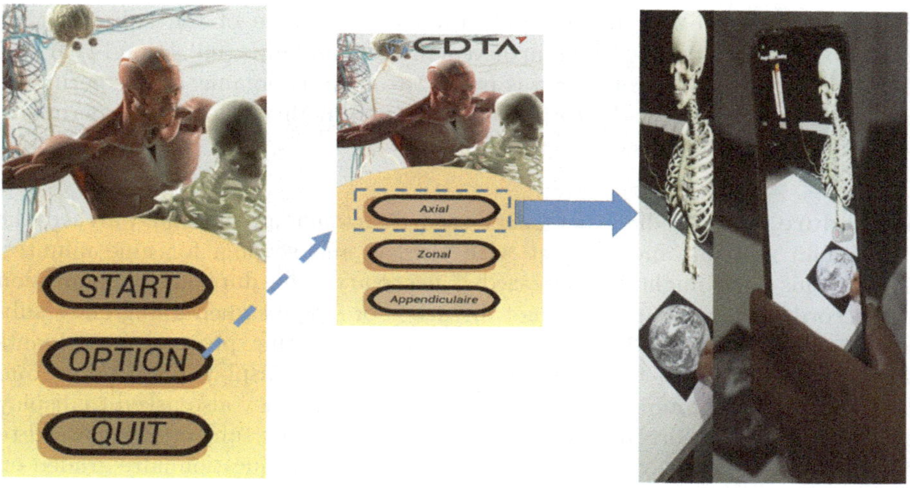

Fig. 9. Selection of a 3D Component.

We use ARCore SDK to support the tracking module in augmented reality, managing 3D user tracking for appropriate display of the skeletal structure of the human body in the real world.

3.1 Experimental Evaluation

This step is crucial to measure the usability and effectiveness of our application. We invited potential users (also called subjects) to interact with the interface we developed and provide feedback on their experience. Usability is a key element to measure the performance of classic and modern Human-Machine Interfaces (HMIs). We chose a well-known usability evaluation method called "System Usability Scale" (SUS) [22] to collect qualitative and quantitative data and identify interaction, ergonomics and user-friendliness issues. This method also provides the strengths and weaknesses of our interface, make improvements and offer an optimized application.

SUS method provide a set of scores which classify the performance of the system or application evaluated.

- If SUS score is 80.3% or higher, it means that users like the app and would recommend it to their friends.
- If SUS value of 68 % or higher, indicates that although the app is good, some improvements are still needed.
- A SUS value below 68% indicates that the app should be prioritized to be accurate.

Participants. We tested our application on a sample of ten subjects (N= 10) who are students (one (01) fourth-year student and nine (09) second-year students) in medicine. A professor was present to help us organize the experiment. Some subjects already had ideas about augmented reality before the experiment, while others had no idea.

Procedure. The medical personnel used their smartphone and watched the 3D human skeleton and interact with clavicle and scapula for nine minutes, though this time might be increased if necessary. The duration of immersion was recorded. Participants were not disturbed or swayed when filling out a subjective questionnaire following the event. In the testing process, participants were required to scan the surrounding environment, display 3D skeleton and then visualize and interact with clavicle and scapula. They also asked to display the 3D data of the corresponding two bones. Finally, the subjects were asked to complete a short survey about their experiences and a questionnaire graded on a seven-point Likert scale when the trials were finished.

- (Q1) How easy was it for you to use the Anatomy app navigation and other features?
- (Q2) I think I will need help from a technician to be able to use this application.
- (Q3) How satisfied are you with the app?
- (Q4) How well does the application meet your specific needs and goals?
- (Q5) Is the AR app design aesthetically appealing and appealing?
- (Q6) In your opinion, does Anatomy offer better accessibility and flexibility for medical learning?

According to the results of questionnaire, *SUS* score was calculated throuigh the following steps:

- Step 1: First, the response should be checked so that it matches the following rating scale: (Strongly disagree, Disagree, Neutral, Agree and Strongly agree).
- Step 2: We calculated an average score. To get the average for each question, all the participants' choices for that question must be added together and then divided by the number of participants.
- Step 3: We calculated the converted score where it is divided into two parts:
  - For questions number 1, 3, 5, 7, 9, 11, 13 and 15 (odd number), the converted score is the average score minus 1.
  - For statements number 2, 4, 6, 8, 10, 12 and 14 (even number), the converted score is 5 minus the average score.
- Step 4: We multiplied the sum of the converted scores by 2.5 to obtain the overall SUS value.

We also calculated the standard deviation to assess the dispersion of the values of the data set from its average value. This approach was applied for each question to know the percentage of difference between the answers, whether they are large or small. The evolution of the corresponding scores is given in Figs. 10.

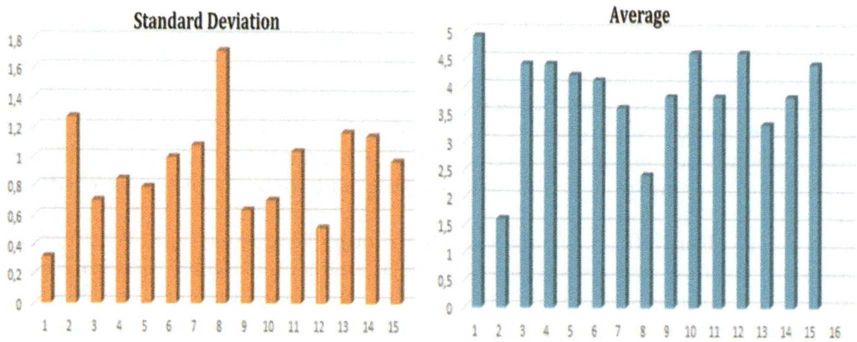

Fig. 10. Mean scores and standard deviation obtained for each question.

In our case, we obtained a SUS value of 84.75% (> 80%). According to SUS standards, we estimate that our application has a high degree of usability. The results of the subjective evaluation shows that our application is more fascinating, easy to understand and use. According to the subjects' opinions, the learning process helps students to visualize and interact with clavicle and scapula bones. Finally, we got suggestions for improving the application and adding additional features.

4 Conclusion

The development of a human anatomy learning system using augmented reality offers a novel, immersive learning experience for medical students and professionals, specifically in Algeria. By overlaying precise virtual anatomical models onto the real world, the system enhances interactive learning and accessibility, allowing users to study anatomy at their own place via their mobile devices. Our preliminary tests at the University of Algiers 1 showed that students found the application user-friendly and effective for visualizing and interacting with human bones. Although the system performed well, further improvements and broader implementation in educational settings are planned. During the academic year 2024/2025, we intend to use our application in a realistic working environment and measure the effectiveness of our application with a large number of undergraduate students from Faculty of Medicine Algiers 1.

Acknowledgments. This work is supported in parts by the DGRSDT (Direction Générale de la Recherche Scientifique et du Développement Technologique), and CDTA (Centre de Développement des Technologies Avancées) via PNR (Programmes Nationaux de Recherche), Algeria-(Project code: 13/132/DFPR/ATRSSV/22).

Disclosure of Interests. The authors declare that they have no conflict of interest.

References

1. Guerroudji, M.A., Amara, K., Zenati, N.: Augmented reality aid in diagnostic assistance for breast cancer detection. Multimed. Tools Appl. 1–14 (2024)
2. Amara, K., et al.: Augmented Reality localisation using 6 DoF phantom head Pose Estimation-based generalisable Deep Learning model. In: 2024 8th International Conference on Image and Signal Processing and their Applications (ISPA), pp. 1–6. IEEE (2024, April)
3. Amara, K., Kerdjidj, O., Atalla, S., Guerroudji, M.A., Mansoor, W., Zenati, N.: Enhancing brain tumour aid diagnosis with augmented reality rendering-based 6 dof object pose estimation. In: 2023 6th International Conference on Signal Processing and Information Security (ICSPIS), pp. 11–16. IEEE (2023, November)
4. Guerroudji, M.A., Amara, K., Lichouri, M., Zenati, N., Masmoudi, M.: A 3D visualization-based augmented reality application for brain tumor segmentation. Comput. Animat. Virtual Worlds **35**(1), e2223 (2024)
5. Jamali, S.S., Shiratuddin, M.F., Wong, K.W., Oskam, C.L.: Utilising mobile-augmented reality for learning human anatomy. Procedia. Soc. Behav. Sci. **197**, 659–668 (2015)
6. Guerroudji, M.A., Amara, K., Aouam, D., Zenati, N., Djekoune, O., Masmoudi, M.: Brain tumor segmentation on MRI using a GVF snake model. In: 2022 7th International Conference on Image and Signal Processing and their Applications (ISPA), pp. 1–5. IEEE (2022, May)
7. Layona, R., Yulianto, B., Tunardi, Y.: Web based augmented reality for human body anatomy learning. Procedia Comput. Sci. **135**, 457–464 (2018)
8. Chaker, R., Gallot, M., Binay, M., Hoyek, N.: User experience of a 3D interactive human anatomy learning tool. Educ. Technol. Soc. **24**(2), 136–150 (2021)
9. Bakar, W.A.W.A., Man, M., Solehan, M.A., Sabri, I.A.A.: GAAR: gross anatomy using augmented reality mobile application. Int. J. Adv. Comput. Sci. Appl. **12**(5) (2021)
10. Kurniawan, M.H., Witjaksono, G.: Human anatomy learning systems using augmented reality on mobile application. Procedia Comput. Sci. **135**, 80–88 (2018)
11. Guerroudji, M.A., Amara, K., Zenati, N., Ameur, Z.: Advancing brain tumor detection: a cutting-edge machine learning approach leveraging CAD systems. In: 2024 8th International Conference on Image and Signal Processing and their Applications (ISPA), pp. 1–6. IEEE (2024, April)
12. Andayani, U., Siregar, B., Pulungan, S.H., Syahputra, M.F., Muchtar, M.A., Arisandi, D.: A visualisation of 3D lung anatomy with augmented reality as interactive medical learning. In: Journal of Physics: Conference Series (Vol. 1235, No. 1, p. 012095). IOP Publishing (2019, June)
13. Guerroudji, M.A., Amara, K., Zenati, N.: Advancing mammography breast mass detection through diffusion segmentation. Multimed. Tools Appl. 1–14 (2024)
14. Guerroudji, M.A., Amara, K., Zenati, N.: HoloBrain: 3D low-cost mobile augmented reality rendering of brain tumour using the GVF snake model segmentation. Comput. Methods Biomech. Biomed. Eng. Imaging Vis. **11**(7), 2302051 (2024)
15. Amara, K., Guerroudji, M.A., Kerdjidj, O., Zenati, N., Ramzan, N.: HOLOTumour: 6DoF Phantom Head pose estimation based deep learning and brain tumour segmentation for AR visualisation and interaction. IEEE Sens. J. (2023)
16. Hossain, M.F., Barman, S., Biswas, N., Haque, A.B.: Augmented reality in medical education: AR Bones. In: 2021 International Conference on Computing, Communication, and Intelligent Systems (ICCCIS), pp. 348–353. IEEE (2021, February)

17. Loucif, S., Al-Rajab, M., Salem, R., Hesham, A., Mahely, D., Ajlouni, M.A.: Learning human anatomy using ARA mobile application. Int. J. Comput. Digit. Syst. **8**(6), 589 (2019)
18. Basurco-Reyes, D., Nuñez-Lopez, A., Sierra-Liñan, F., Zapata-Paulini, J., Cabanillas-Carbonell, M.: Mobile Application with Augmented Reality Focused on the Study of Human Anatomy (2022)
19. Blueprints. https://docs.unrealengine.com/4.27/enUS/ProgrammingAndScripting/Blueprints/GettingStarted/. Accessed 20 Apr 2023
20. Guerroudji, M.A., Hadjadj, Z., Lichouri, M., Amara, K., Zenati, N.: Efficient machine learning-based approach for brain tumor detection using the CAD system. IETE J. Res. **70**(4), 3664–3678 (2024)
21. Duncan-Vaidya, E.A., Stevenson, E.L.: The effectiveness of an augmented reality head-mounted display in learning skull anatomy at a community college. Anat. Sci. Educ. **14**(2), 221–231 (2021)
22. https://www.usability.gov/how-to-andtools/methods/system-usability-scale.html. Accessed 08 June 2023

MobileViT-Based Approach for Arrhythmia Classification Using ECG Scalograms

Hadjer Bechinia[1](✉), Djamel Benmerzoug[1], and Nawres Khlifa[2]

[1] LIRE Laboratory, University Abdelhamid Mehri Constantine 2, Constantine, Algeria
{Hadjer.Bechinia,Djamel.Benmerzoug}@univ-constantine2.dz

[2] Higher Institute of Medical Technologies of Tunis, Research Laboratory of Biophysics and Medical Technologies, University of Tunis el Manar, Tunis, Tunisia
nawres.khlifa@istmt.utm.tn

Abstract. Arrhythmias represent a significant category of cardiovascular disorders. Thus, making early diagnosis and accurate classification is essential for effective patient treatment. Electrocardiogram (ECG) signals, which monitor the heart's electrical activity, serve as the primary method for detecting arrhythmias. In this paper, we present a novel approach to arrhythmia classification by leveraging the image classification capabilities of MobileViT model. Our methodology involves converting ECG signals from the MIT-BIH Arrhythmia Database into scalograms using the Continuous Wavelet Transform (CWT), thereby capturing both time and frequency information. These ECG scalograms are then fed into the MobileViT model for classification into five distinct arrhythmia classes. To enhance the model's generalization capabilities, we employ the mixup data augmentation technique during the training. Our approach achieved an impressive accuracy of 98.97%, demonstrating the effectiveness of our approach in improving the automated early detection and classification of arrhythmias.

Keywords: ECG signals · Arrythmia Classification · Deep Learning · Computer Vision · Continuous wavelet transform

1 Introduction

Arrhythmias represent a significant category of cardiovascular disorders. They can occur independently or alongside other cardiovascular diseases [1]. Arrhythmia is a condition characterized by abnormal heart rhythms caused by irregular electrical activity and conduction within the heart. It can manifest as irregular heartbeats (heart tremor), tachycardia (a resting heart rate exceeding 100 beats per minute), or bradycardia (a resting heart rate below 60 beats per minute) [2]. Due to the high mortality rates associated with heart disease, early diagnosis and precise differentiation of arrhythmias are crucial for effective patient treatment.

S. Chikhi et al. (Eds.): MISC 2024, LNNS 1267, pp. 240–253, 2025.
https://doi.org/10.1007/978-3-031-82112-7_18

The most common method for detecting arrhythmias involves recording an electrocardiogram (ECG), which monitors the heart's electrical activity over time using electrodes placed on the skin. ECG leads, capturing the heart's electrical potential from various angles and positions, can indicate disease states through abnormalities in waveforms or rhythms [1].

An ECG of a single cardiac cycle includes three waves, two intervals, and two segments. These three significant waves represent the heart's distinct electrical activities during one heartbeat cycle, the P wave (atrial depolarization), the QRS complex wave (ventricular depolarization), and the T wave (repolarization). The durations of the two intervals, the PR interval and the QT interval, indicate the time required for the heart to complete the respective electrical changes [3]. Figure 1 illustrates the components of a standard heartbeat.

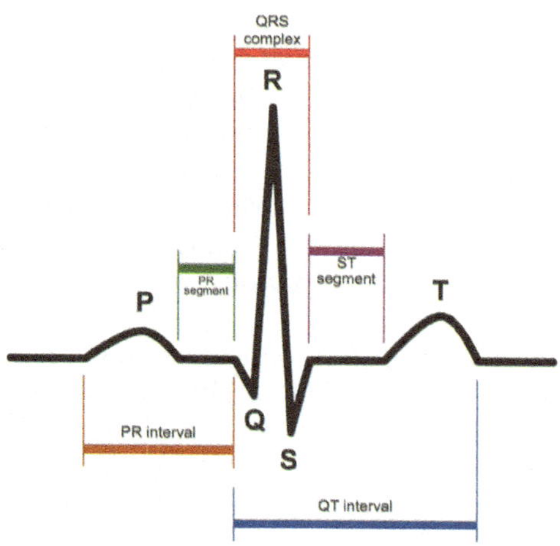

Fig. 1. Normal ECG Signal Waveform

As a result, numerous cardiac abnormalities exhibit characteristic patterns on ECG signals, making their identification crucial for diagnosing heart conditions. In 2019, cardiovascular diseases were responsible for 32% of all deaths worldwide according to the World Health Organization (WHO), highlighting their significant global impact. Therefore, early detection of patients at risk, continuous monitoring of those diagnosed, and a deeper understanding of the underlying disease mechanisms are crucial for improving diagnosis and treatment strategies. [4].

Over the past few years, Deep Neural Networks (DNNs) have become increasingly prevalent for classification and prediction tasks across various domains. Recently, significant advancements in DNNs have markedly improved classification accuracy in numerous medical applications such as [5] and [6]. Modern Computer-Aided Diagnostic Systems (CADS) utilize DNNs to detect arrhythmias from ECG signals, reducing the cost of continuous heart monitoring and enhancing prediction quality [7]. Among the various recent deep learning models, the Vision Transformer (ViT) stands out as a highly effective model for image classification. Leveraging the power of transformer architecture, originally designed for natural language processing, ViT divides images into patches and processes them as sequences, enabling the model to capture intricate patterns and relationships within the data. This approach has led to remarkable performance improvements, often surpassing traditional convolutional neural networks (CNNs) in benchmark tests. ViT's ability to handle large datasets and its versatility in fine-tuning make it a promising choice for a wide range of image classification tasks [8]. However, the Vision Transformer (ViT) model can be computationally intensive and time-consuming to train. This is where MobileViT comes into play, offering a more efficient alternative without sacrificing performance. MobileViT combines the strengths of both MobileNets and ViTs, providing a lightweight yet powerful model for image classification. It retains the transformer-based architecture's ability to capture complex patterns while being optimized for efficiency. This makes MobileViT a superior choice over traditional ViT for efficient image classification [9].

In our previous work [10], we focused on ECG signals presented as grayscale images and utilized lightweight classification models to effectively analyze these signals. Building on this foundation, we want to explore a different representation of ECG signals by applying the Continuous Wavelet Transform (CWT) to capture both time and frequency information. By transforming ECG signals into CWT images, we highlight subtle variations and patterns crucial for accurate classification.

Based on what been said, in this paper, we aim to leverage the powerful image classification capabilities of the lightweight MobileViT model to accurately identify and differentiate between various arrhythmias. The ECG signals are first converted into images using the Continuous Wavelet Transform (CWT) method and then utilizing MobileViT for classification. The CWT is applied to ECG signals to capture both time and frequency information, providing a comprehensive representation of the signals' features. This transformation helps in highlighting the subtle variations and patterns within the ECG signals that are crucial for accurate classification.

In the rest of the paper, we conduct a thorough review of existing literature, analyzing previous research efforts and deep learning techniques focused on diagnosing ECG arrhythmias both as signals and images in Sect. 2. In Sect. 3, we present a detailed description of our proposed MobileViT-based approach. Section 4 is devoted to presenting the results of our experiments and discussing our findings. Finally, in Sect. 5, we conclude the paper and outline potential future directions.

2 Related Works

In this section, we present a comprehensive review of the literature related to the detection and classification of ECG arrhythmias, with a particular focus on the different representations of ECG signals, especially as images, and the application of deep learning models.

First, in [11], the authors introduced a 34-layer CNN for ECG classification, transforming one-dimensional ECG signals into images using CWT. They used signals from three databases, Arrhythmia Database (ARR DB), Congestive Heart Failure Database (CHF DB) and Normal Rhythm Database (NR DB) and compared various preprocessing and data split methods, finding CWT and cross-validation to be the most effective. By combining their CNN with support vector machines (SVM), they achieved a 99.21% accuracy.

In [12] introduced a method to tackle imbalanced classification by using an improved data augmentation technique combining variational auto-encoder (VAE) and auxiliary classifier generative adversarial network (ACGAN). Each ECG beat is transformed into a 2-dimensional grayscale which is then classified using CNN model. The approach achieved 98.45% accuracy on the MIT-BIH arrhythmia database.

[13] introduced a deep learning network for ECG signal classification, using 5,655 ECG recordings. Signals were converted into RGB images with the CWT and classified using a modified AlexNet. The proposed method achieved an accuracy of 98.82% outperforming existing techniques.

Moreover, [14], introduced a novel ECG rhythm classification bypassing R-peak detection, addressing the challenge of arrhythmias like ventricular flutter, which lack distinct R-peaks. The method segmented ECG signals using a sliding window, converted each segment into a scalogram, and feeded them into a pre-trained classifier to generate beat score vectors. These vectors are concatenated into an n-beat score map (n-BSM), which is then used for rhythm classification. Results on the MIT-BIH Arrhythmia Database showed improved performance, including a 97.56% F1 score for ventricular flutter. The method was also validated on the MIT-BIH Malignant Ventricular Ectopy Database (VFDB) and Chapman-Shaoxing databases, achieving average accuracies of 99.28% and 88.83%, respectively.

[15] Classified five ECG arrhythmia classes using a 2D deep CNN, where time-domain ECG signals are transformed into time-frequency spectrograms via short-time Fourier transform (STFT) and then classified using the 2D-CNN. The approach achieved an average accuracy of 99.00% on the MIT-BIH arrhythmia database, significantly outperforming a 1D-CNN model which reached 90.93% accuracy.

The paper of [16] proposed an approach based on the combination of CNN and long short-term memory (LSTM) networks for ECG heartbeat classification. The CNN extracts essential features directly from raw ECG heartbeats, while the LSTM network performs the final classification. The proposed model achieved high classification performance on the MIT-BIH arrhythmia database, with an accuracy of 98.60% outperforming other methods in the literature.

Last, in [17], a deep learning model utilizing time-frequency representations of small ECG segments to enhance computational simplicity and speed without sacrificing performance was proposed for arrhythmia detection. Their 2D-CNN model with CWT-RGB achieved an average accuracy of 99.65%.

Despite the promising results from previous studies, several limitations were noted. For instance, the study by [11] was validated on a small dataset, raising concerns about generalizability. [12] did not capture time features in their classification, using only sequence ECG images, which could impact accuracy.[13] faced data imbalance, while [15] also validated their model on a limited dataset. [16] encountered similar data imbalance challenges, and [17] reported validation on a small dataset as well. Table 1 resumes the presented related works with their proposed approaches, results and limitations.

Table 1. Summary of the related works' Approaches

Ref	Database	Preprocessing	DL model	Input ECG	Results	Limitation
[11]	ARR BD, CHF DB, NR DB	CWT	CNN-SVM	Scalograms	99.21%	Validation on small dataset
[12]	MIT-BIH arrhythmia BD	/	CNN	Gray scale images	98.45%	No time features captured
[13]	MIT-BIH arrhythmia database	CWT	Modified AlexNet	Scalograms	98.82%	Data imbalance
[14]	VFDB	n-BSM	CNN	Scalograms, n-BSM images	99.28%	No pre-training on beat-labeled DBs, high computational costs
[15]	MIT-BIH arrhythmia DB	STFT	CNN	Scalograms	99.00%	Validation on Small dataset
[16]	MIT-BIH arrhythmia database	/	CNN-LSTM	1D signal	98.60%	Data imbalance
[17]	MIT-BIH arrhythmia BD	CWT	2D CNN	Scalograms	99.65%	Validation on Small dataset

3 Methods and Materials

To tackle the aforementioned challenges, we propose a deep learning approach that leverages a larger dataset along with a data augmentation technique. This method utilizes ECG signals from the MIT-BIH Arrhythmia Database to classify arrhythmias through the analysis of ECG scalogram images. The ECG signals are converted to scalograms using the Continuous Wavelet Transform (CWT) with the Mexican hat wavelet to capture the time-frequency characteristics of the signals. These ECG scalograms, representing beats of the five classes, are then fed into the MobileViT model for the classification of arrhythmia classes. During training, the mixup data augmentation method is employed to enhance the model's generalization capabilities. This technique generates new training samples by combining pairs of existing samples and their labels, thereby improving the robustness and performance of the model in classifying the five classes of ECG scalogram images. Figure 2 depicts the overall workflow of the proposed approach.

3.1 Data Acquisition

In this work, ECG recordings featuring five types of arrhythmic beats, normal (N), left bundle branch block (L), right bundle branch block (R), atrial premature (A), and premature ventricular contraction (V) were sourced from the

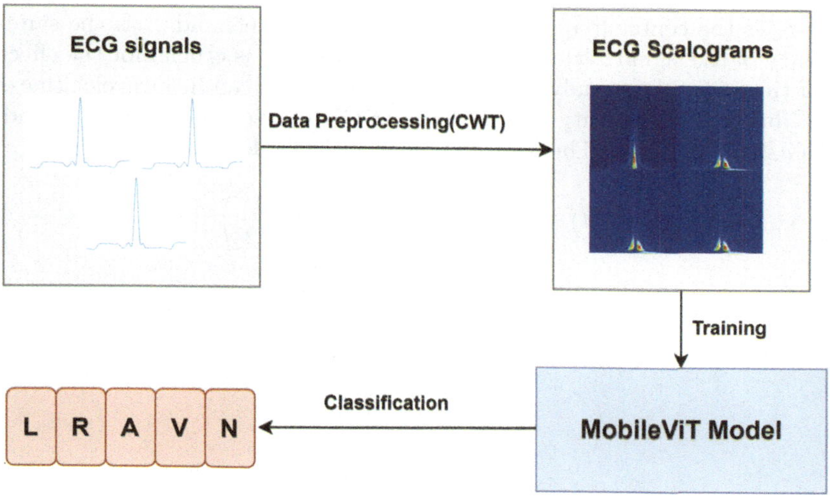

Fig. 2. Proposed Approach Workflow

MIT-BIH Arrhythmia Database which is readily available through the PhysioNet platform[1]. This database comprises 48 half-hour, two-lead ECG recordings sampled at 360 Hz from 47 patients. The MIT-BIH Arrhythmia Database is a well-established and extensively used resource in the field of cardiology for the study of arrhythmias. To ensure a balanced dataset and prevent significant data imbalance, all signals for the L, R, A, and V beats were selected, while only 10,000 N beats were randomly chosen.

3.2 Signal to Image Conversion with CWT

The ECG signal consists of various frequency components, in this study, we transform the ECG signal to the time-frequency domain to enhance feature extraction. The CWT is the most used tool for time-frequency analysis. It employs a family of wavelet functions to decompose a signal into the time-frequency domain. Fundamentally, the CWT is expressed by the following equation [18]:

$$\mathcal{CWT}(x(t)) = \frac{1}{\sqrt{|a|}} \int_{-\infty}^{\infty} x(t)\psi\left(\frac{t-b}{a}\right) dt \tag{1}$$

where a is the scale parameter, b is the translation parameter, and $\psi(t)$ is the wavelet function. The scale can be converted to frequency using the formula:

$$F = \frac{F_c \cdot f_s}{a} \tag{2}$$

[1] https://archive.physionet.org/physiobank/database/mitdb/.

where F_c is the center frequency of the mother wavelet, and f_s is the sampling frequency of the signal $x(t)$. The choice of the wavelet is crucial for the effectiveness of time-frequency analysis. In our work, the Mexican hat wavelet (mexh) is chosen due to its similarity to the shape of QRS waves and its widespread use in ECG signal analysis. The Mexican hat wavelet is defined as [19]:

$$\psi(t) = \frac{2}{\sqrt{3\sqrt{\pi}}} \left(1 - t^2\right) \exp\left(-\frac{t^2}{2}\right) \tag{3}$$

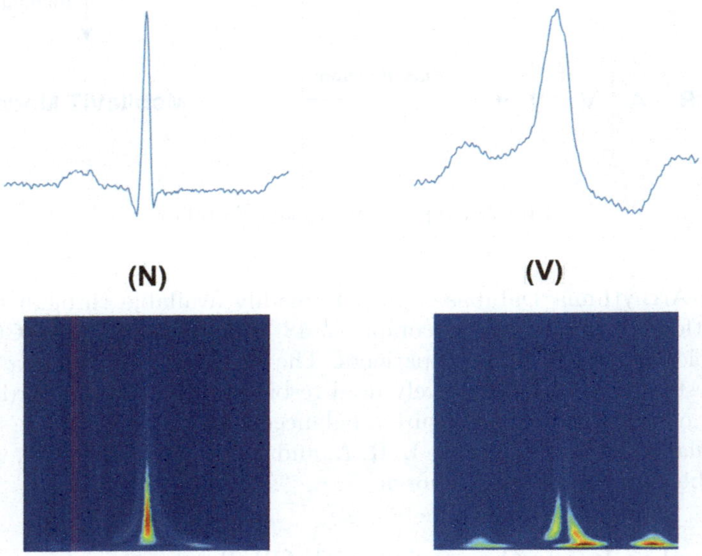

Fig. 3. ECG Scalogram Samples

By applying different scale factors in the CWT, the wavelet coefficients of the signal at various scales are obtained. As depicted in Fig. 3, these wavelet coefficients can be viewed as a 2D scalogram of the ECG signal in the time-frequency domain. The x-axis represents time, the y-axis represents frequency, and the color intensity represents the magnitude or power of the wavelet coefficients. All signals have 256 sampling points and are sampled at a frequency of 360 Hz, and saved as 128 × 128 pixels. Figure 3 illustrates the one-dimensional signal depicting both a normal beat and a V class beat, alongside their time-frequency scalogram transformation derived following the CWT process.

3.3 Mixup Method

To enhance the classification performance, we employed the MixUp method. MixUp, introduced by [20], is a data augmentation technique that creates new training examples by taking convex combinations of pairs of examples and their corresponding labels. Formally, given two training examples (x_i, y_i) and (x_j, y_j), Mixup generates a new example $(\tilde{x}, \tilde{y})$ as follows:

$$\tilde{x} = \lambda x_i + (1 - \lambda)x_j \tag{4}$$

$$\tilde{y} = \lambda y_i + (1 - \lambda)y_j \tag{5}$$

where $\lambda \in [0, 1]$ is a random value drawn from a Beta distribution. By incorporating Mixup into the training process, the model is encouraged to learn more robust and generalized features, thereby improving its performance on the classification task.

For our ECG classification, Mixup is particularly important as it helps the model generalize better to variations in ECG patterns, thus improving the accuracy and robustness of arrhythmia detection. By generating new training samples, Mixup reduces overfitting and enhances the model's ability to distinguish between different types of beats. We chose Mixup because it is straightforward, easy to implement, and not computationally demanding during training, yet it provides good results.

3.4 MobileViT Model for Classification

MobileViT is a lightweight vision transformer model designed to efficiently learn visual representations by combining the strengths of convolutional neural networks (CNNs) and transformers. Unlike standard Vision Transformers (ViTs), which require more parameters and computational resources due to their lack of spatial inductive bias, MobileViT integrates convolution-like properties to enhance performance while maintaining efficiency. The MobileViT architecture includes standard convolutional layers followed by transformer blocks, enabling it to model both local and global information effectively. Additionally, MobileViT uses inverted residual blocks from MobileNetV2, which help reduce the number of parameters and computational cost while maintaining high accuracy. By unfolding the input tensor into non-overlapping patches and applying transformers, MobileViT preserves the spatial order of pixels, ensuring robust feature extraction. This design allows MobileViT to achieve high accuracy with fewer parameters and faster processing times compared to traditional ViTs, making it suitable for mobile and resource-constrained applications [9].

The MobileViT XXS variant is employed, known for its compact size and efficient performance. Its architecture used in our work starts with a convolutional stem followed by a series of inverted residual blocks and MobileViT blocks. The inverted residual blocks use point-wise and depth-wise convolutions to enhance feature extraction, while MobileViT blocks apply transformers to capture long-range dependencies without losing the spatial structure. The architecture includes stages with progressively deeper layers, transitioning through different levels of feature representation. The final stage features a global average pooling layer followed by a dense layer for classification. Figure 4 illustrate the flowchart of the proposed approach including the used MobileViT Block [9] and the inverted residual Block (MV2)[2], while Table 2 resumes the implemented MobileViT XXS architecture.

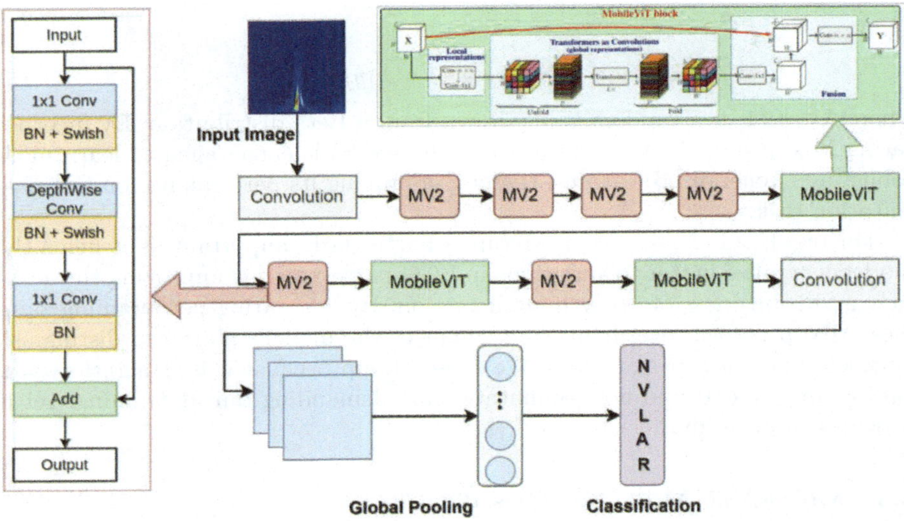

Fig. 4. Flowchart of the Proposed Approach

4 Experimentation Results and Discussion

In this section, we will present the experimentation results and discussion of our model's performance. To evaluate the effectiveness of our proposed approach, we will utilize four key metrics: accuracy, recall, precision, and F1 score. These metrics provide a comprehensive assessment of our model's ability to correctly classify ECG scalogram images across the five arrhythmia classes. To ensure optimal performance, checkpoints were used to save the model with the best performance, and early stopping was implemented, ending the training process at epoch 20. This indicated that the model had reached its optimal performance, balancing high accuracy with minimal overfitting. For the first epochs, the model exhibited a consistent decrease in training loss and stable validation loss. Training accuracy steadily improved, reaching 96.65%, while validation accuracy remained high, fluctuating between 98.49% and 98.69%. For the following epochs, the training loss continued to decline from 0.2419 to 0.2357, and the validation loss remained low, reaching 0.0304. Despite minor fluctuations, training accuracy generally remained high, indicating a well-tuned model with high

[2] https://github.com/nikhilroxtomar/MobileViT-Implementation/tree/main.

Table 2. MobileViT Structure

Layer	Output Size	Output Stride	Number of Repeat	Channels
Input	$128 \times 128 \times 3$	-	-	3
Convolution	$64 \times 64 \times 16$	2	1	16
BatchNorm	$64 \times 64 \times 16$	-	1	16
Activation	$64 \times 64 \times 16$	-	1	16
MV2 Block	$64 \times 64 \times 16$	1	1	16
MV2 Block	$32 \times 32 \times 24$	2	1	24
MV2 Block	$32 \times 32 \times 24$	1	2	24
MV2 Block	$16 \times 16 \times 48$	2	1	48
MobileViT	$16 \times 16 \times 64$	-	1	64
MV2 Block	$8 \times 8 \times 80$	2	1	80
MobileViT	$8 \times 8 \times 96$	-	1	96
MV2 Block	$4 \times 4 \times 160$	2	1	160
MobileViT	$4 \times 4 \times 320$	-	1	320
Convolution	$4 \times 4 \times 320$	1	1	320
BatchNorm	$4 \times 4 \times 320$	-	1	320
Activation	$4 \times 4 \times 320$	-	1	320
Global Pooling	320	-	-	320
Dense	5	-	1	5

performance and minimal overfitting. In the last five epochs, the training loss continued to decrease from 0.2322 in epoch 16 to 0.2210 in epoch 20, suggesting the model was still refining its learning even towards the end. The validation loss remained low and relatively stable, ending at 0.0269, which indicates consistent performance on unseen data. Training accuracy saw a slight increase from 96.92% in epoch 16 to 97.16% in epoch 20, while validation accuracy improved slightly from 98.63% to 99.29%, demonstrating the model's increasing ability to generalize well to the validation set. Figure 5 shows the training and validation accuracies and losses curves for the proposed model.

In the evaluation phase, the overall accuracy of the model is **98.97%**, demonstrating its robust performance. The classification report shown in Table 3 provides a detailed evaluation of our MobielViT model's performance across the five classes: N, L, A, R, and V. The model demonstrates exceptional precision, recall, and F1-scores across all classes, with values predominantly at 0.98 and 0.99. Specifically, the precision for classes A and V is 0.98, while it is 0.99 for classes L, N, and R. The recall values follow a similar pattern, showing 0.94 for class A and 0.99 for the other classes, indicating a high true positive rate. The F1-scores, which balance precision and recall, are consistently high, ranging from 0.96 for class A to 0.99 for classes L, N, and R. The macro and weighted averages for precision, recall, and F1-score are also remarkably high, reflecting

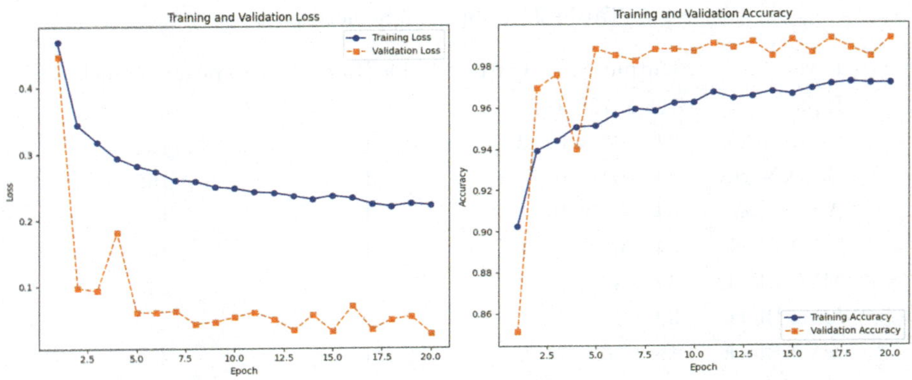

Fig. 5. Training and Validation accuracies' and losses

the model's balanced performance across all classes, with minimal bias towards any class. These results indicate that the MobileViT model is highly effective in classifying the given data, showcasing strong generalization capabilities and minimal overfitting.

Table 3. Classification Report Of the Proposed MobileViT

Class	Precision	Recall	F1-Score	Support
A	0.98	0.94	0.96	447
L	0.99	0.99	0.99	1822
N	0.99	0.99	0.99	2000
R	0.99	0.99	0.99	1444
V	0.98	0.99	0.98	1302
Accuracy	**0.99**			
Macro Avg	0.99	0.98	0.99	7015
Weighted Avg	0.99	0.99	0.99	7015

Additionally, the provided confusion matrix depicted in Fig. 6 provides a detailed breakdown of the performance of our MobileViT model across the five different classes. For class A, the model correctly predicted 421 instances, with minor misclassifications into classes L, N, R, and V, amounting to a total of 26 incorrect predictions. Class L demonstrates a strong performance with 1811 correct predictions, with only 11 instances misclassified into other classes, showcasing the model's high precision and recall for this class. The N class has 1989 correct predictions and minimal misclassifications into classes A, L, and V, indicating robust performance with only 11 errors. For class R, the model achieved 1436 correct predictions, with very few misclassifications, primarily into class V, totaling 8 errors. Finally, class V shows 1286 correct predictions with a small number of instances misclassified into other classes, resulting in 16 errors. Overall, the confusion matrix illustrates the model's high accuracy and ability to correctly classify instances with minimal misclassifications across all classes.

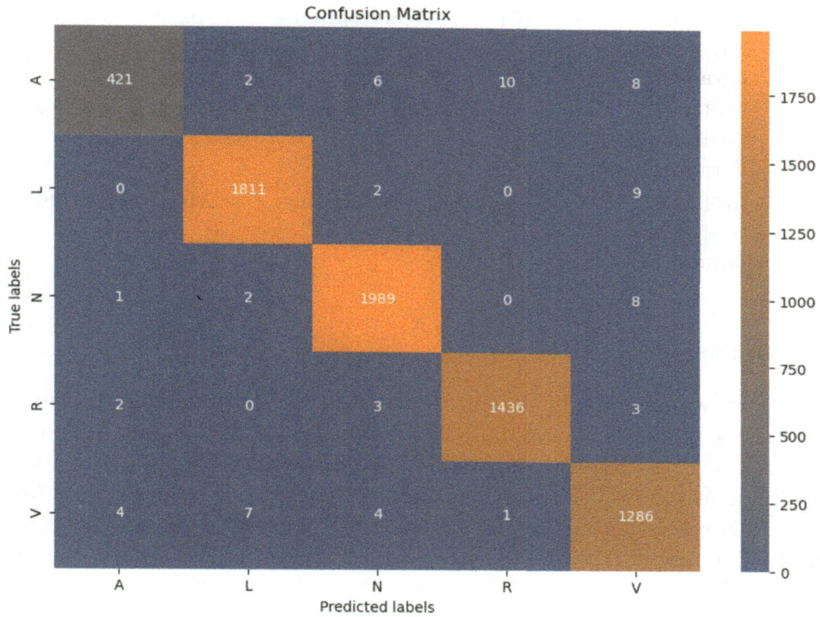

Fig. 6. Confusion Matrix of the Proposed Model

Table 4 provides a comparison of the performance of different deep learning models for ECG classification using various preprocessing techniques and input data representations. Models utilizing the CWT for preprocessing generally demonstrate improved performance, with the proposed MobileViT model achieving the highest accuracy of 98.97%. This model benefits from advanced preprocessing with CWT and data augmentation using Mixup, surpassing other methods. For instance, the 2D CNN model by [12], which used grayscale images, reached an accuracy of 98.45%, while the work by [13] using CWT-based ECG scalograms with the AlexNet model achieved an accuracy of 98.82%. Additionally, the hybrid model by [16], which used CNN-LSTM on ECG signals, achieved an accuracy of 98.60%. Notably, our MobileViT model is lightweight compared to these models, maintaining a good balance between high classification accuracy and computational efficiency. Overall, the comparison suggests that sophisticated models like MobileViT significantly enhance classification accuracy, positioning it as a leading approach in ECG signal classification.

Table 4. Comparison of MobileViT model with Related Works

Reference	DataSet	Preprocessing	Input Data	DL model	Accuracy
[12]	MIT-BIH Arrhythmia database	/	Grayscale images	2D CNN	98.45%
[13]	MIT-BIH Arrhythmia database	CWT	Scalograms	AlexNet	98.67%
[16]	MIT-BIH Arrhythmia database	/	1D Signal	CNN- LSTM	98.60%
Our work	MIT-BIH Arrhythmia dataset	CWT	Scalograms	**MobileViT**	**98.97%**

5 Conclusion

In this paper, we introduced a novel approach for arrhythmia classification using MobileViT, a model that combines the strengths of CNNs and Vision Transformer architectures. By transforming ECG signals into images through CWT and applying MobileViT for classification, we achieved an impressive accuracy of 98.97%. This result highlights the effectiveness of integrating advanced preprocessing techniques with state-of-the-art deep learning models. Given the promising potential of MobileViT to enhance arrhythmia detection accuracy, exploring its use for real-time ECG monitoring is a compelling direction for future research. Additionally, we plan to extend our work by validating the model on larger datasets and experimenting with additional data augmentation techniques. We also aim to explore more advanced deep learning models to further improve classification performance and robustness.

References

1. Murat, F., Yildirim, O., Talo, M., Baloglu, U.B., Demir, Y., Rajendra Acharya, U.: Application of deep learning techniques for heartbeats detection using ECG signals-analysis and review. Comput. Biol. Med. **120**, 103726 (2020)
2. Ma, S., Cui, J., Chen, C.-L., Chen, X., Ma, Y.: An effective data enhancement method for classification of ECG arrhythmia. Measurement **203**, 111978 (2022)
3. Liu, X., Wang, H., Li, Z., Qin, L.: Deep learning in ECG diagnosis: a review. Knowl.-Based Syst. **227**, 107187 (2021)
4. Merdjanovska, E., Rashkovska, A.: Comprehensive survey of computational ECG analysis: databases, methods and applications. Expert Syst. Appl. **203**, 117206 (2022)
5. Haddad, Z., Yaya, B.M., Zgolli, H., Sidibé, D., Tabia, H., Khlifa, N.: Retinal pathologies detection in oct images based on bilinear convolutional neural network. In: 2023 International Conference on Innovations in Intelligent Systems and Applications (INISTA), pp. 1–7. IEEE (2023)

6. Gueddena, Y., et al.: A new intelligent system based deep learning to detect DME and AMD in oct images. Int. Ophthalmol. **44**(1), 191 (2024)
7. Ebrahimi, Z., Loni, M., Daneshtalab, M., Gharehbaghi, A.: A review on deep learning methods for ECG arrhythmia classification. Expert Syst. Appl. X **7**, 100033 (2020)
8. Dosovitskiy, A., et al.: An image is worth 16 × 16 words: transformers for image recognition at scale. arXiv preprint arXiv:2010.11929 (2020)
9. Mehta, S., Rastegari, M.: Mobilevit: light-weight, general-purpose, and mobile-friendly vision transformer. arXiv preprint arXiv:2110.02178 (2021)
10. Bechinia, H., Benmerzoug, D., Khlifa, N.: Approach based lightweight custom convolutional neural network and fine-tuned mobilenet-v2 for ECG arrhythmia signals classification. IEEE Access **12**, 40827–40841 (2024)
11. Ozaltin, O., Yeniay, O.: A novel proposed CNN-SVM architecture for ECG scalograms classification. Soft. Comput. **27**(8), 4639–4658 (2023)
12. Du, C., Liu, P.X., Zheng, M.: Classification of imbalanced electrocardiosignal data using convolutional neural network. Comput. Methods Programs Biomed. **214**, 106483 (2022)
13. Aarthy, S.T., Mazher Iqbal, J.L.: Modified parametric-based alexnet structure to classify ECG signals for cardiovascular diseases. Measur. Sens. **27**, 100816 (2023)
14. Lee, J., Shin, M.: Using beat score maps with successive segmentation for ECG classification without r-peak detection. Biomed. Signal Process. Control **91**, 105982 (2024)
15. Huang, J., Chen, B., Yao, B., He, W.: ECG arrhythmia classification using STFT-based spectrogram and convolutional neural network. IEEE Access **7**, 92871–92880 (2019)
16. Harrane, S., Belkhiri, M.: Classification of ECG heartbeats using deep neural networks. Res. Biomed. Eng. **37**, 737–747 (2021)
17. Mohonta, S.C., Motin, M.A., Kumar, D.K.: Electrocardiogram based arrhythmia classification using wavelet transform with deep learning model. Sens. Bio-Sens. Res. **37**, 100502 (2022)
18. Sun, X., Liu, P., He, Z., Han, Y., Bochao, S.: Automatic classification of electrocardiogram signals based on transfer learning and continuous wavelet transform. Eco. Inform. **69**, 101628 (2022)
19. Wang, T., Changhua, L., Sun, Y., Yang, M., Liu, C., Chunsheng, O.: Automatic ECG classification using continuous wavelet transform and convolutional neural network. Entropy **23**(1), 119 (2021)
20. Zhang, H., Cisse, M., Dauphin, Y.N., Lopez-Paz, D.: Mixup: beyond empirical risk minimization. arXiv preprint arXiv:1710.09412 (2017)

A Combined MaskRCNN-Watershed Model for Colorectal Gland Segmentation from Histological Images

Ferialle Lahreche[1](✉), Slimane Oulad-Naoui[1], Ballou Mansoura[1], Asma Harrouzi[1], and Abdelouahab Moussaoui[2]

[1] Lab. des Mathématiques et Sciences Appliquées, Université de Ghardaia, Bounoura, Algeria
{lahrache.ferialle,ouladnaoui,ballou.mansoura, harrouzi.asma}@univ-ghardaia.dz

[2] University Ferhat Abbas, Setif, Algeria

Abstract. The World Health Organization (WHO) reports ColoRectal Cancer (CRC) as the third most common cancer in Algeria and also the third most common cause of cancer-related deaths worldwide. ColoRectal Adenocarcinoma (CRA) is a common form of CRC where histopathology analysis remains the optimal method for distinguishing between different types of colon cancer, by examining slides of the patient tissue specimens under a microscope. However, this technique is very demanding and time consuming particularly for low-grade cancers, and may lead to the risk of misdiagnosis. Consequently, computers systems are becoming increasingly valuable in alleviating the diagnostic burden for medical professionals. But, medical images are complex in nature, which make their analysis and segmentation a very challenging task. In this work, we tackle CRA image segmentation by combining an object detection deep learning architecture with a traditional segmentation algorithm. Indeed, we adapt the Mask R-CNN architecture to predict instance masks for each gland from histological images. Next, we apply the watershed algorithm to handle the overlapping predicted gland masks. This approach yields a satisfactory result on the GlaS dataset with an F1 score of 0.87, a Dice coefficient of 0.86, and an IoU of 0.78. Despite the promising results, we faced, when conducting this study, two main challenge: the heavy computational requirements of the Mask R-CNN model and the limited availability of histological CRC images.

Keywords: Colorectal Cancer · Medical Image Segmentation · Mask-RCNN · Watershed Algorithm

This work is supported by the Directorate-General for Scientific Research and Technological Development (DGRSDT) - Algeria, and performed under the PRFU Project: C00L07UN470120230001.

S. Chikhi et al. (Eds.): MISC 2024, LNNS 1267, pp. 254–266, 2025.
https://doi.org/10.1007/978-3-031-82112-7_19

1 Introduction

Cancer is one of the most lethal, prevalent, and feared diseases. Its mere mention provokes a sense of imminent extinction. This reputation has a detriment effect on both the patient psyche and the physician's attitude to convince it about the success rate of the treatment scheme and its effectiveness. Among the various types of cancer, colorectal cancer (CRC) is a leading cause of mortality, ranking as the third most frequent reason of cancer-related death worldwide [1]. Intestinal glands, or colonic crypts, are crucial histological structures in the epithelial layer of the colon, responsible for absorbing water and nutrients, secreting mucus, and serving as a niche for epithelial cell regeneration. The epithelial layer continuously regenerates due to the harsh environment, with this renewal process requiring coordination between cell proliferation, differentiation, and apoptosis. Disruptions in this process can lead to colorectal adenocarcinoma [2]. Adenocarcinoma represents the most common CRC type. The early and accurate diagnosis is critical for successful treatment, with colon tissue samples being the gold standard. Pathologists meticulously analyze these biopsies for cancerous features, including glandular morphology, which plays a vital role in tumor grading and prognosis. The morphology of intestinal glands, including architectural appearance and gland formation, is a key criterion for CRC grading.

Automated image processing methods are a promising solution to colon tissue analysis, particularly gland segmentation techniques that determine precisely the location and shape of the gland. Accurate segmentation of glands in colon histology images is crucial for computer-aided diagnosis systems to effectively quantify morphological features associated with cancer progression. Over the years, segmentation methods have evolved from traditional techniques based on color (Hematoxylin and Eosin staining) and shape analysis [9]. However, these methods are time-consuming, challenging, and typically effective only for benign glands with static shapes. To overcome these limitations, more sophisticated approaches utilize machine learning (ML) algorithms, such as clustering based on the color or the shape [3], and the watershed method [5] have been developed. These methods marked significant improvements but still faced challenges in dealing with the complexity and variability of malignant glands. Recently, with the availability of histological image datasets [2] and advancements in deep learning (DL) architectures, particularly convolutional neural networks (CNNs), there has been a substantial progress in gland segmentation [6]. CNNs have demonstrated remarkable capability in feature detection and extraction, enabling more precise and efficient analysis of complex histopathological images.

In this paper, we aim to enhance gland segmentation in colorectal cancer histological images. We tackle CRA image segmentation by adapting an object detection deep learning architecture which we associate with a traditional segmentation algorithm. To this end, we combine the Mask R-CNN [7] model and the Watershed algorithm [4]. The Mask R-CNN is employed to segment the glands and generate instance masks for each one, leveraging its powerful object detection and instance segmentation capabilities. This enables a precise identification of intricate features and diverse shapes of glands within histopathological images. Subsequently, the Watershed algorithm is employed to refine the initial results and address the issue of separating overlapping glands. This process enhances boundary delineation by considering the topographical features of the image, effectively separating closely packed or overlapping glands. The integration of Mask R-CNN and the Watershed algorithm is intended to facilitate more precise and reliable gland segmentation, thereby significantly aiding pathologists in cancer diagnosis and grading.

The remainder of this paper is organized as follows: In Sect. 2, we review the main techniques used for the segmentation and detection of colorectal cancer. Then, in Sect. 3, we outline our proposed approach. Section 4 describes the experiment, the employed dataset, and the environment setup. In Sect. 5 we present and discuss the obtained results. We conclude in Sect. 6 with an outline of possible further extensions to the present work.

2 Related Work

In the sequel, we succinctly review the main techniques used for colorectal cancer segmentation and detection. We categorize these works into three classes: classical techniques, ML-based techniques, and DL-based techniques.

2.1 Classical Techniques

The initial efforts in gland segmentation from histological images involved extracting features characteristic of gland structures. These features typically focused on identifying substructures like cell nuclei or lumens, or on intensity variations associated with hematoxylin staining. For instance, Mukherjee et al. [8] propose a three-step method for analyzing gland morphology in Hematoxylin and Eosin (H&E) stained histology images. First, they apply an edge-preserving filter to smooth the image while maintaining the integrity of the epithelial layer (EL) boundaries. These boundaries hold crucial information about gland morphology. Next, they segment the EL using informative morphological scale space, a technique that analyzes the image at various scales to extract features of different sizes. The informative aspect likely refers to selecting scales most suitable for capturing the EL structures. Finally, they fill the regions within the segmented EL to obtain the complete gland. The method was evaluated on the Glas dataset, achieving a moderate F1 score of 0.68 and a good Dice coefficient of 0.75, indicating successful segmentation with good overlap between the automatically identified glands and the actual glands.

Ben Cheikh et al. [9] utilized color classification techniques to pinpoint the locations of cell nuclei. They employed advanced morphological operations on segmented nuclei objects to analyze their spatial arrangement, thereby discerning glandular regions. These methods were integrated to generate glandular structures. Their approach was tested on a Warwick-QU dataset as part of the Glas2015, resulting in an F-score of 0.86 and an object dice of 0.91. Notably, this approach offers scalability across various histology stains, such as immunohistochemistry and fluorescence, as well as different types of glandular tissues like prostatic glands and mammary glands.

These classical techniques achieved promising results, laying the groundwork for more advanced segmentation methods. Their reliance on handcrafted features, however, often limited their robustness to image variations and complex gland shapes.

2.2 Machine Learning-Based Techniques

Classical methods for gland segmentation have laid the groundwork for more sophisticated approaches. Recent advancements in this field have seen a significant shift towards ML-based methodologies, capable of discerning discriminative features from data such as: intensity variations, textures, and spatial relationships, which can then be used to differentiate gland pixels from background pixels, facilitating accurate segmentation.

Khatun et al. [11] introduced an algorithm for automatic segmentation of glands in colon histology. The method relies on local intensity and texture features, utilizing images from the GlaS dataset. The images are partitioned into patches of varying window sizes, from which the intensity and texture-based features are extracted. Classification of these patches into different labels is achieved through the application of a *Random Forest classifier*. Notably, the proposed algorithm employs a multilevel Random Forest technique hierarchically, ensuring both speed and accuracy. The algorithm demonstrates an average accuracy ranging from 69% to 84%, contingent on the specific test image utilized, affirming its effectiveness in automated gland segmentation for practical clinical applications.

Alfonso et al. [5] proposed a method for gland segmentation that utilizes Hematoxylin and Eosin (H&E) staining information. The method first separates the H&E image into its constituent channels and then focuses on the hematoxylin channel. Smoothing and thresholding are applied to this channel to detect likely gland candidate regions based on their characteristic white coloration. Further refinement steps are employed to achieve a more precise set of candidate glands. Surrounding these candidates, the method segments the nuclei using a watershed algorithm. Finally, a feature extraction step is performed to obtain information about both the candidate glands and their neighboring nuclei. These features encompass shape, color, and spatial relationships. The authors validated their method on a dataset of 1330 fields of view obtained from H&E whole slide images of patients with various conditions. The dataset was manually annotated for gland locations, resulting in 2372 labeled glands. Their method achieved an F1-score of approximately 72%.

Rathore et al. [3] developed a ML technique for distinguishing between benign and malignant tumors. They utilized the RMC and GlaS datasets and employed a gland segmentation algorithm that involved several steps. These steps included clustering tissue components using K-means, extracting features from both the overall image and individual glands, and employing the Radial Basis Function

(RBF) kernel of SVM in the initial layer of detection. Subsequently, a second layer of SVM classification utilized Linear, RBF, and sigmoid kernels. The final classification was determined through majority voting based on predictions from these kernels. Cross-evaluates the intersection of two data sets, detection accuracy achieved 94.5% (Train RMC - Test GlaS) and 93.7% (Train GlaS - Test RMC), while classification accuracy reached 95% for both Train RMC - Test GlaS and Train GlaS - Test RMC scenarios. However, it is noted that the segmentation method was specifically designed for glandular regions and may not be suitable for accurately separating benign and malignant tissue across broader tissue regions.

2.3 Deep Learning-Based Techniques

Unlike traditional methods, DL models particularly CNNs has significantly advanced gland segmentation by automatically learning complex features from images such as: spatial relationships, textures, and intricate gland boundaries. These features are then leveraged to differentiate gland pixels from background pixels, leading to highly accurate segmentation.

The approach proposed by Kosov et al. [12] for gland segmentation employs a two-step strategy that effectively integrates traditional image processing techniques with DL. The first step constructs a probability map for gland positions based on shape and intensity features. A grid is overlaid on the image, and at each grid point, a weighted graph is created to analyze local image properties. This analysis leads to the identification of potential gland contours. However, this reliance on shape analysis might limit the effectiveness for segmenting malignant glands with irregular shapes. The second step employs a pre-trained CNN to classify these candidate regions, refining the segmentation by removing those unlikely to be glands. While this method achieves good results with an accuracy of 0.82 and a Dice score of 0.79 on the Glas dataset, it may be particularly suited for benign glands due to their well-defined shapes.

Mei et al. [13] introduced the Dense Contour-Imbalance Aware (DCIA) framework for precise object segmentation in images. Leveraging DenseNet for feature extraction and Focal Loss to address challenging edges, DCIA significantly outperformed existing methods. The framework polished its predictions with two refinements, leading to superior performance in visual quality and Hausdorff distance metrics. Compared to a similar method, DCIA demonstrated accuracy improvements of up to 12%, promising precise object separation even for intricate outlines. These results highlight DCIA's potential for advanced image segmentation tasks.

Dabass et al. [14] introduced a modified U-Net architecture, This work presents a comprehensive methodology for gland segmentation in medical imaging, characterized by meticulous pre-processing steps including manual ground truth delineation, standardized patch selection, stain normalization, data augmentation, and quality control. At the core of the method lies a sophisticated symmetrical Encoder-Decoder Network Architecture. This architecture not only boasts a symmetrical design but also integrates Atrous-Residual Units for robust feature extraction, Attention Units for gland-specific emphasis, and Transitional Atrous Units to address resolution issues. A weighted cross-entropy loss function guides the training process, ensuring a balanced ratio of foreground and background classes and enhancing segmentation accuracy. Post-processing steps involving thresholding and morphological operations refine the final segmentation. This methodology achieves impressive results, with an F-score of 0.93 and Object-dice coefficient of 0.94 on the Glas dataset, and an F-score of 0.91 and object-dice coefficient of 0.94 on the CREG dataset, underscoring its robustness and effectiveness across varied datasets and scenarios.

Wang et al. [6] introduce Topology-Aware Network (TA-Net), that aims to improve gland segmentation using a specialized NN. TA-Net has two main parts: an encoder and two decoder branches. It uses a framework called SegNet for the encoder, which is like the backbone of the network which helps in capturing important features. The decoder branches focus on two tasks: separating glands from the background (INST segmentation) and understanding the arrangement of glands (TOP estimation).To enhance accuracy, the network considers two aspects: the shape of glands and specific markers within them. This aids to prevent errors like over/under segmentation. Next, the network applies the Watershed algorithm along with some adjustments to refine the results. Evaluation on the GlaS and CRAG datasets, TA-Net achieved significant improvements in accuracy, with F1-scores of 0.9 and 0.84, and object Dice scores of 0.9 and 0.89 respectively.

Sun et al. [15]. introduced a new approach for image segmentation using diffusion models. Their method incorporates a ResNet and Feature Pyramid Network (FPN) within an image dncoder to extract multi-scale features from input images. A diffusion model decoder then applies noise progressively to the image in a forward process, learning to reverse this to reconstruct the original image with conditional encoding and Instance-aware Filters. During this reconstruction, segmentation information is integrated into the decoder. Additionally, a separate mask branch with a dynamic mask head utilizes multi-scale data to predict precise instance masks, addressing potential detail loss during diffusion. The training combines two loss functions with a balancing parameter for optimization. Their method achieves notable results on medical image datasets: an F1-score of 0.98 and object Dice score of 0.93 on the Glas dataset, and an F1-score of 0.85 and object Dice score of 0.83 on the CRAG dataset. However, challenges remain in handling image quality issues such as artifacts and noise.

3 Methodology

Our approach consists of two stages architecture. It associates as Fig. 1 depicts an object detection DL architecture with a traditional post-processing techniques. We leverage transfer learning to tackle gland segmentation in CRC images. A pre-trained Mask R-CNN model with a ResNet-50 backbone serves as the foundation. This model was originally trained for the detection and classification of daily objects [7]. To fine-tune the model we freeze the bounding box layers, and we essentially transfer this valuable knowledge to the domain of gland segmentation. we trained the model on different experiments as we will show. This focuses learning on predicting segmentation masks through fine-tuning with the SGD optimizer.

During the training phase we achieved satisfying results but despite the robustness of the Mask R-CNN model, some of its gland outlines sometimes overlapped in the images, a problem that occurred more frequently in malignant glands than in benign ones. This is because malignant glands often have irregular, dynamic shapes, whereas benign glands have more static, regular shapes. To fix this and get perfectly separated glands, we applied some additional techniques. First, we got rid of any tiny objects the model might have mistakenly identified as glands. This cleaning step reduced errors and made the overall gland identification more accurate. Next, we used the erosion/dilation techniques to slightly shrink the outlined gland areas, which helped separate glands that were practically touching, and to bring the glands back to their original size. Finally, we use the Watershed algorithm, along with the distance transform, to further improve the separation. This last step consider the image as a landscape with hills (glands) and valleys (background). The distance transform calculates how each point is far from the nearest valley. By finding the highest points in this distance map, we could identify the best places to create boundaries and separate the overlapping glands.

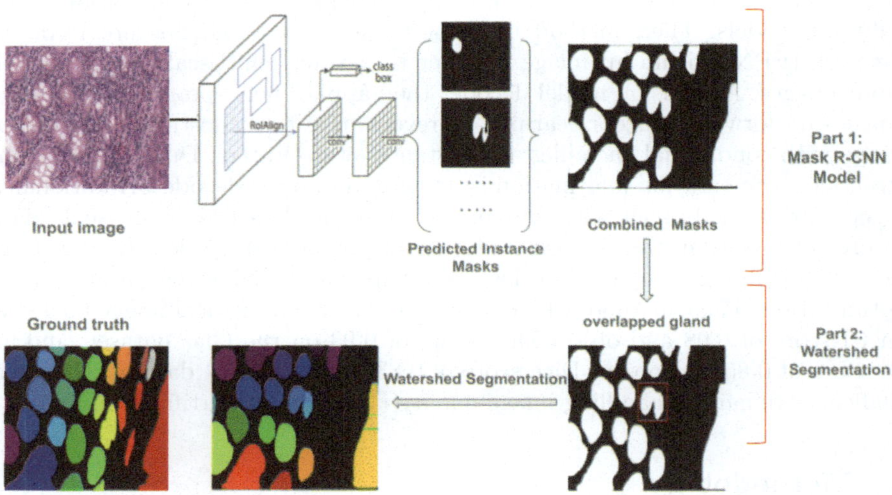

Fig. 1. The architecture of our approach

4 Experiment

4.1 Dataset

In this study, we utilized a dataset obtained from the GlaS challenge [2], The dataset consists of 165 histological images derived from 16 H&E stained histological sections of stage T3 or T4 colorectal adenocarcinoma. Each section originates from a different patient. The histological sections were digitized into whole-slide images (WSIs) using a Zeiss MIRAX MIDI Slide Scanner, with a pixel resolution of 0.465 μm, which were later rescaled to a pixel resolution of 0.620 μm, equivalent to a 20× objective magnification. The dataset contains masks for each image, delineating the glandular structures, created by an expert pathologist. These masks serve as valuable ground truth data for automated segmentation algorithms. The dataset is divided into 85 images for training and 80 for testing divided into two parts: test A with 60 images and test B with 20 images. The Test A sub-dataset contains a mix of benign and malignant glands, including healthy, normal-shaped glands and various stages of abnormal growths, contributing to more balance. While, the Test B sub-dataset consists entirely of abnormal glands, with no glands of normal shape. Table 1 shows the different grades of the glands and the percentage of each grade in the subtest A and B.

Table 1. Percentage of each grade type in Test A and Test B Subdatasets

Grade Type	test A(%)	test B(%)
Adenomatous	**20.00**	**20.00**
Healthy	**35.00**	**0.00**
Moderately Differentiated	**25.00**	**40.00**
Moderately-to-Poorly Differentiated	**6.67**	**20.00**
Poorly Differentiated	**13.33**	**20.00**

We showcase in Fig. 2 examples from the Glas dataset, featuring both malignant and benign gland images along with their corresponding masks. It is noteworthy that the images maintain dimensions of 775 × 522 pixels.

4.2 Data Augmentation

In general, medical images are available only in limited numbers due to different reasons, and colorectal images are not an exception. As one can notice, the dataset size is small and does not justify recourse to the DL approach. To raise the dataset size and enhance the generalization capability of our model, we apply data augmentation. This process involves applying a series of geometric transformations that preserve sense of the data to both the images and their corresponding masks. These transformations included horizontal flipping, vertical flipping, and rotation by 120 degrees. This operation slightly augment the train part size of the dataset to 340 images in lieu of 85.

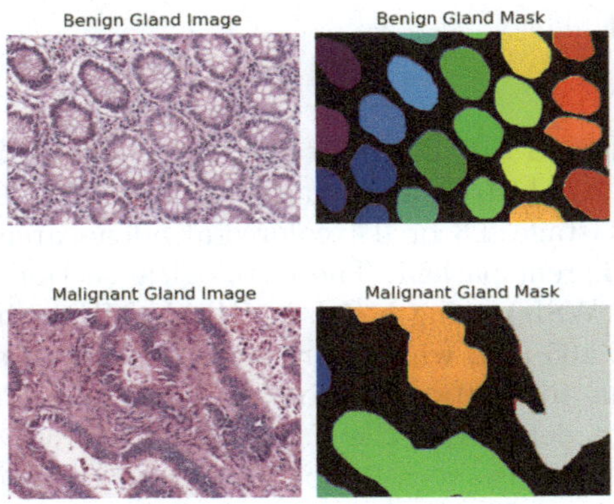

Fig. 2. Histological images and their corresponding masks for benign and malignant gland cancer types.

4.3 Environment

We write our code with the Anaconda platform. It includes a Jupyter Notebook, a user-friendly interface for writing and running Python code with the most common libraries. Experiment is performed with a machine having an Intel(R) Xeon (R) Silver 4216 CPU processor equipped with 128 GB of RAM running Ubuntu 22.04 LTS. Whereas tests are performed on an Intel(R) Core(TM) i5-6200U CPU, 8 GB of RAM, and Windows 10 operating system.

5 Result and Discussion

In this part, we present the results obtained from our model and discuss these findings. We divide this section into two parts: first, the results obtained from experimenting the Mask R-CNN model on Test A and Test B datasets, and second, the results of combining the Mask R-CNN model with post-processing techniques, including morphological methods and the watershed algorithm. Despite its optimization from region-based CNN, we found a huge computation overhead with the Mask R-CNN model training process. So due to this computing resource limitation, we trained our model for 15 epochs with a batch size of 16 only.

5.1 Results

Part 1: Mask R-CNN Model Testing

In this part, we evaluate the performance of the Mask R-CNN model using standard metrics such as Intersection over Union (IoU), Dice coefficient, Precision, Recall, and F1-score. The model was tested on two datasets, Test A and Test B. The result obtained are presented in Table 2 as follows.

Table 2. Mask R-CNN Model Testing Results

Metric	Test A	Test B
Average IoU	**0.77**	**0.73**
Average Dice Coefficient (object dice)	**0.86**	**0.83**
Average Precision	**0.83**	**0.77**
Average Recall	**0.91**	**0.93**
Average F1 Score	**0.87**	**0.84**

The Mask R-CNN model gives good results in predicting instance segmentation. It shows particularly strong performance in Test A as shown in Fig. 3, likely because this test set is balanced, containing both malignant and benign glands. In contrast, Test B is composed of glands with more difficult shapes, primarily malignant, which presents greater challenges for the model.

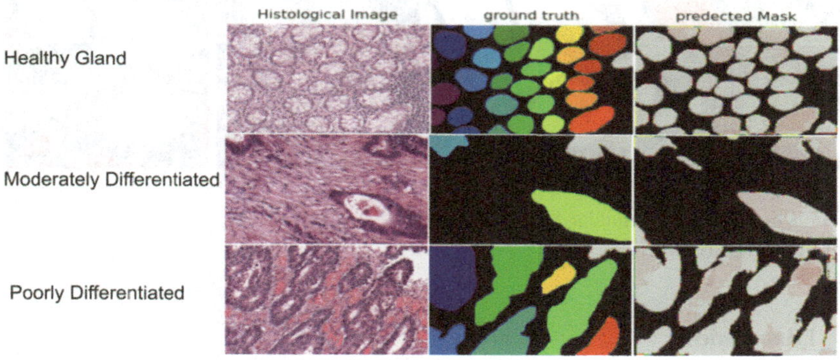

Fig. 3. Predicted Masks of Various Gland Grades from Mask R-CNN (Healthy, Moderately Differentiated, Poorly Differentiated)

Part 2: Post-Processing with Morphological Methods and Watershed Algorithm

While the Mask R-CNN model provided good segmentation results, we observed that it had difficulties in accurately predicting some small objects identified as glands and in separating overlapping glands, especially the malignant glands. To address these issues, we applied additional post-processing steps, including morphological operations (erosion and dilation) and the watershed algorithm. These techniques aimed to improve the segmentation masks produced by the Mask R-CNN model. The results after applying the post-processing steps are summarized in the Table 3 and shown in Fig. 4 below:

Table 3. The Testing Results After the Post-processing

Metric	Test A	Test B
Average IoU	**0.78**	**0.73**
Average Dice Coefficient (object dice)	**0.87**	**0.83**
Average Precision	**0.84**	**0.77**
Average Recall	**0.92**	**0.93**
Average F1 Score	**0.88**	**0.84**

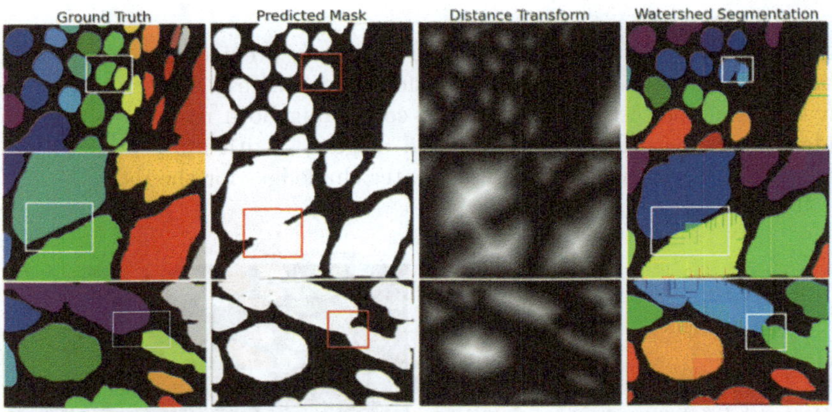

Fig. 4. The predicted masks from mask R-CNN and the Results after applying the Watershed Algorithm

5.2 Discussion

After presenting the results, this section will discuss these findings and evaluate the strengths and challenges of our approach.

In the first part, the Mask R-CNN model gives good results in predicting instance segmentation. It shows particularly strong performance in Test A, with a Dice coefficient of 0.86, F1 score of 0.87, and IoU of 0.77. This is likely because Test A is balanced, containing both malignant and benign glands. In contrast, Test B is composed of glands with more difficult shapes, primarily malignant, which presents greater challenges for segmentation. Although the Mask R-CNN model provides a solid starting point for gland segmentation, applying the watershed algorithm to handle overlapping glands and employing morphological methods significantly improves performance. This combination of the Mask R-CNN and the watershed algorithm enhances segmentation accuracy, especially in Test A. Specifically, the Dice coefficient and F1 score increased by approximately 1.2% after applying morphological methods, while the IoU increased by approximately 2.7%.

The improvement wasn't as dramatic in Test B because the oddly shaped glands were inherently harder to segment precisely. Overall, the study suggests that combining Mask R-CNN's power with the classic watershed technique is a promising approach for gland segmentation. One limitation of our approach is the long training time required for the Mask R-CNN model, which prevented us from experimenting extensively with different parameters. Additionally, the scarcity of histological datasets with gland annotations, especially those representing colorectal cancer, poses a challenge. Nevertheless, the model demonstrated good results, indicating its potential for effective gland segmentation.

6 Conclusion

In this paper, a combined glands segmentation model from histological images for colorectal cancer detection is presented. It accurately identified their locations among various other tissues, as well as their shapes. The model relies on Mask R-CNN architecture with the watershed algorithm where Mask R-CNN alone produced good results with a Dice coefficient of 0.85. After applying watershed a noticeable improvement was recorded.

Despite the promising results, our study faced some challenges. We found that Mask R-CNN is computationally heavy and requires significant time and powerful hardware to train. Additionally, the limited availability of histological images of CRC posed constraints on generating accurate segmentations.

Looking towards the future, medical image segmentation holds immense promise. By leveraging deep learning architectures for object segmentation in medical images and by combining these with established segmentation methods for evaluation, significant advancements can be made. The inherent complexity of medical images, particularly histological ones with diverse tissue types, necessitates exploring preprocessing techniques to isolate the objects of interest from surrounding tissues. This can significantly aid deep learning techniques in feature extraction, ultimately improving segmentation performance.

References

1. WHO. Colorectal Cancer, 2023. https://www.who.int/fr/news-room/fact-sheets/detail/colorectal-cancer. Accessed 15 Dec 2023
2. Sirinukunwattana, K., et al.: Gland segmentation in colon histology images: the glas challenge contest. Med. Image Anal. **35**, 489–502 (2017)
3. Rathore, S., Iftikhar, M.A., Chaddad, A., Niazi, T., Karasic, T., Bilello, M.: Segmentation and grade prediction of colon cancer digital pathology images across multiple institutions. Cancers **11**(11) (2019)
4. Vincent, L., Soille, P.: Watersheds in digital spaces: an efficient algorithm based on immersion simulations. IEEE Trans. Pattern Anal. Mach. Intell. **13**, 583-598 (1991). https://doi.org/10.1109/34.87344
5. Alfonso, S., Corredor, G., Moncayo, R., Barrera, C.R., Sanchez, A.Y., Toro, P., Romero, E.: A method to detect glands in histological gastric cancer images. In: Romero, E., Lepore, N., Brieva, J. (eds.) 14th International Symposium on Medical

Information Processing and Analysis, vol. 10975, p. 109750X. International Society for Optics and Photonics, SPIE (2018)
6. Wang, H., Xian, M., Vakanski, A., Ta-net: topology-aware network for gland segmentation. In: IEEE/CVF Winter Conference on Applications of Computer Vision, pp. 3241–3249 (2022). automated gland segmentation in colon histopathology images. Informatics in Medicine Unlocked **27**(100784), 2021 (Jan 2022)
7. He, K., Gkioxari, G., Dollár, P., Girshick, R.: Mask r-cnn. In: 2017 IEEE International Conference on Computer Vision (ICCV), pp. 2980–2988 (2017)
8. Paul, A., Mukherjee, D.P.: Gland segmentation from histology images using informative morphological scale space. In: 2016 IEEE International Conference on Image Processing (ICIP), pp. 4121–4125 (2016)
9. Cheikh, B.B., Bertheau, P., Racoceanu, D.: A structure-based approach for colon gland segmentation in digital pathology. In: Gurcan, M.N., Madabhushi, A. (eds.) Medical Imaging 2016: Digital Pathology, vol. 9791, p. 97910J. International Society for Optics and Photonics, SPIE (2016)
10. Lamang, L.D., Kim, B.W.: Deep learning approaches to colorectal cancer diagnosis: a review. Appl. Sci. **11**(22) (2021)
11. Khatun, R., Chatterjee, S.: Machine learning approach for segmenting glands in colon histology images using local intensity and texture features. In: 2018 IEEE 8th International Advance Computing Conference (IACC), pp. 314–320 (2018)
12. Kosov, A., Khvostikov, A., Krylov, A.: Adaptive method of glands segmentation on histological images. In: Proceedings of the 30th International Conference on Computer Graphics and Machine Vision (GraphiCon 2020). Part 2, pp. 39–1, December 2020
13. Mei, L., Guo, X., Huang, X., Weng, Y., Liu, S., Lei, C.: Dense contour-imbalance aware framework for colon gland instance segmentation. Biomed. Signal Process. Control **60**, 101988 (2020)
14. Dabass, M., Dabass, J.: An Atrous Convolved Hybrid Seg-Net Model with residual and attention mechanism for gland detection and segmentation in histopathological images. Comput. Biol. Med. **155**, 106690 (2023)
15. Sun, M., Wang, J., Gong, Q., Huang, W.: Enhancing gland segmentation in colon histology images using an instance-aware diffusion model. Comput. Biol. Med. **166**, 107527 (2023). Epub ahead of print

CASF: Conflict Areas Survivor Framework

Nardjes Bouchemal(✉), Mounira Bouzahzah, and Saber Ziad Barki

LISI Laboratory of Intelligent Systems and Informatics, Abdelhafid Boussouf University Center of Mila, Mila, Algeria
n.bouchemal.dz@ieee.org, m.bouzahzah@centre-univ-mila.dz

Abstract. The Conflict Areas Survivor Framework (CASF) is an innovative platform, serving as a robust information hub, empowering individuals residing in conflict zones through an array of features. From volunteer coordination to essential supplies locator, healthcare services access, safe areas identification, communication hub, resource mapping, emergency services integration, to crowdsourced intelligence, supply chain management, psychosocial support, and accessibility features, CASF embodies a comprehensive toolset. By fostering collaboration and resilience, CASF facilitates effective response efforts and community support, leveraging technology to navigate challenges and promote well-being in conflict-ridden environments.

This dynamic solution not only addresses immediate needs but also cultivates a network of empowered individuals working collectively for the welfare of their communities, encapsulating a paradigm shift in conflict response mechanisms. CASF is based on a minimum of technologies because of the very restricted environment.

Keywords: Healthcare services access · safe areas identification · crowdsourced intelligence · voice recognition

1 Introduction

Conflict areas around the world are regions where ongoing violent clashes, wars, or tensions significantly disrupt the lives of people. The impact on civilian populations in these areas is profound, leading to widespread suffering, displacement, and loss of life.

In the last three years, over 82.4 million people were forcibly displaced worldwide due to persecution, conflict, violence, or human rights violations, with 55 million internally displaced, according to UNHCR (The United Nations High Commissioner for Refugees) and IDMC (The Internal Displacement Monitoring Centre), reports [9].

The Conflict Areas Survivor Framework (CASF) is a robust, dynamic and adaptive solution that leverages technology to provide crucial support, foster community resilience, and facilitate effective response efforts in conflict-ridden environments.

S. Chikhi et al. (Eds.): MISC 2024, LNNS 1267, pp. 267–278, 2025.
https://doi.org/10.1007/978-3-031-82112-7_20

This innovative framework encourages collaboration between those in need and those willing to give, creating a network of empowered individuals working together for the well-being of their communities.

Recognizing the critical situations faced by the targeted users, CASF prioritizes ease of use in its design. There is no authentication required, acknowledging the challenges users face in conflict areas. The framework operates seamlessly with non-permanent internet connections, understanding connectivity constraints in such environments. Moreover, it accommodates various communication supports, including GPS and SMS, ensuring accessibility in diverse circumstances.

In a ground-breaking feature, the framework incorporates voice activation of the application, allowing users to navigate and initiate actions through verbal commands, [10]. This user-friendly approach aims to empower individuals, even in the most challenging circumstances, by providing them with quick and efficient access to essential resources and services. CASF is not just a framework; it's a lifeline in times of crisis, designed to be robust, accessible, responsive and supportive.

The rest of the paper is organized as follow: after related work section, we present CASF framework details then in section four the most important functionalities are given. Section five provides evaluation of the framework by comparing it with other works.

2 Related Work

The development of the Conflict Areas Survivor Framework (CASF) draws upon a rich history of research and technological advancements in the realm of mobile computing, emergency response frameworks [11] and community resilience platforms [12]. This section reviews relevant literature and existing frameworks that have informed and inspired the design of CASF.

2.1 Emergency Response Frameworks

Emergency response frameworks have evolved significantly, leveraging advancements in mobile and wireless technologies to improve situational awareness and coordination. Frameworks like Sahana [1] and Ushahidi [2] have demonstrated the potential of crowdsourced data in managing disaster response and relief efforts. Sahana, originally developed in the aftermath of the 2004 Indian Ocean tsunami, offers a comprehensive suite of tools for disaster management, including volunteer coordination, missing person registries, and inventory management for relief supplies. Ushahidi, developed in response to the 2008 Kenyan post-election violence, allows users to report incidents via SMS, email, or web forms, which are then mapped to provide real-time situational awareness.

While these frameworks provide valuable functionalities, CASF extends their capabilities by integrating real-time monitoring, voice-activated assistance, and psychosocial support tailored specifically for conflict zones.

2.2 Healthcare Access and Resource Mapping

Access to healthcare and essential resources is critical in conflict and disaster scenarios, [13] Frameworks like OpenMRS [3] and DHIS2 [4] have been instrumental in managing health information in low-resource settings. OpenMRS, an open-source medical record framework, supports the efficient management of patient data and healthcare services. DHIS2, a flexible, web-based open-source information framework, facilitates the collection, analysis, and reporting of health data, widely used in over 60 countries.

CASF builds upon these foundations by incorporating real-time updates on the availability of essential supplies and healthcare services, ensuring that users can locate critical resources quickly and efficiently.

2.3 Community-Based Platforms

Community engagement and resilience are pivotal in conflict-affected areas, [14]. Platforms like FrontlineSMS [5] and CommCare [6] empower local communities by enabling communication and data collection via mobile phones. FrontlineSMS allows organizations to send and receive text messages without internet access, facilitating community coordination and information dissemination. CommCare, used extensively in health interventions, enables community health workers to collect and report data using mobile devices, improving service delivery and tracking.

CASF enhances community resilience by offering a centralized communication hub, resource mapping, and volunteer coordination, fostering a network of empowered individuals working together for mutual support.

2.4 Crowdsourced Intelligence and Safety

The use of crowdsourced data to enhance safety and situational awareness has been explored in various projects, [15]. SafeCity [7], for example, collects anonymous reports of sexual harassment and abuse, mapping them to identify hotspots and inform police and community action.

CASF leverages crowdsourced intelligence to provide real-time information on resource status, safety conditions, and potential hazards, validated and prioritized to enhance the reliability and usefulness of the data.

2.5 Accessibility and Usability

Ensuring accessibility and usability in emergency frameworks is paramount [16]. Research on inclusive design, such as the work by the WHO on the Global Accessibility Reporting Initiative (GARI) [8], highlights the importance of features like voice commands, screen readers, and text-to-speech functionalities to accommodate users with varying needs. CASF integrates these accessibility features, ensuring that even individuals with disabilities can effectively use the framework in critical situations.

3 CASF Framework

3.1 The Framework Description

The Conflict Areas Survivor Framework (CASF) builds on the strengths of existing emergency response, healthcare access, community-based, crowdsourced intelligence, and accessibility-focused frameworks. By integrating these diverse functionalities into a cohesive platform tailored for conflict zones, it offers a comprehensive solution that addresses immediate needs. The architecture is composed essentially of four levels, see Fig. 1: User devices (1), Voice recognition (2), Communication module and the cloud level (4 and 5). The proposed framework has two operating modes:

Real-Time Monitoring
Evaluate the person's condition in real time and create a special record for each person that contains all the basic data such as the condition of the area, the condition of the person, and the geographical area using GPS (4 and 5 Fig. 1). If the framework notices a specific pattern of silence (through sound and motion analysis techniques using artificial intelligence), it asks the person a question. If he does not find a response or interaction from him, he sends a message to the emergency service via SMS (every 10 min, see (3) Fig. 1). This message contains all the information recorded by that person (such as status, location, the last time the person interacted with the framework, and the remaining time until the battery runs out).

Emergency Mode
When the person stops interacting with the framework, the system launches emergency mode. This mode consists of stopping all secondary functions and reducing the screen brightness in order to preserve the battery for the longest possible time. The system also enters a surveillance state, where it analyses the sounds coming from the place along with Turn on Wi-Fi, and if sounds of movement, speech, or one of the rescuers' networks are detected, it plays strong rescue sounds with vibration and flash lighting in order to draw people's attention and know the location of the injured person.

4 CASF Functionalities

This section deals within the different functionalities or services offered by the proposed framework.

4.1 Initial Launch of the Framework

One of the most difficult challenges is the initial launch of the framework to ensure its secure and trustworthy usage. To address this, we propose a strategy where ten trusted users are selected as initial adopters. These users are responsible for onboarding only trusted individuals, such as family members, neighbours, and community leaders, into the framework. This approach aims to establish a network of trust within the user base, safeguarding against potential misuse or unauthorized access to the framework's functionalities.

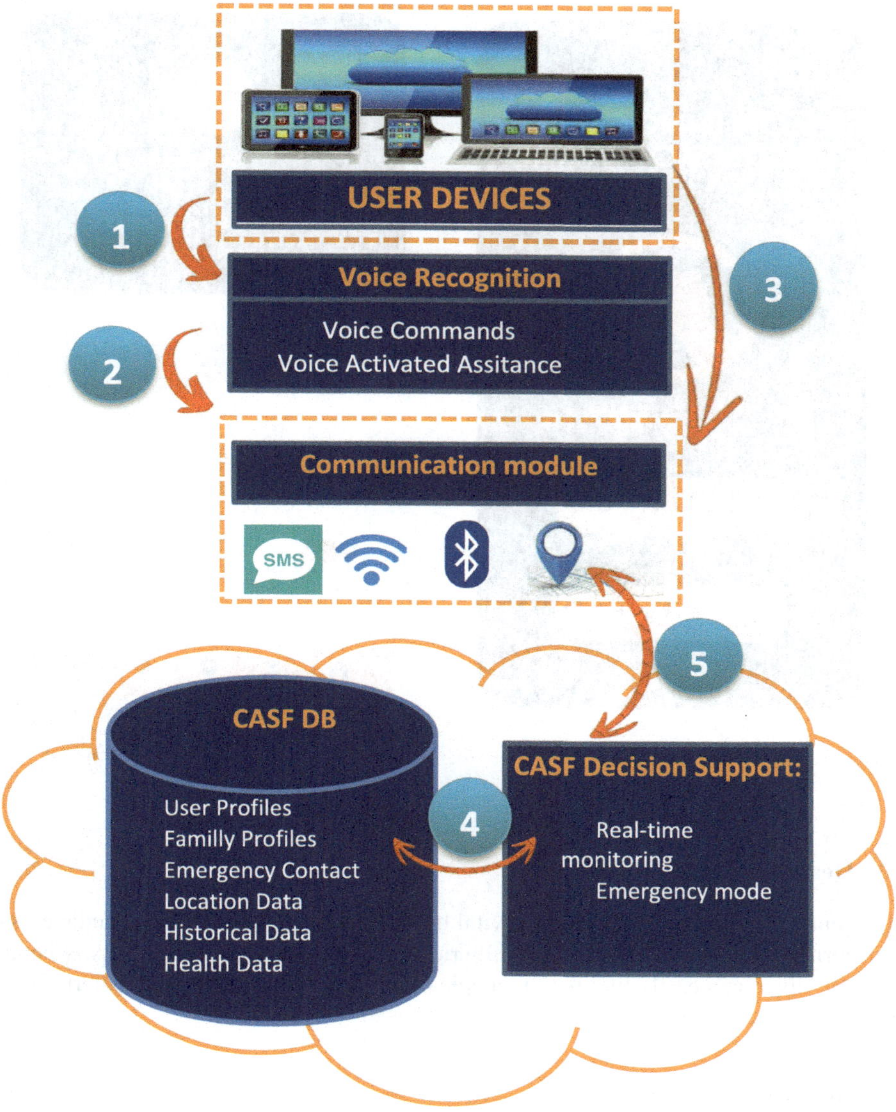

Fig. 1. CASF Global Architecture

4.2 Essential Supplies Locator

Users effortlessly locate and access critical supplies like water, food, and other necessities through an intuitive interface. Essential supplies are meticulously categorized, with real-time updates on their availability, Fig. 2.

Fig. 2. Survivor Framework

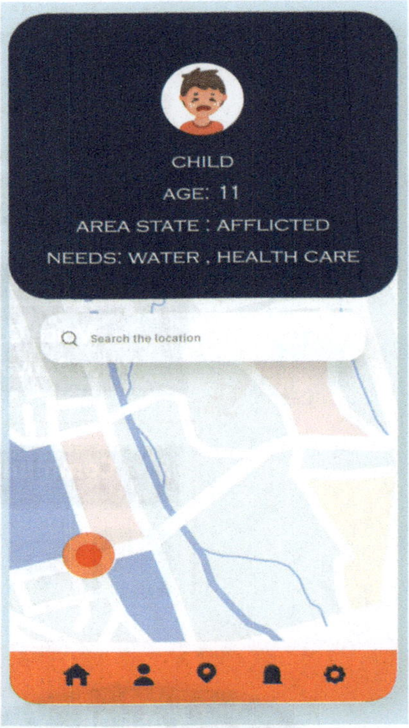

Fig. 3. Healthcare Services Access

4.3 Healthcare Services Access

The framework streamlines access to vital healthcare services, including medications, transportation to hospitals, first aid, surgeries, and health monitoring. Users, whether given or Giver, can easily find nearby medical facilities and receive timely alerts about available services, Fig. 3.

4.4 Resource Mapping

A map-based interface visualizes the distribution of essential supplies, healthcare facilities, safe areas, and potential hazards.

Users can mark locations of available resources, medical facilities, and safe zones, promoting efficient resource allocation, Fig. 4.

4.5 Volunteer Coordination

The framework features a volunteer registry for two types of users:

Given: Individuals seeking essential supplies such as water, food and other necessities. They also require health services, including:

First Aid: Immediate medical assistance for injuries and emergencies.
Transportation: Facilitation of transportation to medical facilities or safe areas.
Surgery: Access to surgical procedures if necessary.
Health Monitoring: Ongoing monitoring of health conditions for individuals requiring special care.

Giver: Individuals willing to contribute, offering: **Essential Needs**: Donations of water, food, and other necessities. **Health Services:** First aid, transportation, surgery, and health monitoring.

4.6 Educational Resources

The platform provides educational materials on first aid, disaster preparedness, and safety measures. Tutorials and guides empower citizens to contribute effectively to their own and others' well-being during crises, Fig. 5.

4.7 Crowdsourced Intelligence

Citizens contribute real-time information on resource status, safety conditions, and potential hazards. The framework validates and prioritizes crowdsourced data to enhance situational awareness.

4.8 Voice-Activated Assistance

The framework includes advanced voice recognition capabilities, ensuring ease of use for users in critical situations.

No authentication is required, recognizing the challenges faced by users. It is operable with non-permanent internet connections, acknowledging connectivity constraints. It is also compatible with various communication supports such as GPS and SMS.

Voice activation of the application allows users to initiate actions through verbal commands, enhancing usability in challenging environments.

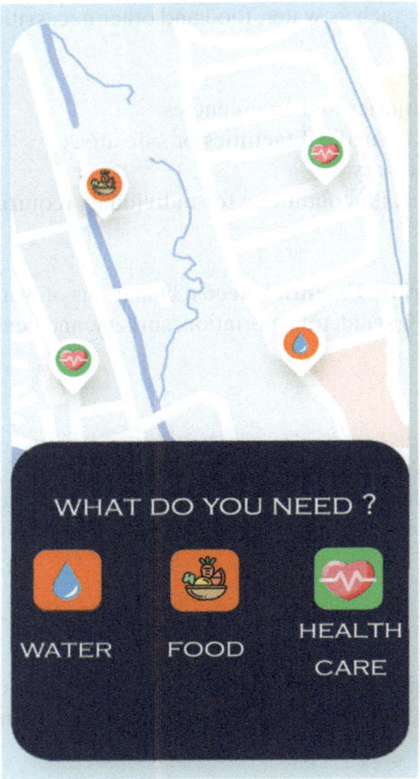

Fig. 4. Resource Mapping

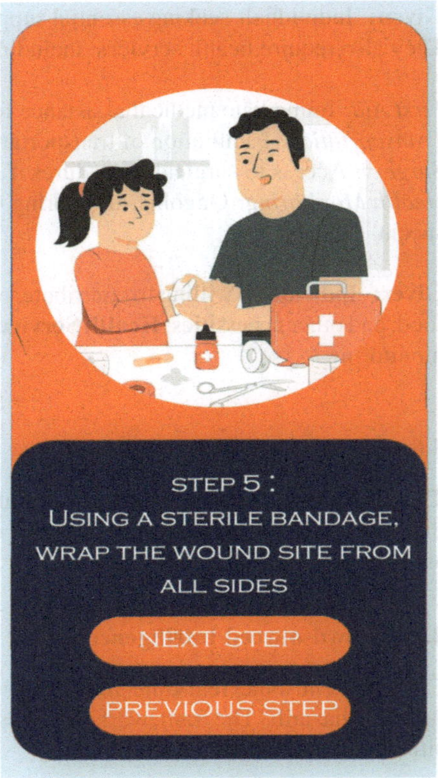

Fig. 5. Educational Resources

5 CASF Evaluation

Table 1 is a comparative table of the platforms discussed in the "Related Work" section, highlighting their key features and how they relate to the CASF Framework. Also, we propose a comparative graph to more understand the performance of the platform, see Fig. 6.

In a comparative analysis of response times across various platforms (Fig. 7), CASF stands out with a remarkably quick response time of just 1 h, showcasing its efficiency and reliability in urgent situations. Following CASF, Frontline SMS also performs well with a response time of 2 h. Platforms such as Ushahidi and CommCare show moderate efficiency with a 3-h response time. Sahana and GARI, each with a 4-h response time, perform at an average level. OpenMRS and SafeCity, both with 5-h response times, and DHIS2, with the slowest response time of 6 h, indicate areas needing significant improvement. Overall, CASF's superior performance highlights its optimization for rapid response, while other platforms may need to streamline their processes to enhance efficiency.

Table 1. Comparative table of different platforms

Platform	Purpose	Key Features	Limitations	Comparison to CASF
Sahana	Disaster management	Volunteer coordination, missing person registries, inventory management	Primarily focused on natural disasters	CASF includes voice-activated assistance, real-time monitoring, and psychosocial support
Ushahidi	Incident reporting and mapping	Crowdsourced incident reporting, mapping	Limited to incident reporting and mapping	CASF expands with healthcare access, resource mapping, and emergency mode features
OpenMRS	Health information management	Patient data management, healthcare services	Focused on healthcare data management	CASF integrates healthcare services with resource and safe area identification
DHIS2	Health data collection and analysis	Health data collection, analysis, and reporting	Primarily for health data management	CASF combines health services with communication hub and volunteer coordination
Frontline SMS	Community communication	SMS-based communication	Limited to SMS communication	CASF uses multiple communication supports (GPS, SMS) and real-time monitoring
CommCare	Community health worker support	Data collection, reporting via mobile devices	Focused on health interventions	CASF includes educational resources and psychosocial support, enhancing overall resilience

(continued)

Table 1. (*continued*)

Platform	Purpose	Key Features	Limitations	Comparison to CASF
SafeCity	Safety reporting	Anonymous reporting, hotspot mapping	Focused on harassment and abuse reporting	CASF provides comprehensive safety and resource information in conflict zones
GARI	Accessibility reporting	Voice commands, screen readers, text-to-speech	Primarily a reporting initiative	CASF integrates accessibility features directly into the platform for ease of use

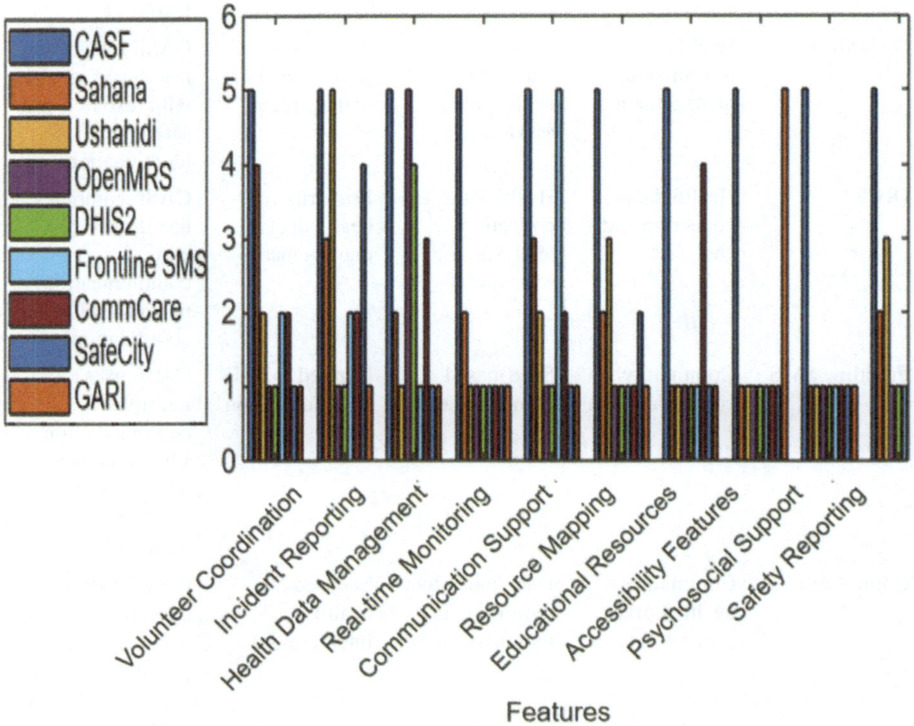

Fig. 6. Comparison of CASF with Other Platforms

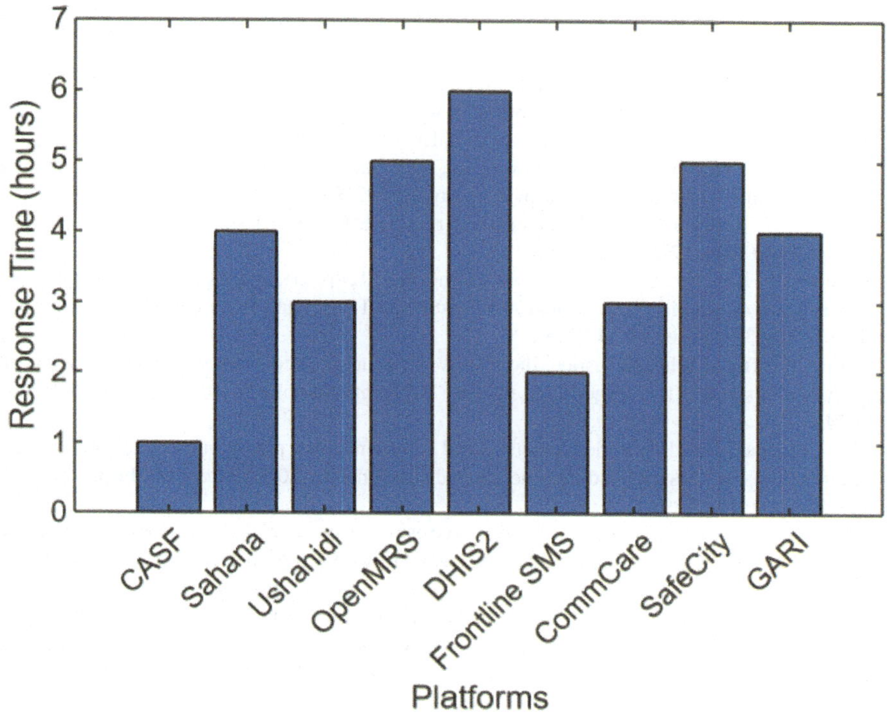

Fig. 7. Comparison of CASF Response Time with other Platforms

6 Conclusion

In conclusion, the Conflict Areas Survivor Framework (CASF) stands as a beacon of hope amidst the challenges faced by conflict-affected communities. Through innovative technology, collaborative partnerships, and unwavering dedication, CASF embodies the spirit of resilience and solidarity, empowering individuals to overcome adversity and rebuild their lives. As we navigate the complexities of humanitarian crises, let us remain steadfast in our commitment to serving those in need, fostering trust, and promoting peace and prosperity for all. Together, we can forge a brighter future, where compassion and solidarity prevail, and no one is left behind.

References

1. Sahana Foundation: Sahana Software Foundation (n.d.). Retrieved from https://sahanafoundation.org. Accessed 1 Aug. 2024
2. Ushahidi: Ushahidi (n.d.). Retrieved from https://www.ushahidi.com. Accessed 1 Aug. 2024
3. OpenMRS: OpenMRS (n.d.). Retrieved from https://openmrs.org. Accessed 10 July 2024
4. DHIS2: DHIS2 Overview (n.d.). Retrieved from https://dhis2.org. Accessed 12 July 2024
5. FrontlineSMS: FrontlineSMS (n.d.). Retrieved from https://www.frontlinesms.com. Accessed 1 July 2024

6. CommCare. (n.d.). *CommCare.* Retrieved from https://www.dimagi.com/commcare/ (Accessed July 1, 2024)
7. SafeCity: SafeCity (n.d.). Retrieved from https://safecity.in. Accessed 1 July 2024
8. (GARI). Retrieved from [gari.info]. https://www.gari.info
9. World Health Organization: Global accessibility reporting initiative (n.d.)
10. Azhiimah, A., Khotimah, K., Sumbawati, M., Santosa, A.: Automatic control based on voice commands and Arduino (2020). https://doi.org/10.2991/aer.k.201124.006
11. Mustafa, N.: Introduction to the emergency response framework (ERF) (2024). https://doi.org/10.13140/RG.2.2.21461.84965
12. Khujumi, I., Kour, D.: Psychological wellbeing, social support, and resilience among population in areas of conflict. Educational Administration Theory and Practices 30 (2024). https://doi.org/10.53555/kuey.v30i4.2795
13. Usman, F.: Towards health justice: the effects of Lebanon's economic crisis on access to primary health services by refugees in Lebanon (2024). https://doi.org/10.13140/RG.2.2.10584.10241
14. Lyles, C., et al.: Building a client resource and communication platform for community-based organizations: co-design to address health and social needs (2023) (Preprint). https://doi.org/10.2196/preprints.53939
15. Herhkovitz, S.: Crowdsourced intelligence (Crosint): using crowds for national security. Int. J. Intell. Secur. Public Affairs **22**, 42–55 (2020). https://doi.org/10.1080/23800992.2020.1744824
16. Vo, A., Bhaskar, R., Chi, T.Y., Faddoul, G.: Identifying socially vulnerable regions with persistent low accessibility to emergency care through a spatial decision framework. J. Decis. Syst.Decis. Syst. **29**(4), 201–222 (2020). https://doi.org/10.1080/12460125.2020.1796300

Author Index

A
Abid, Malika 29
Ait Abderrahim, Imène 212
Ait Mohammed, Linda 205
Alim-Ferhat, Fatiha 205
Allal, Imed 182
Amara, Kahina 226
Ameur, Imane 43
Ameur, Mazene 43
Ameur, Mohamed El Amine 43

B
Barki, Saber Ziad 267
Baziz, Meriem 182
Bechinia, Hadjer 240
Benbelkacem, Samir 226
Benguenane, Messaoud 3
Benkacimi, Ali 226
Benkaddour, Mohammed Kamel 29
Benmerzoug, Djamel 240
Bennouar, Djamal 112
Benouis, Mohamed 29
Benrached, Youcef 212
Bensalem, Fatima Zohra 102
Boucetta, Anfal Yousra 182
Bouchemal, Nardjes 267
Boudabia, Saad 102
Boudine, Leila 226
Bouzahzah, Mounira 267

C
Chaib, Noureddine 3
Chaoui, Allaoua 69
Chouikh, Asma 69

D
Dagha, Houssem Eddine 43
Daksi, Mohamed Yacine 143
Dib, Mehdi 143
Djerida, Achraf 127
Draoui, Ahmed 102

F
Fatima, Boumahdi 171

G
Gacem, Rayane 226
Guerroudji, Mohamed Amine 226
Guidoum, Arsalane Chouaib 156

H
Hamdad, Leila 182
Hamdi, Skander 87
Harous, Saad 14
Harrouzi, Asma 254
Hebal, Sara 14
Henider, Amel 156
Houha, Amel 53

I
Issolah, Imene 226

K
Kerkouche, ELHillali 69
Khalfaoui, Khaled 69
Khemis, Sabrina 87
Khlifa, Nawres 240

L
Lahreche, Ferialle 254
Laiourate, Kheira 102
Louail, Lemia 14

M
Mahdjoub, Youcef 102
Mansoura, Ballou 254
Masmoudi, Mostefa 226
Matougui, Zakaria 143
Mechta, Djamila 14
Moussaoui, Abdelouahab 87, 254

N
Neciri, Said 3

S. Chikhi et al. (Eds.): MISC 2024, LNNS 1267, pp. 279–280, 2025.
https://doi.org/10.1007/978-3-031-82112-7

O
Omar, Mokhtari 171
Ouahrani, Leila 112
Oulad-Naoui, Slimane 254
Oussama, Nedjah 171

T
Talbi, Farid 205
Tari, Abdelkamel 53
Taright, Samya 226
Toubaline, Nesrine 226
Touia, Fatima-Zohra 226

Y
Yasmine, Mancer 171

Z
Zenati, Nadia 226

The manufacturer's authorised representative in the EU is Springer Nature Customer Service Centre GmbH, Europaplatz 3, 69115 Heidelberg, Germany. If you have any concerns regarding our products, please contact ProductSafety@springernature.com

Printed and bound by CPI Group (UK) Ltd, Croydon, CR0 4YY
30/09/2026
02225561-0004